Conservation of Bio- and Geo-Diversity and Landscape Changes

Conservation of Bio- and Geo-Diversity and Landscape Changes

Guest Editors

Alexandru-Ionuţ Petrişor
Adrian Ursu
Ilinca-Valentina Stoica

MDPI

Basel • Beijing • Wuhan • Barcelona • Belgrade • Novi Sad • Cluj • Manchester

Guest Editors

Alexandru-Ionuţ Petrişor
Doctoral School of Urban Planning
Ion Mincu University of Architecture and Urbanism
Bucharest
Romania

Adrian Ursu
Faculty of Geography and Geology
University "Alexandru Ioan Cuza"
Iași
Romania

Ilinca-Valentina Stoica
Faculty of Geography and Interdisciplinary Centre for Advanced Research on Territorial Dynamics
University of Bucharest
Bucharest
Romania

Editorial Office
MDPI AG
Grosspeteranlage 5
4052 Basel, Switzerland

This is a reprint of the Special Issue, published open access by the journal *Land* (ISSN 2073-445X), freely accessible at: https://www.mdpi.com/journal/land/special_issues/bio_and_geo_diversity.

For citation purposes, cite each article independently as indicated on the article page online and as indicated below:

Lastname, A.A.; Lastname, B.B. Article Title. *Journal Name* **Year**, *Volume Number*, Page Range.

ISBN 978-3-7258-6716-5 (Hbk)
ISBN 978-3-7258-6717-2 (PDF)
https://doi.org/10.3390/books978-3-7258-6717-2

Contents

About the Editors

Alexandru-Ionuţ Petrişor

Alexandru-Ionuţ Petrişor (born 1975) earned a BS in Ecology (1997) and a PhD in Geography (2011) from the University of Bucharest, a MSPH in Biostatistics (2000) and a PhD in Ecology (2004) from the University of South Carolina, a certificate degree in Project Management (2012) from Bucharest Academy of Economic Studies, and a Habilitation in Urban Planning (2014) and the highest research rank in Biology (2013) and Geography (2017) from the Romanian Ministry of Education. He is a Professor and the Director of the Doctoral School of Urban Planning at Ion Mincu University of Architecture and Urbanism in Bucharest, a Professor at the Technical University of Moldova, and a Senior Researcher, Rank I at the National Institute for Research and Development in Constructions, Urbanism and Sustainable Spatial Development URBAN-INCERC in Bucharest, Romania. His work in over 80 research grants has yielded over 800 publications, 550 presentations, and environmental studies for 40 spatial plans, recognized by memberships including Sigma Xi, the Society for Urban Ecology and International Association for Landscape Ecology, and awards in international invention shows, including the Geneva International Exhibition of Inventions. He is a member of many editorial boards and the Chief Editor of *Urbanism Architecture Constructions* (Scopus). His research interests include different urban/landscape ecology topics and predatory publishing.

Adrian Ursu

Adrian Ursu, PhD, is an Associate Professor in the Department of Geography within the Faculty of Geography and Geology at Alexandru Ioan Cuza University of Iași, Romania. His research interests include Environmental Science, Remote Sensing, GIS, Nature Conservation, Environmental Geography, Land Use Change, Risk Assessment, and Heritage Studies. He has contributed to major international initiatives such as the Corine Land Cover 2000 and Corine Land Cover 2006 projects for Romania, funded by the European Environment Agency. He has also participated in numerous national research projects focused on Risk Assessment, Sustainable Development, Land Degradation, GIS applications for Natura 2000 sites, and Noise Pollution Modelling in urban environments. Dr. Ursu serves as Editor-in-Chief of the journal *Present Environment and Sustainable Development* (ESCI–Clarivate) and has been Guest Editor for two Special Issues in *Remote Sensing* (Q1 IF) and *Land* (Q2 IF). Since 2019, he has coordinated the International Conference Present Environment and Sustainable Development, which celebrated its 20th edition in 2025. He is also one of the coordinators of the International Symposium on Geographical Information Systems, initiated in 1991 and organized in partnership with four academic institutions from Romania and the Republic of Moldova.

Ilinca-Valentina Stoica

Ilinca-Valentina Stoica holds a PhD in Geography (2010) and pursued postdoctoral studies focused on integrated rural development in restrictive environments (2010–2013), both at the University of Bucharest. Currently, she is working at the University of Bucharest in the Faculty of Geography, Human and Economic Geography Department and within the Interdisciplinary Centre for Advanced Research on Territorial Dynamics (CICADIT). She has been involved in various scientific projects focused on spatial planning, rural development, and the urban–rural interface, among others, at different scales of analysis (local, national or international). Her current research interests include small towns, urban expansion processes, post-socialist dynamics, socio-economic disparities, and local development in mountain areas.

Preface

The reprint includes articles dealing with the relationship between biological diversity (including geo- and eco-diversity) and land cover and use changes, including the factors able to affect this relationship and possibilities to control them. The articles included in this collection are grouped around several interconnected themes: urbanization, biodiversity loss, habitat quality, and the role of planning and technology in environmental management. Overall, the key findings indicate that landscape structure and historical land use are often more significant drivers of biodiversity than broad climatic variables. At the same time, the findings indicate alarming trends, such as a generalized loss of urban green infrastructure, fueled by poorly regulated urban sprawl and real estate development that consumes agricultural and natural lands; long-term, diffuse threats, such as atmospheric nitrogen deposition, which are becoming primary drivers of biodiversity loss and are overriding traditional landscape-based ecological factors; and natural stressors like drought. Advanced remote sensing technologies are indispensable tools for monitoring large-scale transformations. Altogether, the studies underscore the critical role of effective, science-informed human management. While targeted interventions can rapidly restore ecosystem functions, policies can also yield unintended negative consequences, highlighting the complexity of managing interconnected natural systems.

Alexandru-Ionuţ Petrişor, Adrian Ursu, and Ilinca-Valentina Stoica
Guest Editors

land

MDPI

Editorial

Editorial: Conservation of Bio- and Geodiversity and Landscape Changes

Alexandru-Ionut Petrisor [1,2,3,*], Adrian Ursu [4] and Ilinca-Valentina Stoica [5]

1 Doctoral School of Urban Planning, Ion Mincu University of Architecture and Urbanism, Academiei 18-20, 010014 Bucharest, Romania
2 Department of Architecture, Faculty of Architecture and Urban Planning, Technical University of Moldova, Ştefan cel Mare şi Sfânt Blvd., 168, 2004 Chișinău, Moldova
3 National Institute for Research and Development in Constructions, Urbanism and Sustainable Spatial Development, Şos. Pantelimon, 266, 021652 Bucharest, Romania
4 Faculty of Geography and Geology, University "Alexandru Ioan Cuza", 700506 Iași, Romania
5 Faculty of Geography and Interdisciplinary Centre for Advanced Research on Territorial Dynamics, University of Bucharest, 030018 Bucharest, Romania
* Correspondence: alexandru_petrisor@yahoo.com

1. Introduction

The Rio Convention on Biological Diversity has established biodiversity as a key concept governing the development policies of most countries towards assuming responsibility for the environment [1]. Many scientists tie biodiversity (as well as geo- and ecodiversity) to the ecological infrastructure of our planet [2]. The Convention includes recommended measures dealing with the conservation of biodiversity, as well as measures that are needed to ensure that future development is achieved in a way that does not affect biodiversity [3]. At the same time, as part of "global changes" [4], land use and land cover changes are a major threat to biodiversity. This Special Issue deals with the relationship between the two, and with the factors that can affect it, such as its drivers and possibilities of controlling it.

We initially thought that the relationship between the conservation of bio- and geodiversity and landscape changes could be best described by a collection of contributions focused on both pure and applied research, as well as comparative studies. Examples of possible topics included the impact of land use and land cover change on bio-, geo-, and ecodiversity, including the drivers and means of controlling or mitigating it; the impacts of land use and land cover changes on natural protected areas, i.e., its drivers and means of controlling or mitigating it; spatial planning tools for preventing land use and land cover changes within and around natural protected areas at different spatial scales; natural hazards and bio-, geo-, and ecodiversity and human influences on their relationship; and local development challenges in natural protected areas. The result is a collection of contributions analyzing direct human activity, policy decisions, and climatic factors that are changing ecosystems significantly worldwide.

2. Synopsis of Contributions

The contributions forming this collection can be grouped around several interconnected themes: urbanization, biodiversity loss, habitat quality, and the role of planning and technology in environmental management.

2.1. Impact of Urbanization on Biodiversity and Habitat Quality

A major recurring finding revealed by the contributions is that urbanization acts as a major driver of biodiversity loss and habitat degradation [5]. This finding is common

 https://doi.org/10.3390/land15020203

to studies dealing with avian and plant diversity. In more detail, in North African urban environments, bird diversity was found to decrease significantly along an urbanization gradient, highlighting that urban green spaces are critical biodiversity hotspots that need to be interconnected [5]. Similarly, in the Yangtze River Floodplain, habitat fragmentation caused by human activities like aquaculture negatively affects goose population trends [6]. In Sabah, Borneo, the conversion of intact forests into industrial oil palm plantations has caused a severe and statistically significant decline in plant functional diversity. This decline proves that direct land use change, driven by agricultural expansion and built-up development, is causing measurable degradation of ecosystems, confirming findings from smaller field studies and showcasing the potential of satellite observation for monitoring biodiversity in critical hotspots [7]. Furthermore, built-up development drivers such as population density and GDP are generally negatively correlated with habitat quality in regions like the Hung River Valley [8].

In addition, quantitative analyses of Polish and Romanian cities show generalized loss and fragmentation of urban green infrastructure between 2006 and 2018. The decline of urban green infrastructure in post-socialist Europe is driven by mechanisms of urban sprawl and weak planning enforcement. The loss is attributed to "derogatory planning", where economic interests lead to exemptions from urban regulations, resulting in built-up densification at the expense of nature [9].

2.2. Drivers of Species Composition and Population Trends

Some articles identified specific landscape and historical factors that determine how species are distributed, including historical continuity, environmental heterogeneity, and landscape attributes. Historical continuity is illustrated by a study carried out in two Central European national parks: the Podyjí and Thayatal National Parks (Czech Republic/Austria). The study revealed a fundamental shift in the primary drivers determining plant species richness, showing that historical land cover (dating back to the 1950s) had a larger impact on current plant species composition than present-day land cover. Even in highly protected areas, insidious, long-term environmental pressures are dominant drivers of ecological change, altering the fundamental drivers of biodiversity. This finding indicates that the widespread, chronic issue of nitrogen deposition has become a more powerful influence on biodiversity than the inherent physical characteristics of the landscape. These results suggest that even well-preserved areas are not immune to large-scale, diffuse environmental threats that require broader policy solutions [10]. In the same study, the "river phenomenon", characterized by deep valleys with high geomorphological and meso-climatic diversity—which is a primary driver of species richness in these regions—illustrates environmental heterogeneity [8].

A 24-year study carried in Shengjin Lake in China's Yangtze Floodplain found that an increasing population trend was driven primarily by habitat and landscape attributes, not climate. For wintering geese, the availability of suitable habitat and landscape attributes (such as connectivity and patch size) are the key forces affecting population trends, while climatic factors have a much weaker influence. Larger herbivores can modify vegetation in a way that creates more suitable foraging conditions for smaller geese. These findings contrast with the decreasing population trends reported in other regional wetlands, suggesting that well-intentioned policies can have unforeseen negative effects and indicating the critical role of landscape structure and history in determining biodiversity. The physical configuration and historical context of a landscape appear to be primary determinants of species population trends and community composition, often outweighing the immediate effects of climatic factors. Altogether, these findings highlight the need for adaptive management strategies that consider complex ecological interactions [6].

https://doi.org/10.3390/land1502020

2.3. Coastal and Wetland Degradation

The studies dealing with this topic, particularly with land reclamation and hydrological changes and drought, emphasize the vulnerability of wetlands and coastal ecosystems to human intervention and climate extremes. A study carried out in China's northern Yellow Sea shows that land reclamation for industry and aquaculture is the main cause of a 63% decline in Suaeda salsa saltmarshes over the last three decades [11].

Hydrological changes and drought are illustrated by a study on wetland birds' assemblages in Nuntași and Tuzla Lakes (Danube Delta Biosphere Reserve). The findings indicate that an ecosystem's resilience to natural stressors like drought is critically linked to human management. Extreme weather, such as the severe droughts in Romania in 2013 and 2020, can lead to the total drying of lakes, causing piscivorous birds like pelicans and swans to abandon the area while creating temporary feeding grounds for waders and gulls in the mudflats. Inadequate oversight can amplify the impacts of climate, while targeted interventions can provide rapid and effective remedies. Altogether, these results underline the interplay of climate stress and human management [12].

2.4. Socio-Economic Sustainability and Planning Challenges

Two studies carried out in Romania—one dealing with residential satisfaction and another with niche agriculture—show that socio-economic needs often conflict with environmental preservation, but some niche solutions exist. The rapid, developer-led expansion of residential areas in Bucharest and its surroundings, often on former agricultural or industrial land and offering modern dwellings at affordable prices, has created a paradox of residential satisfaction. While residents are pleased with the modernity of their new dwellings, they are highly dissatisfied with the inadequate infrastructure, as the new developments fall short in providing essential facilities like centralized sewerage, public transport, and paved roads. Residents often hold local authorities responsible for allowing development without first establishing the necessary utility networks [13]. Lavender farming is identified as a sustainable economic and ecological solution for small Romanian farms. It offers high profits per hectare, supports beekeeping and biodiversity, and requires low maintenance, although it faces barriers such as a lack of labor and dedicated markets [14].

2.5. Methodologies: Remote Sensing and Geo-Statistical Approaches

So far, we have analyzed the theoretical contributions of the collected studies. However, this presentation would be incomplete without looking at their methodologies. From this standpoint, a cross-cutting theme is the utility of technology in monitoring environmental changes. Satellite remote sensing proved to be a critical tool for quantifying the studied impacts over large and often inaccessible areas. More exactly, remote sensing (using Landsat and Sentinel-2) allows for "wall-to-wall" mapping of plant functional diversity and the monitoring of rapid land cover changes over large, inaccessible regions like Sabah or the Yellow Sea [7]. Similarly, the use of analytical models like InVEST for assessing habitat quality and GWR (Geographically Weighted Regression) for identifying its drivers provided a scientific basis for developing sustainable land use policies [8].

In a nutshell, we can think of a regional ecosystem as a complex puzzle. Urbanization and land reclamation are like removing pieces from the edge, causing the overall picture of biodiversity to shrink [5]. Planning and green infrastructure act as the glue between pieces; if the glue is weak (derogatory planning), the puzzle is fragmented [5]. Finally, technologies like remote sensing are the magnifying glasses that allow scientists to see exactly where pieces are missing or fading before the whole picture is lost [7,12].

https://doi.org/10.3390/land15020203

3. Conclusions

Overall, the key findings of all the studies included in this Special Issue indicate that landscape structure and historical land use are often more significant drivers of biodiversity than broad climatic variables. Advanced remote sensing technologies are proving indispensable for monitoring these large-scale transformations, from the loss of functional diversity in Bornean forests to hydrological stress in Romanian wetlands. In post-socialist European cities, a generalized loss of urban green infrastructure is underway, fueled by poorly regulated urban sprawl and real estate development that consumes agricultural and natural lands. Concurrently, studies from natural protected areas in Central Europe highlight the emergence of long-term, diffuse threats, such as atmospheric nitrogen deposition, which are becoming primary drivers of biodiversity loss and are overriding traditional landscape-based ecological factors. These studies underscore the critical role of effective, science-informed human management. Case studies demonstrate that while inadequate oversight can exacerbate natural stressors like drought, targeted interventions can rapidly restore ecosystem functions. However, policies can also yield unintended negative consequences, which highlights the complexity of managing interconnected natural systems.

Author Contributions: Writing—original draft preparation, A.-I.P., A.U. and I.-V.S.; writing—review and editing, A.-I.P., A.U. and I.-V.S. All authors have read and agreed to the published version of the manuscript.

Funding: This research received no external funding.

Data Availability Statement: Data sharing is not applicable to this article.

Acknowledgments: The authors would like to thank the team at *Land* for helping with and coordinating the Special Issue.

Conflicts of Interest: The authors declare no competing interests.

References

1. Raustiala, K.; Victor, D.G. Biodiversity Since Rio: The Future of the Convention on Biological Diversity. *Environ. Sci. Pol. Sust. Dev.* **1996**, *38*, 16–45. [CrossRef]
2. Perschke, M.J.; Harris, L.R.; Sink, K.J.; Lombard, A.T. Ecological Infrastructure as a framework for mapping ecosystem services for place-based conservation and management. *J. Nat. Cons.* **2023**, *73*, 126389. [CrossRef]
3. Biermann, F. Curtain down and Nothing Settled: Global Sustainability Governance after the 'Rio+20' Earth Summit. *Environ. Plan. C Gov. Pol.* **2013**, *31*, 1099–1114. [CrossRef]
4. Dale, V.H.; Efroymnson, R.A.; Kline, K.L. The land use—Climate change—Energy nexus. *Landsc. Ecol.* **2011**, *26*, 755–773. [CrossRef]
5. Aouissi, H.A.; Petrişor, A.-I.; Ababsa, M.; Boştenaru-Dan, M.; Tourki, M.; Bouslama, Z. Influence of Land Use on Avian Diversity in North African Urban Environments. *Land* **2021**, *10*, 434. [CrossRef]
6. Chen, S.; Zhang, Y.; Borzée, A.; Liang, T.; Zhang, M.; Shi, H.; Chen, B.; Xu, W.; Song, Y.; Mao, L. Landscape Attributes Best Explain the Population Trend of Wintering Greater White-Fronted Goose (*Anser albifrons*) in the Yangtze River Floodplain. *Land* **2021**, *10*, 865. [CrossRef]
7. Hauser, L.T.; Timmermans, J.; Soudzilovskaia, N.A.; van Bodegom, P.M. Linking Land Use and Plant Functional Diversity Patterns in Sabah, Borneo, through Large-Scale Spatially Continuous Sentinel-2 Inference. *Land* **2022**, *11*, 572. [CrossRef]
8. Fan, X.; Gu, X.; Yu, H.; Long, A.; Tiando, D.S.; Ou, S.; Li, J.; Rong, Y.; Tang, G.; Zheng, Y.; et al. The Spatial and Temporal Evolution and Drivers of Habitat Quality in the Hung River Valley. *Land* **2021**, *10*, 1369. [CrossRef]
9. Petrișor, A.-I.; Mierzejewska, L.; Mitrea, A. Mechanisms of Change in Urban Green Infrastructure—Evidence from Romania and Poland. *Land* **2022**, *11*, 592. [CrossRef]
10. Němec, R.; Vymazalová, M.; Skokanová, H. The Impact of Fine-Scale Present and Historical Land Cover on Plant Diversity in Central European National Parks with Heterogeneous Landscapes. *Land* **2022**, *11*, 814. [CrossRef]
11. Zhang, J.; Zhang, Y.; Lloyd, H.; Zhang, Z.; Li, D. Rapid Reclamation and Degradation of *Suaeda salsa* Saltmarsh along Coastal China's Northern Yellow Sea. *Land* **2021**, *10*, 835. [CrossRef]

https://doi.org/10.3390/land15020203

2. Jitariu, V.; Dorosencu, A.; Ichim, P.; Ion, C. Severe Drought Monitoring by Remote Sensing Methods and Its Impact on Wetlands Birds Assemblages in Nuntași and Tuzla Lakes (Danube Delta Biosphere Reserve). *Land* **2022**, *11*, 672. [CrossRef]
3. Preda, M.; Vijulie, I.; Lequeux-Dincă, A.-I.; Jurchescu, M.; Mareci, A.; Preda, A. How Do the New Residential Areas in Bucharest Satisfy Population Demands, and Where Do They Fall Short? *Land* **2022**, *11*, 855. [CrossRef]
4. Vijulie, I.; Lequeux-Dincă, A.-I.; Preda, M.; Mareci, A.; Matei, E. Could Lavender Farming Go from a Niche Crop to a Suitable Solution for Romanian Small Farms? *Land* **2022**, *11*, 662. [CrossRef]

https://doi.org/10.3390/land15020203

land

MDPI

Article

Influence of Land Use on Avian Diversity in North African Urban Environments

Hani Amir Aouissi [1,2], Alexandru-Ionuţ Petrişor [3,4,*], Mostefa Ababsa [1], Maria Boştenaru-Dan [5], Mahmoud Tourki and Zihad Bouslama [2,7]

[1] Scientific and Technical Research Center on Arid Regions (CRSTRA), Biskra 07000, Algeria; aouissi.amir@gmail.com (H.A.A.); mustapha.ababsa@gmail.com (M.A.)
[2] Faculty of Sciences, Department of Biology, Badji-Mokhtar Annaba University, Annaba 23000, Algeria; ihadb@yahoo.fr
[3] Doctoral School of Urban Planning, Ion Mincu University of Architecture and Urbanism, 010014 Bucharest, Romania
[4] National Institute for Research and Development in Tourism, 50741 Bucharest, Romania
[5] Department for the Management of Research, Ion Mincu University of Architecture and Urbanism, 010014 Bucharest, Romania; Maria.Bostenaru-Dan@alumni.uni-karlsruhe.de
[6] Soil and Sustainable Development Laboratory, Faculty of Earth Sciences, Badji-Mokhtar Annaba University, Annaba 23000, Algeria; mahmoud.tourki23@gmail.com
[7] Environmental Research Center (CRE), Badji-Mokhtar Annaba University, Annaba 23000, Algeria
* Correspondence: alexandru.petrisor@uauim.ro; Tel.: +40-213-077-191

Abstract: Land cover and use changes are important to study for their impact on ecosystem services and ultimately on sustainability. In urban environments, a particularly important research question addresses the relationship between urbanization-related changes and biodiversity, subject to controversies in the literature. Birds are an important ecological group, and useful for answering this question. The present study builds upon the hypothesis according to which avian diversity decreases with urbanization. In order to answer it, a sample of 4245 observations from 650 sites in Annaba, Algeria, obtained through the point abundance index method, were investigated by computing Shannon-Wiener's diversity index and the species richness, mapping them, and analyzing the results statistically. The findings confirm the study hypothesis and are relevant for planning, as they stress the role of urban green spaces as biodiversity hotspots, and plead for the need of connecting them. From a planning perspective, the results emphasize the need for interconnecting the green infrastructure through avian corridors. Moreover, the results fill in an important lack of data on the biodiversity of the region, and are relevant for other similar Mediterranean areas. Future studies could use the findings to compare with data from other countries and continents.

Keywords: urbanization; diversity indices; Annaba; geo-statistical approaches; urban ecology; bird abundance

1. Introduction

The crucial importance of studying land cover and use changes (LCUC) results from their being part of the "global changes" [1], constituting a major component [2]. They transform natural into man-dominated systems [3], threatening biodiversity [4], affecting the quality of water, land and air resources, ecosystem processes, functions, services, equilibrium, and resilience, and making an important impact on the climate system [5–9]. For these reasons, studying LCUC makes an important contribution to the sustainable development debate [6,10]. LCUC are more intense when associated with land fragmentation [11], and constitute an important threat for coastal areas [12].

Some authors consider that modernization [13] particularly migration to urban centers [14], is an important driver of LCUC. A supporting argument is the fact that an important LCUC, urbanization, is now considered a major driving force of biodiversity loss and biological homogenization [15], generating a disproportionate share of environmental impacts compared to the total area affected by it [16]. The effects include habitat fragmentation [16–20], which in its turn influences species and biogeochemical cycles [18]. Urban sprawl is also a main threat to non-urbanized areas [16,21]. The Habitat II conference introduced the concept of "ecological footprint" to measure how much the new constructions, including infrastructure, affect the environment [22].

Over time, urban ecologists studied the ecology in the city, ecology of the city, and urban sustainability [23]; researches have classified the nature of cities in four categories: remains of natural systems, extensions of natural systems, landscaped or managed areas, and spontaneous, invasive, or ruderal species [24,25]. As a consequence of fragmentation, urban ecosystems are characterized by a built matrix embedding natural corridors and small, fragmented patches, and processes including succession and invasion [26–29].

While the overall influence of fragmentation on the urban ecosystem has been studied from a plethora of perspectives, its direct connection with biodiversity is still debated in the literature. Some authors associate fragmentation with a low biodiversity of small isolated patches [30,31], especially because species within these patches are more exposed to anthropogenic impacts [31,32] and have a reduced areal [33]. As a consequence, the size of patches is considered a good predictor of species richness [31,34]. Other authors found out that urban sprawl reduces species richness, but the abundances of some species might peak due to edge effects [27,34]. Moreover, even rare and endangered species can be preserved in urbanized habitats [26,35] when there is enough connectivity to provide corridors for certain species [34]. On an intermediate position, some authors consider that urban biodiversity depends on the spatial structure (understood as size of habitats, and distance between them) [19]. In addition, the relationship of different species with humans (hemerophobic, hemerodiaphoric, or hemerophile) plays an important role [36]. Roden [37] approaches the same issue as Mollov [36], differentiating between various species in the city and also highlighting the need for avian biodiversity. More specifically, for birds it is not so important to have continuous green spaces (assured by connecting green areas and through the presence of trees alongside of roads) but, because of their specific movement ways, and according to Angold et al. [38], they require diversity of green spaces. For this reason, it is important to phrase land use policies which provide for the need for biodiversity.

While it is impossible to study exhaustively all species, ecologists focus on some key groups. Birds are among the most studied groups in urban ecology [34,39], and avian species are known as excellent bio-indicators [40]. Moreover, the presence of birds in cities contributes to the soundscapes of cities, investigated in some studies [41–43]. The cited literature show that natural sounds are preferred by city inhabitants. The way the avian biodiversity contributes to the soundscape of the city will be subject to a future study, similar to [44], conducted in Europe by a network on the topic of COST TUD Action TD-0804 Soundscapes of European Cities and Landscapes. Moreover, bird species richness is highly variable across the urban-to-rural gradient [45]. The characteristics of rural individuals are different compared to their urban peers, both at species and population levels [46]. In this context, urban agriculture [47] deserves a special attention.

Ornithological studies investigating urban environments are scarce in North Africa, especially in Algeria, and even more precisely in the north of Algeria. Information is still incomplete, fragmentary, or even lacking, especially for some urban-adapted and urban-exploiter species [48,49]. In the area of Annaba, despite its ecological importance and localization (detailed in the Materials and Methods section), there are too few studies dealing with the avian diversity. The most noteworthy urban studies carried out in the area focused only on single species or groups, considering the phenotype of the Collared Dove (*Streptopelia decaocto*) [50], focusing on the health status of the Feral Pigeon (*Columba livia*) [51] or on the *Columbidae* [52,53].

The aim of this study is to investigate the controversial relationship between biodiversity and urbanization by analyzing the avian diversity in Annaba, Algeria along an urbanization gradient, using a novel geo-statistical approach. We hypothesize that diversity decreases towards the urban environment, along with the reduction of vegetated areas. Unlike the previous ones, this study investigates all avian species encountered during the field sampling.

2. Materials and Methods

This study relies on a novel geo-statistical approach, combining geospatial techniques with statistics. In a nutshell, field observations were mapped, combined with remote sensing data on land cover and use, and spatial analyses carried out based on the statistical processing of data.

2.1. Case Study: Annaba, Algeria

The study was carried out in and around the City of Annaba from the Wilaya of Annaba in Algeria. Figure 1 shows the study location in an international and national context, and also the location of sampling points.

Annaba city is the fourth largest city after Algiers, Oran, and Constantine. It is located between the latitude of 7°42′ and 7°48′ east, and the longitude of 36°50′ and 37°57′ north in the north east of Algeria at 100 km from the Algerian–Tunisian border in the east. The city is at 600 km from the capital Algiers, covers 1412 km^2, and its climate is typically Mediterranean, with an average annual temperature of 18 °C and an annual rainfall ranging from 650 to 1000 mm with a peak in winter and deficit in summer [54]. Annaba City is bordered in the north and west by the Massif of Edough, the Mediterranean Sea to the east and the alluvial plain of the Seybouse Wadi to the south. The average population in 2020 is estimated to 715,370 inhabitants throughout the region, including over 358,000 in the city of Annaba [55].

Annaba is one of the oldest cities in Algeria, founded in 1295 BC. The city was named successively Ubon, Hippo Regius, Hippone, Bouna, Bled El Aneb, Bône, and finally Annaba [56]. The antique city was place to the passage of several civilizations: Phoenicians, Punic, Numidian, Roman, and Byzantine. Nevertheless, the Roman remains represent the Roman phase that we see today: cultural buildings, villas, theaters, fountains, etc. [57]. The Arab-Muslim period (7th and 8th century) left as mark a part of the historic downtown of Annaba "Place d´arme" represented by capitols of the composite gallery, porticoes with triple section, wall niches, etc. [58]. In the French Colonial period, the neoclassical style made its appearance [47,59] to the point where the architectural and urban production in Algeria merges with that of France. Until contemporary times, from the 1970s, there was a production of collective buildings under the framework of a more collective architecture characterized by a modern style poor in decoration. This is found especially in the main hotels of the city, along with a post-modern style used especially for shopping centers, and dominated by arches and colors, as well as glass buildings (clinics, banks, administrations, etc.) [60].

Over time, the city of Annaba has experienced increased sprawl, green spaces making way to buildings. The city is mainly made up of buildings with very low vegetation in the surrounding area, and the current trend is to suppress vegetation in order to build homes, commercial places, etc., mainly for economic reasons [61]. There are several predominantly wild forest areas in the periphery, such as those around "La basilique de Saint Augustin", the "19 May 1945" Stadium), the Sidi-Achour road, and especially the mountainous region of "Séraïdi". However, there are fully open parks in the downtown: Cours de la revolution, Placette Alexis Lambert, and others delimited: Boukhtouta Hocine Garden, The Squares of Edough, as well as cemeteries, etc. The most important site from an ecological viewpoint is the Christian Cemetery, which is a closed space inaccessible to the vast majority of population with an area of 6.5 ha, and the largest park in an urban environment, characterized by a substantial avian diversity [62].

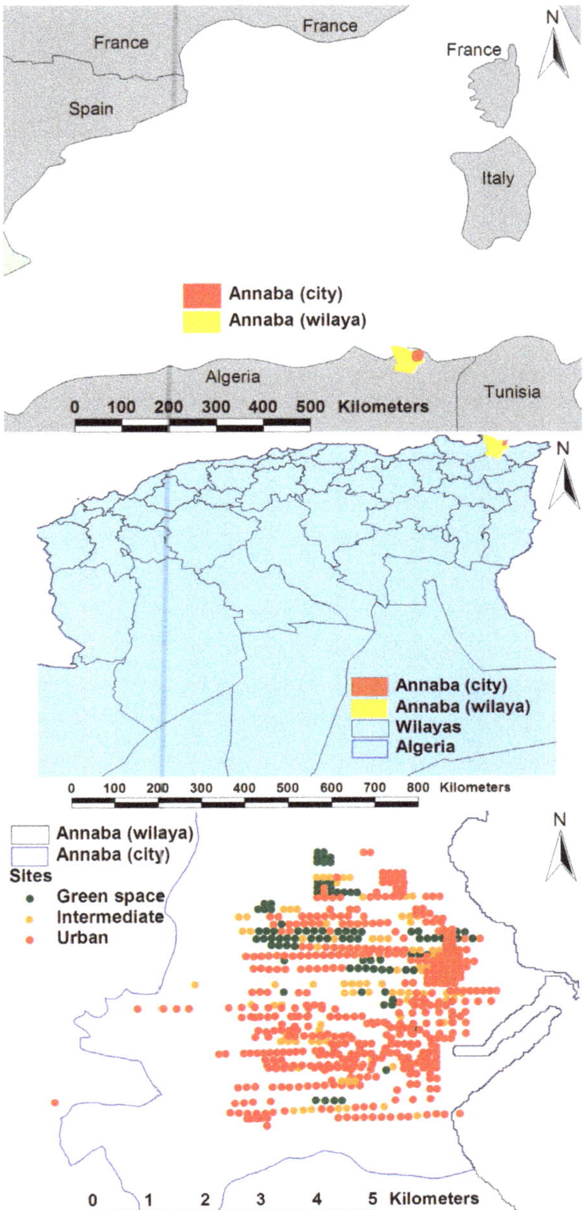

Figure 1. Location of the case study in an international (**top**) and national context (**middle**), and of the sampling sites (**bottom**).

2.2. Data

The data used in this research comes from two sources: field observations and remote sensing data.

Field observations were acquired during the period of 2017–2018 from 650 sites, classified, following the methodology of Aouissi et al. [63,64], based on the land cover and use data, as "green space"(characterized by the dominance of vegetation, including parks, cemeteries, and wooded green spaces, 92 sites), "intermediate" (sites where the presence of trees and greenery was noted, but with adjacent built areas, 92 sites), and "urban" (where the built up area is predominant, and without green spaces, 466 sites). There were 4245 raw observations, each line including the name of species, number of individuals, coordinates of site, date of observation, type of site, presence of vegetation (%), and whether the observation was made during the reproductive season or not.

We used the method of IPA ("indice ponctuel d'abondance"/point abundance index) [65]. It consists, for an observer, of staying motionless for 15 minutes and recording all individual birds heard or seen [66]. The sampling was performed twice by a single observer (H.A.A) on all 650 sites, previously identified cartographically using GPS. The first sampling was carried out at the beginning of spring and during summer (between 7 February and 1 September), in order to catch the presence of sedentary and breeding migratory species during the breeding season. The second sampling was carried out later on in the season (between 5 September and 1 February), outside of the breeding season. Data was collected on each site twice a day: early in the morning (approximately between 5:00 and 7:00 GMT + 1 in summer or between 6:00 and 8:00 GMT + 1 in winter) and before sunset (approximately between 17:00 and 19:00 GMT + 1 in winter or 18:00 to 20:00 GMT + 1 in summer). In order to avoid overlapping, a minimum distance of 100 m was maintained between any two sites.

Spatial data was derived from satellite imagery, overlaying the distribution of sites with the image and determining the land cover and use for each site according to the classification presented above. The satellite imagery and geographical data, including the administrative limits, was obtained from the Cartography Department of the National Hydraulic Basins Agency (ABH CSM) "Agence de Bassin Hydrographique Constantinois-Seybousse-Mellegue" [67] by M.T.

2.3. Data Processing and Analysis

Data was analyzed using the General Linear Models to find the influence of different factors, including year and month of collection, presence of vegetation, and whether the observation was made during the reproductive season or not, on the number of individuals from each species.

Following the statistical analysis, it was found that since only 323 of the total 4245 observations were from 2018, the year did not significantly influence the avian diversity, and the model could not run. As a result, during the next step data was aggregated at the site level, combining all observations for the same species from both years. The resulting data set had 2738 observations, each line containing the site, species, and number of individuals. This data set was used to compute for each site separately two indices of diversity: the species richness, defined as the total number of species per site, and Shannon-Wiener's informational entropy index, computed using Equation (1):

$$H = \sum_{n}^{i=1} (-1) \times p_i \times \ln(p_i) \tag{1}$$

where n is the total number of species, and p_i is the share of individuals from species i from the total number of individuals per site.

The choice of the two indices was motivated by their ability to compare the diversity of different sites [68] or of the same site across time [69] and visualize the distribution in a spatial perspective, when used in conjunction with geo-statistical approaches [70].

The results were analyzed statistically, in order to determine the relationship between land cover and use and biodiversity, using the analysis of variance (ANOVA) followed by three post-hoc comparison tests: Tukey, Bonferroni, and Scheffé. The statistics were computed using Microsoft Excel 2003 (Microsoft, Spring Valley, CA, USA).

The geospatial analyses examined the spatial distribution of biodiversity, assessed using the species richness and Shannon-Wiener's informational entropy index, against land cover and use. For this purpose, ArcGIS10.X (Redlands, CA, USA) was used for the spatial interpolation of the two biodiversity values via simple kriging [71], preferred due to its intuitiveness and broad use. The land cover surfaces were also derived by interpolation based on assigning proximities with the Spatial Analyst extension of ArcView GIS 3.X (Redlands, CA, USA).

3. Results

Overall, all results are attempting to answer the question whether avian biodiversity depends on land cover and use, testing its variation across a gradient of urbanization.

3.1. Overall View of the Sample

The analysis of the 4245 observations revealed the presence of 28 bird species. Table 1 presents the species identified, showing the total number of individuals, its average numerical distribution per site, and number of sites where the species was found.

Table 1. Species found in Annaba, Algeria during 2017–2018. The table shows the total and average number per site and the number of sites where each species was present.

Species	Total No.	Average No. per Site	No. of Sites Present
Apus apus	108	3.72	29
Bubulcus ibis	10	1.11	9
Chloris chloris	117	1.65	71
Chroicocephalus ridibundus	167	4.64	36
Columba livia	4424	8.27	535
Columba oenas	1	1.00	1
Columba palumbus	1	1.00	1
Cyanistes caeruleus	86	1.62	53
Cyanistes teneriffae	67	2.23	30
Delichon urbicum	66	2.20	30
Erithacus rubecula	50	1.19	42
Falco tinnunculus	62	1.29	48
Fringilla coelebs	161	1.96	82
Goeland leucopée	102	3.09	33
Hirundo rustica	329	7.65	43
Muscicapa striata	112	1.33	84
Parus major	84	1.50	56
Passer domesticus	1763	5.23	337
Phoenicurus ochruros	113	1.36	83
Phylloscopus collybita	132	1.94	68
Pycnonotus barbatus	116	1.38	84
Serinus serinus	114	1.90	60
Spilopelia senegalensis	117	2.72	43
Streptopelia decaocto	5745	9.28	619
Streptopelia turtur	74	3.36	22
Sturnus vulgaris	158	6.32	25
Sylvia atricapilla	189	2.08	91
Turdus merula	517	4.17	124

The view of the entire sample indicates a diverse presence, with clear outliers: the Feral Pigeon (*Columba livia*) and the Collared Dove (*Streptopelia decaocto*) were the most numerous, and also present in most sites in large numbers. The Stock Dove (*Columba oenas*) and the Common Wood Pigeon (*Columba palumbus*) were at the opposite extreme, each found only once in a single site (note that these species were completely absent from our surveys during previous counts). The number of individuals per site also depends on the behavior of species, as some of them tend to group (e.g., *Sturnus vulgaris*, *Passer domesticus*), and others to spread out (for example, the predator *Falco tinunculus*).

The results of statistical analysis of the overall data set are presented in Table 2. The table combines two General Linear Models; the first one, labeled "full model", includes all variables, and the second, labeled "prediction model", only those statistically significant. The first model was overall statistically significant, with F = 12.83 and $p < 0.0001$; it had an adjusted R^2 of 0.043524. The second model was overall statistically significant, with F = 13.71 and $p < 0.0001$; it had an adjusted R^2 of 0.043419. The table indicates that the month of observation, land cover and use, and presence of vegetation have a statistically significant influence on the number of individuals from each species.

Table 2. General Linear Models examining the relationship between the number of individuals from each bird species found in Annaba, Algeria during 2017–2018 and the variables with a potential influence on it.

Variable	Full Model					Prediction Model				
	DF	Type III SS	Mean Square	F Value	Pr > F	DF	Type III SS	Mean Square	F Value	Pr > F
Month	11	1021.83	92.89	8.60	<0.0001	11	1175.08	106.83	9.90	<0.0001
Land cover and use	2	724.48	362.24	33.55	<0.0001	2	727.05	363.52	33.68	<0.0001
Presence of vegetation	1	103.29	103.29	9.57	0.0020	1	102.86	102.86	9.53	0.0020
Reproductive season	1	5.02	5.02	0.47	0.4953	—	—	—	—	—

3.2. Relationship Between Biodiversity and Land Cover and Use

The results of the analysis of variance (ANOVA) indicate that there are statistically significant differences between the three types of sites ("green space", "intermediate", and "urban") with respect to species richness (F = 440.77, $p < 0.0001$) and Shannon-Wiener's informational entropy index (F = 177.17, $p < 0.0001$). The results were confirmed by all three post-hoc tests, which found significant differences between the values of the two indicators among each possible pairs of sites. Figure 2 presents the average values of the two indices, indicating that the underlying hypothesis of the study is verified by the statistical analysis.

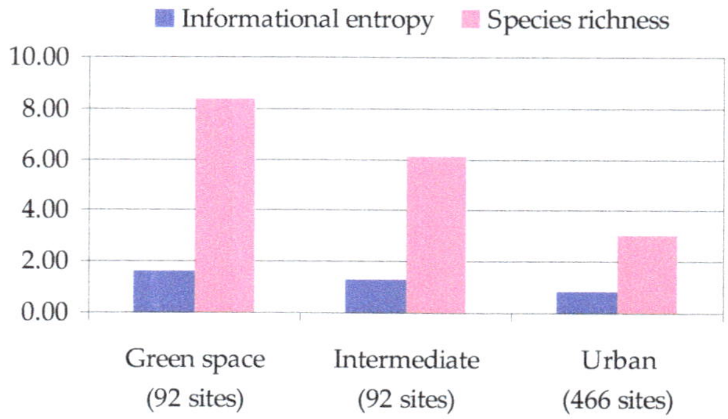

Figure 2. Distribution of Shannon-Wiener's informational entropy index and species richness across the gradient of urbanization in Annaba, Algeria. The image shows that biodiversity tends to increase when moving from the built-up area towards the natural habitats.

The same results were obtained from geospatial analysis. Figure 3 presents the spatial distribution of Shannon-Wiener's informational entropy index and species richness, interpolated from the sampling sites; it can be seen that green spaces and intermediate areas are located in areas with higher diversity.

In Figure 4, the same distribution is overlaid with the land cover and use derived from the spatial interpolation of sampling sites. The image shows that areas with high biodiversity (dark shades) overlap with green spaces and intermediate areas.

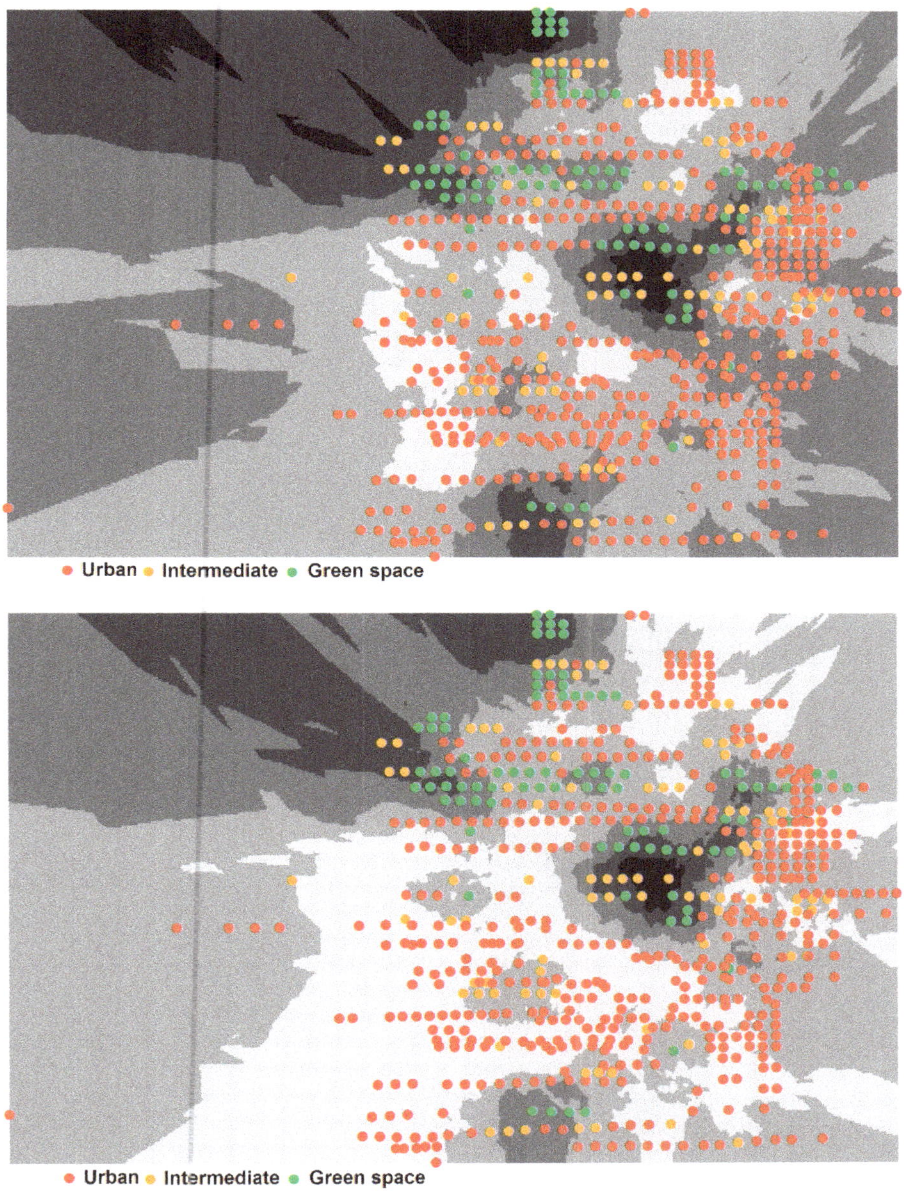

Figure 3. Interpolated spatial distribution of Shannon-Wiener's informational entropy index (**top**) and species richness (**bottom**) in Annaba, Algeria. The darker shades correspond to higher values of the two indices, indicating higher biodiversity.

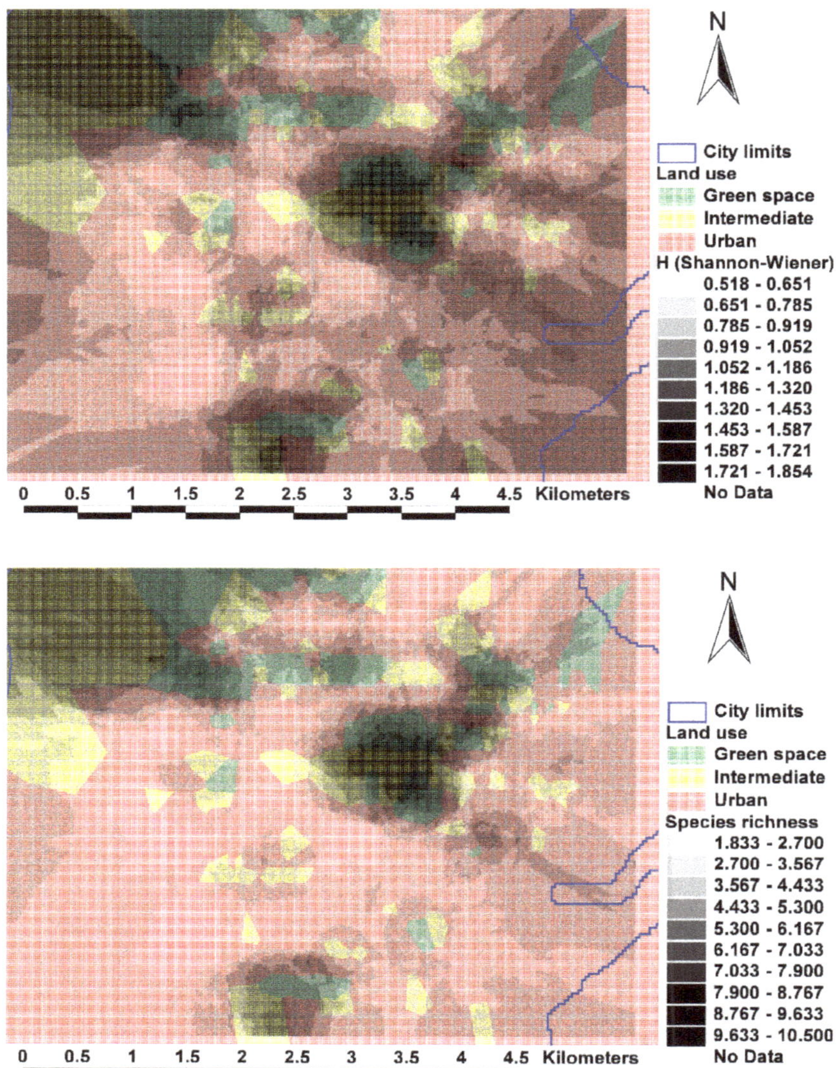

Figure 4. Interpolated spatial distribution of Shannon-Wiener's informational entropy index (**top**) and species richness (**bottom**) and interpolated land cover and use in Annaba, Algeria. The darker shades correspond to higher values of the two indices, indicating higher biodiversity.

4. Discussion

4.1. Significance of the Findings and Comparison with Similar Studies

This study aimed to test the hypothesis according to which biodiversity is inversely proportional with the degree of urbanization. The statistical analyses confirmed the hypothesis, showing than the biodiversity of green spaces is significantly greater than the one of intermediate sites, which in its turn is significantly greater than the one of urban sites. These results bring additional evidence to the fact that biodiversity is negatively influenced by urbanization [30,31], leaning the balance of the existing controversies in this direction. At the same time, spatial analyses cast a new light on the potential implications, especially in relationship to planning, discussed in the next section. Last but not least, the results have a special importance from a methodological perspective, indicating the importance of combining statistical approaches with geo-spatial techniques in order to get a detailed picture of the spatial distribution of biodiversity.

Several similar studies were carried out in other regions. In a study carried out in Mar del Plata City, southeastern Buenos Aires Province, Argentina [72], 39 species were counted; the most abundant was the Feral Pigeon, and there was no relationship between the level of urbanization and stability of bird species richness. Another study considered three Swiss cities: Zurich, Lucerne, and Lugano [73]; in 96 sampling points, 63 species were found, generally outside of the cities, while in highly urbanized areas, only few were found, namely *Columba livia* and *Passer domesticus*, which is very similar to our results. The study showed that bird species richness and diversity were negatively affected by increasing sealed area or buildings, and positively influenced by increasing vegetation structures, in particular trees. A bird count was carried out in Algeria using the same method in a much less urbanized area (Ziban region); 42 species were found, most of them from the Turdidae family [74].

The relationship between avian biodiversity and urbanization in the Mediterranean climate has been addressed by Di Pietro et al. [75] and Vignoli et al. [76], both investigating it in the metropolitan area of Rome. The first study, focusing on breeding birds, showed that the effects of urbanization are little investigated in other species-area relationship studies in urban ecology. The second study, based on 69 breeding bird species, compared the urban–rural gradient between the core of metropolis and surrounding area, subdividing the city in cells classified according to landscape, and the highest heterogeneity was found for the rural gradient. The authors explicitly called for studies in other cities in order to validate their findings. We consider these studies relevant for comparison due to being carried out in the same climate region as in Algeria and taking into account the reference to the green planting map of Rome. The 2013 Le Notre Landscape Forum in Rome also dealt with the urban periphery, which is worthy considering in the context of urban sprawl [77]. Vignoli et al. [76] looked specifically at this periphery as a heritage of urban villas in a city inhabited for millennia. The decrease of biodiversity was observed at point count and landscape scale. The study showed that different species exhibit different trends in the rural-urban gradient. Although the agricultural land was richer in species, this is also true for urban villas, now part of the city, which are remnants of the former countryside and have large gardens. This sustains our findings regarding the relationship between the green spaces and avian biodiversity, even if the green spaces are disconnected. The conclusion of Vignoli et al. [76] is that green areas in the city (such as the gardens of the villas) provide avian biodiversity to the inner-city areas. In Rome, they are replacing parks (for example the villa Borghese garden), with implications calling for the proper conservation of these areas.

Our study is one of only few such studies that have been conducted in Northern Africa. Provided that Algeria in general remains very little studied, and the study region specifically, this study fills in an important gap in the knowledge of North African avian biodiversity, urban environment in general, and the biodiversity related to this type of habitats is often overlooked. In the majority of cases, studies were carried out only outside of urbanized areas until recently, when an interest in the city started developing, as evidenced by the study of Belabed-Zediri [78], focused on water birds from urban areas. The idea that cities are regarded as deserts from an ornithological point of view tends to be abandoned over time [72,79].

In our study, even if the results show a decrease in bird diversity along the urbanization gradient, they also show an important richness of green spaces within the urban areas, and the presence of same species found outside the city. A high concentration of diversity and abundance in the environments with significant vegetation should be noted, testifying to the importance of green spaces in urban environments, further discussed in the next section in relationship to planning.

4.2. The Planning Perspective

The planning consequences rely on previous studies. Planners are considered responsible for developing a strategy for urban green infrastructure [30,68,80]; even taking it into account in planning and management can increase urban sustainability [81–87], by turning the vicious circle determined by its fragmentation into a virtuous one [88] when creating corridors to link natural patches. The strategy must be built on the premise that conservation should protect networks instead of parcels [30]; this is vital to maintaining ecosystem services [26]. The possible solutions include restoring connectivity [89] or buffering natural areas [32], and linking protected areas or parks [17]. Most important, the strict enforcement of planning provision makes an important contribution to the conservation of biodiversity, in opposition to any deviation from it by the means of derogatory planning [12,90].

In the particular case of this study, the lesson according to which biodiversity decreases along the urbanization gradient implies a need for solutions aimed at connecting the biodiversity hotspots, i.e., green spaces. Provided that the "intermediate" sites also show a high biodiversity, significantly higher than the urban ones, they can be used as links between the green spaces, ensuring the movement of species across sites.

In Northern Europe a decline was observed in the sparrow population, despite many environmental restoration activities (e.g., unsealing the ground through permeable plates in parking places etc.) which is highlighted by the literature [91]. This underlines that the urbanization influence on the decline of avian biodiversity is not reversible if the area undergoes ecological restoration afterwards [73], and thus the efforts to preserve the initial ecological status have to be done. This was also present in the COP21 prescriptions, and caused citizen participation movements in Northern Europe against transport infrastructure extension around cities, which can lead to the loss of biodiversity in cities and the agricultural land around them [92]. The conflicts in urban management between stakeholders are not to be decoupled from the energy discussion by COP21, but include the debate on the neighborhoods at the city periphery with single family housing. The Algerian urban model follows, as stated earlier, the French Western European urban model with urban sprawl occurring in the periphery through the emergence of new neighborhoods in the 20th century. Instead, the "house with garden" model, also of French influence (see also [93]), should be preserved to provide not only for connections, but also for the diversity of green spaces. Currently low-rise housing and green spaces are endangered by speculative building. Preserving the green areas of the city and recycling the historic building stock also helps moving to a circular economy and countering the urban heat islands, with a positive effect on vegetation and biodiversity.

Few other studies considered the relationship with planning practices. In Canoas, the metropolitan area of Porto Alegre (Brazil), 100 species were found, mostly doves; their number was negatively affected by the increase of urbanization (and noise level), but the share of green areas had a positive effect. Nevertheless, the study found that the presence of particular habitats (wetlands, grasslands, woodlots), patchily distributed in the urban matrix, could buffer the effects of urbanization on birds, and recommended taking them into account in urban planning. The study recommended supporting citizens in maintaining residential vegetation (e.g., private yards), and, hence, keeping native vegetation areas inside the urban area, to increase the green areas and promote biological conservation [94]. In Valdivia (Chile), green spaces had a positive impact on the distribution and diversity of 32 bird species; various categories of green space can have very different effects even exerting a negative influence, such as the municipal green areas. Municipal green spaces, designed and maintained for recreational purposes, were found more homogeneous with respect to land cover and vertical heterogeneity when compared to non-municipal green areas. The authors recommended planners pay a special attention to municipal green areas, where habitat quality for birds can be improved by reducing the impervious surface and creating and conserving a multi-layered vegetation structure, and preserve non-municipal green areas, including wetlands critically threatened by urban development [95]. In the municipalities of Vancouver and Burnaby (British Columbia), 25 common species were found; the results suggested that both local and landscape-scale resources were important in determining the distribution of birds in urban areas. They recommended the integration of parks, reserves, and surrounding residential areas into urban planning to preserve the diversity of resident avifauna and overall species diversity. Also, development on the verge of continuously forested areas was found useful to minimize the impervious surface cover and house size, maintain native tree and berry shrubs, integrate new ponds, and preserve and develop natural freshwater sources. Residential areas near parks were found to be more likely to recruit sensitive nesting species and probably experience frequent use by species from nearby parks [19].

4.3. Future Research Directions

Our study could be extended to explore biodiversity indicators for many dimensions; we could, for example, verify the stability of the explicative model of biodiversity during our period (2017–2018), previously, and for the future periods using several statistical tests such as the Chow Test, etc. However, currently it is difficult to do investigative work, in particular because of the epidemiological crisis due to the COVID-19 pandemic, especially in Algeria [96].

Future research could also explore more explicative factors of biodiversity, such as industry, agricultural activity, demographic growth, ecological indicators, etc. Urban agriculture can be a determining factor. In Algeria [97], but also in the European Union [98], attention is given, apart of green spaces, to animal species in urban agriculture. Therefore, not only recreation areas, but also production areas can connect the green areas of the city with the peri-urban ones.

It would be interesting to extend the study area to several cities, and make comparisons, for example, between semi-arid and arid areas of Algeria; this may help clarifying better the relationship between urbanization and bird communities. At least, especially in North African and Middle-East regions, this kind of studies remain scarce and all additional data can be considered important. A comparison can be made between areas around the Mediterranean, including the European ones.

The debate can also be carried towards the role of individual dwellings versus collective housing. In each case, adequate green public space has to be assured. In cities which do not have historical fortifications, individual homes with gardens proliferate. On the other hand, the spread of the International Style favored the construction of blocks of flats in the green areas. These green spots, rarely mapped, contribute to linking green spaces. A representative project was carried out in Rome, Italy to map green spaces [99]. A similar study can be developed based on an urban analysis.

5. Conclusions

This study has explored the hypothesis according to which biodiversity, and in particular avian diversity, decreases from the urban towards the natural areas. The hypothesis was confirmed by the findings, providing some important insight from a planning perspective. First, urban green spaces are important biodiversity hotspots, and second, connectivity is important to link them and permit the migration of birds from one area to another. From a planning perspective, this study brings additional evidence for the need for connecting the green infrastructure, enabling corridors for the avian fauna. At the same time, the research is relevant for its region, as the North-African biodiversity is insufficiently explored in the literature, and also for the coastal Mediterranean environments. Future studies could make use of the data and results and check whether the patterns are comparable with those from Europe and elsewhere. Last but not least, the study underlines the potential of diversity indices, such as Shannon-Wiener or species richness, for being used in a spatial setting, with outcomes relevant both for planning and scientific purposes.

Author Contributions: Conceptualization, H.A.A., A.-I.P. and M.B.-D.; methodology, H.A.A. and A.-I.P.; software, A.-I.P. and M.T.; validation, A.-I.P., M.B.-D. and Z.B.; formal analysis, A.-I.P. and M.A.; investigation, H.A.A.; resources, H.A.A. and M.T.; data curation, A.-I.P. and M.A.; writing—original draft preparation, H.A.A., A.-I.P. and M.B.-D.; writing—review and editing, H.A.A., A.-I.P. and M.B.-D.; visualization, H.A.A., A.-I.P. and M.B.-D.; supervision, A.-I.P., M.B.-D. and Z.B. All authors have read and agreed to the published version of the manuscript.

Funding: This research was funded by the DGRSDT and MESRS (Algerian Ministry).

Data Availability Statement: The data presented in this study are collected within a research project and can be made available on request from the corresponding author.

Acknowledgments: Many thanks are addressed to Samir Merouana for his assistance in data entry, Anis Aouissi and Aissam Gaagai for their help with GIS, and the DGRSDT and MESRS for their support.

Conflicts of Interest: The authors declare no conflict of interest. The funders had no role in the design of the study; in the collection, analyses, or interpretation of data; in the writing of the manuscript; or in the decision to publish the results.

References

1. Dale, V.H.; Efroymnson, R.A.; Kline, K.L. The land use–Climate change–Energy nexus. *Landsc. Ecol.* **2011**, *26*, 755–773. [CrossRef
2. Lambin, E.F.; Meyfroidt, P. Global land use change, economic globalization, and the looming land scarcity. *Proc. Nat. Acad. Sc. USA* **2011**, *108*, 3465–3472. [CrossRef]
3. Elias, P.; Dekolo, S.; Babatola, O. Land use change modelling in developing countries: Issues and prospects. *Int. J. Geogr. Geol.* **2012**, *1*, 23–41.
4. Hansen, A.J.; Rasker, R.; Maxwell, B.; Rotella, J.J.; Johnson, J.D.; Wright Parmenter, A.; Langner, U.; Cohen, W.B.; Lawrence, R.; Kraska, M.P.V. Ecological Causes and Consequences of Demographic Change in the New West. *BioScience* **2002**, *52*, 151–1[illegible] [CrossRef]
5. Ojima, D.S.; Galvin, K.A.; Turner, B.L., II. The Global Impact of Land-Cover Change. *BioScience* **1994**, *44*, 300–304. [CrossRef]
6. Lambin, E.F.; Rounsevell, M.D.A.; Geist, H.J. Are agricultural land-use models able to predict changes in land-use intensity *Agric. Ecosys. Environ.* **2000**, *82*, 321–331. [CrossRef]
7. Metzger, M.J.; Rounsevell, M.D.A.; Acosta-Michlik, L.; Leemans, R.; Schröter, D. The vulnerability of ecosystem services to land use change. *Agric. Ecosys. Environ.* **2006**, *114*, 69–85. [CrossRef]
8. Li, R.-Q.; Dong, M.; Cui, J.-Y.; Zhang, L.-L.; Cui, Q.-G.; He, W.-M. Quantification of the Impact of Land-Use Changes on Ecosystem Services: A Case Study in Pingbian County, China. *Environ. Monit. Assess.* **2007**, *128*, 503–510. [CrossRef] [PubMed]
9. Pelorosso, R.; Gobattoni, F.; La Rosa, D.; Leone, A. Ecosystem Services based planning and design of Urban Green Infrastructure for sustainable cities. In *Atti della XVIII Conferenza Nazionale SIU. Italia '45-'45. Radici, Condizioni, Prospettive, Venezia 11–13 giugno 2015*; Munarin, S., Carta, M., Eds.; Planum Publisher: Rome, Italy, 2015; pp. 763–769.
10. Reidsma, P.; König, H.; Feng, S.; Bezlepkina, I.; Nesheim, I.; Bonin, M.; Sghaier, M.; Purushothaman, S.; Sieber, S. van Ittersum, M.K.; et al. Methods and tools for integrated assessment of land use policies on sustainable development in developing countries. *Land Use Policy* **2011**, *28*, 604–617. [CrossRef]
11. Müller, D.; Kuemmerle, T.; Rusu, M.; Griffiths, P. Lost in transition: Determinants of post-socialist cropland abandonment in Romania. *J. Land Use Sci.* **2009**, *4*, 109–129. [CrossRef]
12. Petrişor, A.-I.; Hamma, W.; Nguyen, H.D.; Randazzo, G.; Muzirafuti, A.; Stan, M.-I.; Tran, V.T.; Aştefănoaiei, R.; Bui, Q.-T.; Vintilă, D.-F.; et al. Degradation of Coastlines under the Pressure of Urbanization and Tourism: Evidence on the Change of Land Systems from Europe, Asia and Africa. *Land* **2020**, *9*, 275. [CrossRef]
13. Bičík, I.; Janoušek, Z.; Kabrda, J. Czechia: Changes in landscape use in the transformation period. *Rom J. Geogr.* **2015**, *59*, 3–17.
14. Izquierdo, A.E.; De Angelo, C.D.; Aide, T.M. Thirty years of human demography and land-use change in the Atlantic Forest of Misiones, Argentina: An evaluation of the forest transition model. *Ecol. Soc.* **2008**, *13*, 3. [CrossRef]
15. Pauchard, A.; Aguayo, M.; Peña, U.R. Multiple effects of urbanization on the biodiversity of developing countries: The case of a fast-growing metropolitan area (Concepción, Chile). *Biol. Conserv.* **2006**, *127*, 272–281. [CrossRef]
16. Irwin, E.G.; Bockstael, N.E. The evolution of urban sprawl: Evidence of spatial heterogeneity and increasing land fragmentation. *Proc. Nat. Acad. Sci. USA* **2007**, *104*, 20672–20677. [CrossRef] [PubMed]
17. McMahon, E.T. Green Infrastructure. *Plan Comm. J.* **2000**, *37*, 4–7.
18. Luck, M.; Wu, J. A gradient analysis of urban landscape pattern: A case study from the Phoenix metropolitan region, Arizona, USA. *Landsc. Ecol.* **2002**, *17*, 327–339. [CrossRef]
19. Melles, S.; Glenn, S.; Martin, K. Urban bird diversity and landscape complexity: Species—Environment associations along a multiscale habitat gradient. *Ecol. Soc.* **2003**, *7*, 5. [CrossRef]
20. Kabisch, N.; Haase, D. Green spaces of European cities revisited for 1990–2006. *Landsc. Urban Plan.* **2013**, *110*, 113–122. [CrossRef

21. La Greca, P.; La Rosa, D.; Martinico, F.; Privitera, R. Agricultural and green infrastructures: The role of non-urbanised areas for eco-sustainable planning in a metropolitan region. *Environ. Pollut.* **2011**, *159*, 2193–2202. [CrossRef] [PubMed]
22. Boştenaru-Dan, M. Management of innovation: Innovation policy in the urban development. *Urban. Archit. Constr.* **2013**, *4*, 3–18.
23. Wu, J. Urban ecology and sustainability: The state-of-the-science and future directions. *Landsc. Urban Plan.* **2014**, *125*, 209–221. [CrossRef]
24. Qureshi, S.; Breuste, J.H. Prospects of Biodiversity in the Mega-City of Karachi, Pakistan: Potentials, Constraints and Implications. In *Urban Biodiversity and Design*, 1st ed.; Müller, N., Werner, P., Kelcey, J.G., Eds.; Blackwell: Chichister, UK, 2010; pp. 497–517.
25. Breuste, J.H.; Qureshi, S.; Li, J. Scaling down the ecosystem services at local level for urban parks of three megacities. *Hercynia N. F.* **2013**, *46*, 1–20.
26. Niemelä, J. Ecology and urban planning. *Biodivers Conserv.* **1999**, *8*, 119–131. [CrossRef]
27. Grimm, N.B.; Faeth, S.H.; Golubiewski, N.E.; Redman, C.L.; Wu, J.; Bai, X.; Briggs, J.M. Global Change and the Ecology of Cities. *Science* **2008**, *319*, 756–760. [CrossRef] [PubMed]
28. Taylor Lovell, S.; Taylor, J.R. Supplying urban ecosystem services through multifunctional green infrastructure in the United States. *Landsc. Ecol.* **2013**, *28*, 1447–1463. [CrossRef]
29. Andersson, E.; Barthel, S.; Borgström, S.; Colding, J.; Elmqvist, T.; Folke, C.; Gren, Å. Reconnecting Cities to the Biosphere: Stewardship of Green Infrastructure and Urban Ecosystem Services. *Ambio* **2014**, *43*, 445–453. [CrossRef]
30. Benedict, M.A.; McMahon, E.T. *Green Infrastructure: Smart Conservation for the 21st Century*; Sprawl Watch Clearinghouse Monograph Series: Washington, DC, USA, 2001.
31. Gibb, H.; Hochuli, D.F. Habitat fragmentation in an urban environment: Large and small fragments support different arthropod assemblages. *Biol. Conserv.* **2002**, *106*, 91–100. [CrossRef]
32. Marzluff, J.M.; Ewing, K. Restoration of Fragmented Landscapes for the Conservation of Birds: A General Framework and Specific Recommendations for Urbanizing Landscapes. *Restor. Ecol.* **2001**, *9*, 280–292. [CrossRef]
33. McKinney, M.L. Effects of urbanization on species richness: A review of plants and animals. *Urban Ecosyst.* **2008**, *11*, 161–176. [CrossRef]
34. Fernández-Juricic, E.; Jokimäki, J. A habitat island approach to conserving birds in urban landscapes: Case studies from southern and northern Europe. *Biodivers Conserv.* **2001**, *10*, 2023–2043. [CrossRef]
35. McKinney, M.L. Urbanization, Biodiversity, and Conservation. *BioScience* **2002**, *52*, 883–890. [CrossRef]
36. Mollov, I.A. Level of Synanthropy of the Amphibians and Reptiles from the City of Plovdiv (Bulgaria). *Ecol. Balk.* **2014**, *6*, 109–112.
37. Roden, V. *Urbane Biodiversität als Städtebaurechtliches Nachhaltigkeitskonzept*; Mohr Siebeck: Tübingen, Germany, 2017; p. 389.
38. Angold, P.G.; Sadler, J.P.; Hill, M.O.; Pullin, A.; Rushton, S.; Austin, K.; Small, E.; Wood, B.; Wadsworth, R.; Sanderson, R.; et al. Biodiversity in urban habitat patches. *Sci. Total Environ.* **2006**, *360*, 196–204. [CrossRef]
39. Marzluff, J.M. Worldwide urbanization and its effects on birds. In *Avian Ecology and Conservation in an Urbanizing World*; Marzluff, J.M., Bowman, R., Donnelly, R., Eds.; Springer: Boston, MA, USA, 2011; pp. 19–47.
40. Fraixedas, S.; Lindén, A.; Piha, M.; Cabeza, M.; Gregory, R.; Lehikoinen, A. A state-of-the-art review on birds as indicators of biodiversity: Advances, challenges, and future directions. *Ecol. Indic.* **2020**, *118*, 106728. [CrossRef]
41. Liu, J.; Kang, J.; Luo, T.; Behm, H. Landscape effects on soundscape experience in city parks. *Sci. Total Environ.* **2013**, *454–455*, 474–481. [CrossRef]
42. Yang, W.; Kang, J. Soundscape and Sound Preferences in Urban Squares: A Case Study in Sheffield. *J. Urban Des.* **2005**, *10*, 61–80. [CrossRef]
43. Rehan, R.M. The phonic identity of the city urban soundscape for sustainable spaces. *HBRC J.* **2016**, *12*, 337–349. [CrossRef]
44. Yorukoglu, P.N.D.; Ustun Onur, A.Z. Semiotic interpretation of a city soundscape. *Semiotica* **2019**, *226*, 73–87. [CrossRef]
45. Pal, M.; Pop, P.; Mahapatra, A.; Bhagat, R.; Hore, U. Diversity and structure of bird assemblages along urban-rural gradient in Kolkata, India. *Urban For. Urban Green.* **2019**, *38*, 84–96. [CrossRef]
46. Chenchouni, H. Variation in White Stork (Ciconia ciconia) diet along a climatic gradient and across rural-to-urban landscapes in North Africa. *Int. J. Biometeorol.* **2017**, *61*, 549–564. [CrossRef] [PubMed]
47. Panagopoulos, T.; Jankovska, I.; Boştenaru-Dan, M. Urban green infrastructure: The role of urban agriculture in city resilience. *Urban. Archit. Constr.* **2018**, *9*, 55–70.
48. Bendjoudi, D.; Chenchouni, H.; Doumandji, S.; Voisin, J.F. Bird species diversity of the Mitidja Plain (Northern Algeria) with emphasis on the dynamics of invasive and expanding species. *Acrocephalus* **2013**, *34*, 13–26. [CrossRef]
49. Bendjoudi, D.; Voisin, J.F.; Doumandji, S.; Merabet, A.; Benyounes, N.; Chenchouni, H. Rapid increase in numbers and change of land-use in two expanding Columbidae species (Columba palumbus and Streptopelia decaocto) in Algeria. *Avian Res.* **2015**, *6*, 18. [CrossRef]
50. Belabed, A.I.; Aouissi, H.A.; Zediri, H.; Djemadi, I.; Driss, K.; Houhamdi, M.; Bouslama, Z. The effect of urbanization on the phenotype of the Collared Dove (Streptopelia decaocto) in northeastern Algeria. *Bull. Inst. Sci. Rabat. Sect. Sci. Vie* **2014**, *35*, 155–164.
51. Bendjoudi, D.; Marnich, F.; Messaoudi, Z. Premières données sur les parasites chez deux espèces de Columbidés, la tourterelle turque Streptopelia decaocto et le pigeon biset Columba livia. *Rev. Agrobiol.* **2018**, *8*, 809–816.
52. Absi, K.; Farhi, Y.; Halis, Y.; Belhamra, M. A comparison of the reproduction of collared doves Streptopelia decaocto and turtle doves Streptopelia turtur in the Ziban Oases (Biskra, Algeria). *J. Entomol. Zool. Stud.* **2015**, *2*, 1–6.

53. Saâd, N.; Hanane, S.; Khemis, M.D.E.H.; Farhi, K. Landscape composition governs the abundance patterns of native and invasive Columbidae species along an urban–rural gradient and contribute to their partitioning. *Biol. Invasions* **2021**, *1*, 1–15.
54. Ouali, N.; Belabed, B.E.; Chenchouni, H. Modelling environment contamination with heavy metals in flathead grey mullet Mugil cephalus and upper sediments from North African coasts of the Mediterranean Sea. *Sci. Total Environ.* **2018**, *639*, 156–174. [CrossRef]
55. Otmani, H.; Belkessa, R.; Bengoufa, S.; Boukhediche, W.; Djerrai, N.; Abbad, K. Assessment of shoreline dynamics on the Eastern Coast of Algiers (Algeria): A spatiotemporal analysis using in situ measurements and geospatial tools. *Arab. J. Geosci.* **2020**, *13*, 124. [CrossRef]
56. Cote, M.; Camps, G. Annaba. (Hippone-Bône). *Encyclopédie Berbère* **1988**, *5*, 674–685. [CrossRef]
57. Dahmani, S. De Hippone-Bùna à Annaba (Histoire de la Fondation d'une Métropole). Dar El-Houda: Ain M'Lila, Algeria, 2002; pp. 49–60.
58. Salah-Salah, H.; Spiga, S. La médina de Annaba: Histoire, mémoire et identité. *El-Tawassol Sci. Hum. Soc.* **2017**, *50*, 314–326.
59. Malek, C. L'Imaginaire Arabo-Musulman. Editions PUF: Paris, France, 1993.
60. Djouad, F.Z.; Spiga, S. La Production Architecturale à Annaba: Entre Héritage Délaissé Et Contemporanéité Désirée. *Synthèse Rev. Sci. Technol. Synthèse* **2018**, *36*, 49–60.
61. Benlakhlef, B. L'aire métropolitaine, un nouveau champ d'analyse de la ville algérienne. L'exemple d'Annaba. In *Penser la Ville–Approches Comparatives*; Edition Halshs-00382602: Khenchela, Algeria, 2008; p. 259.
62. Boughediri, S.; Benslama, M. Pluie pollinique et inventaire floristique dans la ville d'Annaba (Algérie). *Nat. Technol.* **2020**, *23*, 63–73.
63. Aouissi, H.A.; Belabed, A.I.; Bouslama, Z. Doves' mapping and inventory into the urban sites of Annaba (Northeastern of Algeria). *Adv. Environ. Biol.* **2015**, *12*, 328–338.
64. Aouissi, H.A.; Gasparini, J.; Belabed, A.I.; Bouslama, Z. Impact of greenspaces in city on avian species richness and abundance in Northern Africa. *Comptes. Rendus. Biol.* **2017**, *340*, 394–400. [CrossRef]
65. Frochot, B.; Roché, J. Suivi de populations d'oiseaux nicheurs par la méthode des indices ponctuels d'abondance (IPA). *Alauda* **1990**, *58*, 29–35.
66. Blondel, J.; Frochot, B.; Ferry, C. La méthode des Indices Ponctuels d' Abondance (I.P.A.) ou des relevés d'avifaune par «stations d'écoute». *Alauda* **1970**, *38*, 55–71.
67. ABH CSM. *Cahiers de L'agence des Bassins Hydrographiques no. 7: Bassins de la Seybouse Actualisation*; Ministry of Water Resources: Constantine, Algeria, 1999.
68. Scăunaşu, D.A.; Petrişor, A.-I.; Ivanov, F.M. Diversity of arthropod communities as an indicator of changes produced by the utilization of silvicultural techniques. *Ekologia* **2012**, *31*, 22–32. [CrossRef]
69. Petrişor, A.-I. Study about the Degradation of Lentic Ecosystems in Pantelimon Lakes. In *2000 Proceedings of the Section on Statistics and the Environment*; American Statistical Association: Alexandria, VA, USA, 2000; pp. 68–70.
70. Petrişor, A.-I.; Cheval, S.; Zinevici, V.; Parpală, L. Geostatistical analysis of time series: An exploratory analysis of climatologic and ecological data using a spatial instrument. *J. Appl. Quant. Methods* **2011**, *6*, 80–87.
71. Johnston, K.; Ver Hoef, J.M.; Krivoruchko, K.; Lucas, N.; Ver Hoef, J.M. *Using ArcGIS Geostatistical Analyst*; Environmental Systems Research Institute: Redlands, CA, USA, 2001.
72. Leveau, L.M.; Leveau, C.M. The role of urbanization and seasonality on the temporal variability of bird communities. *Landsc. Urban Plan.* **2012**, *3*, 271–276. [CrossRef]
73. Fontana, S.; Sattler, T.; Bontadina, F.; Moretti, M. How to manage the urban green to improve bird diversity and community structure. *Landsc. Urban Plan.* **2011**, *101*, 278–285. [CrossRef]
74. Farhi, Y.; Absi, K.; Belhamra, M. Composition de l'avifaune des steppes présahariennes des Ziban (Biskra, Algérie). *Courr. Du Savoir.* **2016**, *21*, 121–128.
75. Di Pietro, S.; Mantoni, C.; Fattorini, S. Influence of urbanization on the avian species-area relationship: Insights from the breeding birds of Rome. *Urban Ecosyst.* **2020**. [CrossRef]
76. Vignoli, L.; Scirè, S.; Bologna, M.A. Rural–urban gradient and land use in a millenary metropolis: How urbanization affects avian functional groups and the role of old villas in bird assemblage patterning. *Web Ecol.* **2013**, *13*, 49–67. [CrossRef]
77. Stiles, R.; Bouche-Pillon, S.; Brinkhuijsen, M.; de Vries, J.; Di Carlo, F.; Fetzer, E.; Libbrecht, H.; Maksymiuk, G.; Meeres, S.; Meireles Rodrigues, F. *Rome's Landscape*; LE:NOTRE Landscape Forum 2013: Vienna, Austria, 2014.
78. Belabed-Zediri, H.; Belabed, A.I.; Bouslama, Z. Etude comparative de l'écologie de la reproduction des rallidés entre milieu urbain et milieu naturel dans le nord-est Algérien. *Bull. Soc. Zool. Fr.* **2020**, *145*, 489–508.
79. Leveau, L.M.; Leveau, C.M. Street design in suburban areas and its impact on bird communities: Considering different diversity facets over the year. *Urban For. Urban Green.* **2020**, *48*, 126578. [CrossRef]
80. Spanò, M.; Gentile, F.; Davies, C.; Lafortezza, R. The DPSIR framework in support of green infrastructure planning: A case study in Southern Italy. *Land Use Policy* **2017**, *61*, 242–250. [CrossRef]
81. Hansen, R.; Olafsson, A.S.; van der Jagt, A.P.N.; Rall, E.; Pauleit, S. Planning multifunctional green infrastructure for compact cities: What is the state of practice? *Ecol. Indic.* **2019**, *96*, 99–110. [CrossRef]
82. Cilliers, S.; Cilliers, J.; Lubbe, R.; Siebert, S. Ecosystem services of urban green spaces in African countries—Perspectives and challenges. *Urban Ecosyst.* **2013**, *16*, 681–702. [CrossRef]

3. Badiu, D.-L.; Iojă, I.C.; Pătroescu, M.; Breuste, J.H.; Artmann, M.; Niţă, M.R.; Grădinaru, S.R.; Hossu, C.A.; Onose, D.-A. Is urban green space per capita a valuable target to achieve cities' sustainability goals? Romania as a case study. *Ecol. Indic.* **2016**, *70*, 53–66. [CrossRef]

4. Artmann, M.; Chen, X.; Iojă, I.C.; Hof, A.; Onose, D.-A.; Ponizy, L.; Zavodnik Lamovšek, A.; Breuste, J.H. The role of urban green spaces in care facilities for elderly people across European cities. *Urban For. Urban Green.* **2017**, *27*, 203–213. [CrossRef]

5. Artmann, M.; Kohler, M.; Meinel, G.; Gan, J.; Iojă, I.C. How smart growth and green infrastructure can mutually support each other–A conceptual framework for compact and green cities. *Ecol. Indic.* **2017**, *96*, 10–22. [CrossRef]

6. Grădinaru, S.R.; Ioja, I.C.; Patru-Stupariu, I.; Hersperger, A.M. Are spatial planning objectives reflected in the evolution of urban landscape patterns? A framework for the evaluation of spatial planning outcomes. *Sustainability* **2017**, *9*, 1279. [CrossRef]

7. Turkelboom, F.; Leone, M.; Jacobs, S.; Kelemen, E.; García-Llorente, M.; Baró, F.; Termansen, M.; Barton, D.N.; Berry, P.; Stange, E.; et al. When we cannot have it all: Ecosystem services trade-offs in the context of spatial planning. *Ecosyst. Serv.* **2018**, *29*, 566–578. [CrossRef]

8. Petrişor, A.-I.; Andronache, I.C.; Petrişor, L.E.; Ciobotaru, A.-M.; Peptenatu, D. Assessing the fragmentation of the green infrastructure in Romanian cities using fractal models and numerical taxonomy. *Procedia Environ. Sci.* **2016**, *32*, 110–123. [CrossRef]

9. Niţă, M.R.; Năstase, I.-I.; Badiu, D.-L.; Onose, D.-A.; Gavrilidis, A.-A. Evaluating urban forests connectivity in relation to urban functions in Romanian cities. *Carpathian J. Earth Environ. Sci.* **2018**, *13*, 291–299. [CrossRef]

0. Taşcu-Stavre, M.; Bănică, C. Old and New in Vama Veche and 2 Mai. *Urban. Arhit. Constr.* **2014**, *5*, 73–82.

1. Shaw, L.M.; Chamberlain, D.; Evans, M. The House Sparrow Passer domesticus in urban areas: Reviewing a possible link between post-decline distribution and human socioeconomic status. *J. Ornithol.* **2008**, *149*, 293–299. [CrossRef]

2. Steinhäußer, R.; Siebert, R.; Steinführer, A.; Hellmich, M. National and regional land-use conflicts in Germany from the perspective of stakeholders. *Land Use Policy* **2015**, *49*, 183–194. [CrossRef]

3. Tudora, I. *La curte—Gradina, Cartier si Peisaj Urban in Bucuresti (In the Yard—Urban Garden, Neighborhood and Landscape in Bucharest)*; Curtea Veche: Bucharest, Romania, 2009; 200p.

4. Rodrigues, A.G.; Borges-Martins, M.; Zilio, F. Bird diversity in an urban ecosystem: The role of local habitats in understanding the effects of urbanization. *Iheringia Série Zool.* **2018**, *108*, 1–11. [CrossRef]

5. Silva, C.P.; García, C.E.; Estay, S.A.; Barbosa, O. Bird richness and abundance in response to urban form in a Latin American city: Valdivia, Chile as a case study. *PLoS ONE* **2015**, *10*, e0138120. [CrossRef]

6. Ababsa, M.; Aouissi, H.A. Current State of the Coronavirus (Covid-19) in Algeria. *J. Community Med. Health Care* **2020**, *5*, 1036. [CrossRef]

7. Nori, M.; El Mourid, M.; Giorgi, P.; Nefzaoui, A. Herding in a Shifting Mediterranean Changing Agro-Pastoral Livelihoods in the Mashreq & Maghreb Region. Working Paper. Mediterranean Programme Series, EUI RSCAS, 2009/52. Available online: http://hdl.handle.net/1814/12714 (accessed on 19 February 2021).

8. Grădinaru, S.R.; Triboi, R.; Iojă, C.I.; Artmann, M. Contribution of agricultural activities to urban sustainability: Insights from pastoral practices in Bucharest and its peri-urban area. *Habitat Int.* **2018**, *82*, 62–71. [CrossRef]

9. Accademia di Francia a Roma—Villa Medici. Simon de Dreuille—Villa Medici. Available online: https://www.villamedici.it/residenze/simon-de-dreuille/ (accessed on 19 February 2021).

land

MDPI

Article

Landscape Attributes Best Explain the Population Trend of Wintering Greater White-Fronted Goose (*Anser albifrons*) in the Yangtze River Floodplain

Sheng Chen [1], Yong Zhang [1,*], Amaël Borzée [2], Tao Liang [1], Manyu Zhang [1], Hui Shi [1], Bin Chen [1], Wenbin Xu [3], Yunwei Song [3] and Lingfeng Mao [1]

1 Co-Innovation Center for Sustainable Forestry in Southern China, College of Biology and the Environment, Nanjing Forestry University, Nanjing 210037, China; neo0306@njfu.edu.cn (S.C.); liangtrep@njfu.edu.cn (T.L.); manyuzhang@njfu.edu.cn (M.Z.); shihui@njfu.edu.cn (H.S.); bin.chen@njfu.edu.cn (B.C.); maolingfeng2008@163.com (L.M.)

2 Laboratory of Animal Behaviour and Conservation, College of Biology and the Environment, Nanjing Forestry University, Nanjing 210037, China; amaelborzee@gmail.com

3 Anhui Shengjin Lake National Nature Reserve, Dongzhi 247200, China; wenbinx34@gmail.com (W.X.); songyunwei80@gmail.com (Y.S.)

* Correspondence: yong.zhang@njfu.edu.cn

Abstract: Biodiversity in the Middle and Lower Yangtze Floodplain has critically decreased during the last several decades, driven by numerous determinants. Hence, identification of primary drivers of animal population decline is a priority for conservation. Analyzing long time-series data is a powerful way to assess drivers of declines, but the data are often missing, hampering effective conservation policymaking. In this study, based on twenty-four years (from 1996 to 2019) of annual maximal count data, we investigated the effects of climate and landscape changes on the increasing population trend of the Greater White-Fronted Goose (*Anser albifrons*) at a Ramsar site in the Middle and Lower Yangtze Floodplain, China. Our results showed that the availability of a suitable habitat and landscape attributes are the key driving forces affecting the population trend, while the effects of climate factors are weak. Specifically, increasing the area of suitable habitat and alleviating habitat fragmentation through a fishing ban policy may have provided a more suitable habitat to the geese, contributing to the increasing population trend. However, we also observed that the grazing prohibition policy implemented in 2017 at Shengjin Lake may have potentially negatively affected geese abundance, as grazing by larger herbivores may favor smaller geese species by modifying the vegetation community and structure. Based on our results, we suggest several practical countermeasures to improve the habitat suitability for herbivorous goose species wintering in this region.

Keywords: Anatidae; population trend; landscape changes; recessional grassland; Yangtze wetlands; bird conservation

1. Introduction

Wildlife population trends can be affected by various environmental and anthropogenic factors [1]. Among them, land use and climate change are considered to be the major threats [2,3]. Habitat loss, fragmentation, and degradation severely affect bird species abundance, distribution range, trend, and diversity [3–5]. Poleward shifts have also been documented for bird species under the influence of global warming [6]. However, as long time-series data are often limited to a small number of systems or regions [7], a comprehensive understanding of the causal effects of land use and climate changes on individual bird species is still largely missing, hindering effective conservation.

Known as a valuable natural ecosystem type, wetlands not only provide important ecosystem services globally, such as flood control, water purification, and carbon sequestration [8], but also provide food and shelter for numerous species [9]. However, about 50% of the global wetland area has been lost since 1900 [10], and the loss of wetland began to intensify in recent years [11]. Especially in Asia, human activities in wetlands have greatly increased compared with other regions [12], which has posed a severe threat to waterbirds that rely on wetlands for their life cycle [13].

The middle and lower Yangtze River Floodplain is covered by the largest freshwater lake cluster in East Asia, which plays a critical role in supporting hundreds of migratory wintering waterbird species migrating along the East Asian–Australasian Flyway, such as geese [14]. However, with economic development, the Yangtze lakes have also experienced severe land use changes [12], and as a consequence, abundances and distributions of goose species have also changed [15].

Goose species wintering in the Yangtze floodplain mainly feed on recessional grassland. Hence, a primary determinant of goose population abundance is the size of the available habitat, which is governed by water level fluctuations. According to the individual–area relationship, population size will increase with increasing habitat area [16]. Hence, population abundance might be negatively affected by the reduction in recessional grasslands. Following the optimal foraging theory, forage quantity may also affect goose species abundance by affecting their daily nutrition intake [17,18]. As predicted by the functional response, goose species generally display Type II or Type IV functional responses, indicating that goose abundance will first increase with increasing forage quantity and then level off or decrease when the maximum intake is reached [17,19].

More importantly, in this region, human activities have caused large-scale habitat loss and fragmentation [20], potentially influencing geese population abundance and trends [15]. For example, aquaculture activities such as constructing fishing nets and cofferdams may split a large habitat into several smaller patches, negatively affecting geese abundance [21]. In opposition, larger fields of higher-yielding grasses may favor foraging [22]. In addition, habitat fragmentation may also increase the edge effect and decrease habitat connectivity, causing a decline in goose population [23,24].

Climate factors can also influence abundance of goose species that rely on wetlands, mainly through changes in temperature and precipitation. Wintering birds foraging in relatively warm habitats can reduce their metabolic rate [25] as the warmer temperature may reduce the cost of thermoregulation [26], and hence a warmer area may attract more geese. In addition, both temperature and precipitation are predicted to be positively correlated with grassland primary productivity [27], which will positively affect the number of herbivorous geese [28]. However, a higher precipitation may also result in increasing water levels in wetlands, which may decrease food availability for grazing waterbirds through flooding [29].

Former studies have found that the population abundances of the Anatidae species (e.g., bean goose *Anser fabialis*, swan goose *Anser cygnoides*, and tundra swan *Cygnus columbianus*) generally declined in wetlands along the Yangtze River floodplain [4,15]. The Greater White-Fronted Goose (*Anser albifrons*) is another dominant wintering waterbird species in this region, which strictly forages on recessional grassland. It is also one of the most vulnerable waterbirds affected by human activities along the East Asian–Australasian Flyway [30,31]. However, because of the absence of long time-series population data, the drivers of population trend have still not been studied in the Yangtze River Floodplain, thus hindering effective conservation strategies.

Anhui Shengjin Lake National Nature Reserve, a Ramsar site located in the middle and lower Yangtze floodplain, is one of the wetlands with the highest wintering waterbird density in the region. In addition, rapid economic development has caused greater landscape changes in the past few decades, offering a good opportunity to study the influence of human activities on the population trend of waterbirds. In this study, based on twenty-four years (from 1996 to 2019) of continuous waterbird monitoring, we investigated how landscape change, habitat fragmentation, as well as climate factors have affected the population trends of the Greater White-Fronted Goose at the Shengjin

Lake National Nature Reserve. Specifically, we aim to determine what are the key factors affecting the population trend of the Greater White-Fronted Goose.

2. Materials and Methods

2.1. Study Area

Shengjin Lake National Nature Reserve (30°16′–30°25′ N, 116°59′–117°12′ E) was designated as a Ramsar site in 2015 and lies on the southern bank of the middle and lower Yangtze River in Anhui Province, China (Figure 1). It is one of the wetlands with the highest density of waterbirds in the middle and lower Yangtze floodplain. It is a seasonally inundated, extending to ca 133 km^2 in the wet season in summer. The lake area decreases to ca 34 km^2 in the dry season in winter when the water level recedes, exposing extensive mudflats, grasslands, sedge (*Carex* spp.) meadows, and seasonal wetlands. The climate is characterized by a subtropical monsoon with average annual temperature of around 16.1 °C and average January temperature of about 4.0 °C. Average annual rainfall is about 1600 mm. The lake is connected to the Yangtze River via the Huangpen sluice, which was built in 1965.

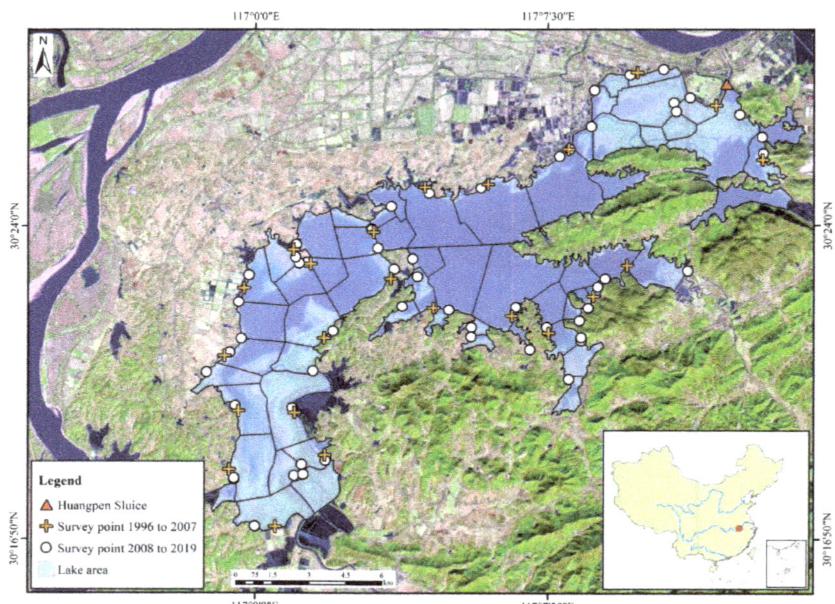

Figure 1. Location of Shengjin Lake National Reserve in China and the waterbird monitoring design.

Owing to the rapid economic development since the 1900s, human activities have severely modified the landscape of Shengjin Lake, potential affecting wintering waterbird abundance. In 2017, to better protect the wintering waterbird, a fishing and grazing prohibition was implemented in the area, and relevant regulations were officially issued in 2018.

To determine the whole range of the lake area, we extracted the water area in summers of 1989, 2002, and 2019 during the high water periods and calculated the total water area separately using ENVI5.3. The result showed that 96.7% of the lake area overlapped between years, indicating only little change in the whole lake area during the past thirty years (Appendix A, Table A1). Therefore, in further analyses, we used the boundary of the Shengjin Lake in 2002 as the whole lake area to quantify the habitat features and landscape matrices of the lake area from 1996 to 2019.

2.2. Count Data of Greater White-Fronted Geese

Greater White-Fronted Geese were surveyed using point counting methodology from 1996 to 2019. Geese were identified and their abundance was recorded through a 18–60× telescope. Most of the surveys were carried out by two teams of two persons in two days.

From 1996 to 2007, the number of overwintering Greater White-Fronted Geese was annually surveyed by the skilled staff of Shengjin Lake National Nature Reserve. Based on the lake morphology, 23 counting points covering the whole lake were set up, and the observers visited all counting points every month during the wintering periods.

From 2008 to 2019, the number of overwintering geese was counted systematically by the authors every 16 days during winter (October–April; for the years 2015–2018, only January counts were available), following the methods described in our former studies [32]. Briefly, Shengjin Lake was divided into 59 survey areas covered by 56 counting points to ensure that the entire lake was completely surveyed (Figure 1). The reason for increasing the number of the counting points is because the aquaculture activities, which accelerated in 2006, used a larger amount of bamboo poles and fishing nets, interrupting the vision of observers.

The maximum annual number of Greater White-Fronted Geese for each winter was further extracted for winters from 1996 to 2019, except for winters from 2015 to 2018, when only January counts were available and used. The maximum numbers of wintering Greater White-Fronted Geese were normally observed in January, suggesting that bias caused by the survey effort between 2015 and 2018 had little effect on the final results.

2.3. Satellite Image Processing

Satellite images used for the analyses were obtained from the United States Geological Survey (USGS; http://www.usgs.gov/; accessed on 20 April 2020), including Thematic Mapper (TM), Enhanced Thematic Mapper Plus (ETM+) images, and Operational Land Imager (OLI) images (with a consistent spatial resolution of 30 m). Images with a cloud cover less than 10% were selected. Most of the images were taken in November and December during the years of 1996–2019 (Appendix A, Table A2). Gap filling was used to deal with the problems of data duplications and loss in ETM+ images owing to the failure of the Scan Line Corrector [33]. After that, image calibration, atmospheric correction, and geometric correction were conducted.

We classified the habitat within the lake area into four categories: water, grassland, mudflat and emergent vegetation. Modified normalized difference water index (MNDWI) [34] was calculated to delineate the water surface area for each image using the formula below:

$$MNDWI = (green - SWIR)/(green + SWIR) \quad (1)$$

where *green* and *SWIR* (short-wave infrared) represent the second and seventh bands of Landsat5 TM and Landsat7 ETM+ images, respectively, and the third and seventh bands of Landsat8 OLI images, respectively.

Mudflat and grassland were classified by manually selecting the empirical threshold value of the normalized difference vegetation index (NDVI) [35], and the emergent vegetation was classified through visual interpretation on this basis. NDVI was calculated following the method below:

$$NDVI = (NIR - red)/(NIR + red) \quad (2)$$

where *red* and *NIR* (near-infrared) represent the third and fourth bands of Landsat5 TM and Landsat7 ETM+ images, respectively, and the fourth and fifth bands of Landsat8 OLI images, respectively.

Finally, patches with an area less than 90 m × 90 m (determined by the satellite images resolution) were aggregated through a minority analysis to facilitate the following analyses. All the above processing was conducted in the software ENVI 5.3. The patch area of each classification and NDVI of grassland were calculated using ArcGIS 10.3.

2.4. Climatic Variables

Waterbirds' population size was generally affected by temperature and precipitation [36]. For example, the abundance of waterbirds decreases with decreasing temperatures during the wintering period [36], and considering that grassland primary productivity is predicted to be positively correlated with both temperature and precipitation [27], this relationship is likely to positively affect the number of herbivorous geese [28]. We thus expected the waterbird population trend will be positively correlated with temperature and precipitation. Hence, 10 climatic variables (Table 1) were selected to correlate with changes in abundance of the Greater White-Fronted Goose: mean annual temperature (MAT), mean temperature of the coldest quarter (MTCQ), mean temperature of the driest quarter (MTDQ), minimum temperature of the coldest month (MTCM), temperature seasonality (TSN), mean annual precipitation (MAP), precipitation of the coldest quarter (PCQ), precipitation of the driest quarter (PDQ), temperature annual range (TAR), and precipitation seasonality (PSN). The air temperature and precipitation of Shengjin Lake were acquired from meteorological stations through China Meteorological Data Service Center (CMDSC) (http://data.cma.cn/; accessed on 8 June 2020).

Table 1. Potential explanatory variables, their descriptions and abbreviations, and data sources used in this study. CMDSC: the China Meteorological Data Service Center.

Categories	Variables	Abbreviations	Sources
Climatic variables CLIM	Mean annual temperature	MAT	CMDSC
	Mean temperature of the coldest quarter	MTCQ	CMDSC
	Mean temperature of the driest quarter	MTDQ	CMDSC
	Min temperature of the coldest month	MTCM	CMDSC
	Mean annual precipitation	MAP	CMDSC
	Precipitation of the coldest quarter	PCQ	CMDSC
	Precipitation of the driest quarter	PDQ	CMDSC
	Temperature annual range	TAR	CMDSC
	Temperature seasonality	TSN	CMDSC
	Precipitation seasonality	PSN	CMDSC
Ecological variables ECOL	Mudflat area	MA	Image processing
	Grassland area	GA	Image processing
	Emergent vegetation area	EVA	Image processing
	NDVI	NDVI	Image processing
	NDVI coefficient of variation	NDVICV	Image processing
Landscape metrics LAND	Largest patch index of mudflat	MLPI	Image processing
	Patch density of mudflat	MPD	Image processing
	Connectance index of mudflat	MCONNECT	Image processing
	Largest patch index of grassland	GLPI	Image processing
	Patch density of grassland	GPD	Image processing
	Connectance index of grassland	GCONNECT	Image processing
	Largest patch index of emergent vegetation	EVLPI	Image processing
	Patch density of emergent vegetation	EVPD	Image processing
	Connectance index of emergent vegetation	EVCONNECT	Image processing
	Largest patch index of lake area	LALPI	Image processing
	Patch density of lake area	LAPD	Image processing
	Connectance index of lake area	LACONNECT	Image processing
	Simpson's diversity index of lake area	LASIMP	Image processing
	Simpson's evenness index of lake area	LASIMPEVE	Image processing

2.5. Ecological Variables

Bird abundance is predicted increase with increases in patch size [17]. In addition, vegetation biomass and waterfowl abundance are often positive related [37]. Hence, according to the diet and habitat usage of the Greater White-Fronted Goose, we refer to five ecological variables (Table 1) that potentially affect the Greater White-Fronted Goose population size: mudflat area (MA), grassland area (GA), emergent vegetation area (EVA), NDVI, and coefficient of variation of NDVI (NDVICV) of grassland.

2.6. Landscape Metrics

Landscape metrics at both class (to quantify the attributes of a specific landscape class) and landscape level (to quantify the entire landscape mosaic) were calculated as shown in previous studies where species abundance is affected by landscape attributes at both levels [38]. To do this, landscape metrics were selected according to the ecological characteristics of the studied species [39,40]. In total, five landscape metrics were included (Table 1): largest patch index (LPI), patch density (PD), connectance Index (CONNECT), Simpson's diversity index (SIDI), and Simpson's evenness index (SIEI), covering the dominance, fragmentation, connectivity, and diversity of the landscape [41]. The variations in bird habitats (mudflat, grassland, and emergent vegetation) were determined as class-level metrics (LPI, PD, CONNECT). Landscape-level indexes (LPI, PD, CONNECT, SIDI, and SIEI) were measured based on the results of all classifications to reflect the changes in the landscape in the lake area. The software FRAGSTATS 4.2 [42] was applied to compute all metrics.

2.7. Statistical Analyses

Population trends and annual indices were calculated for the Greater White-Fronted Goose based on the abundance data in the R-package rtrim [43]. Rtrim is a commonly used tool in bird monitoring, using a Poisson general log-linear model. Rtrim produces annual growth rate and annual abundance indices, including their standard errors.

All explanatory variables were grouped into three variable sets. Before analyses, all independent and dependent variables were standardized using the Z-score method to interpret parameter estimates on a comparable scale. To account for the risk of multicollinearity, Pearson correlation coefficients (r) were calculated for all possible pairs of independent variables belonging to the same variable set. Then, univariate linear regression analyses were applied with all dependent variables tested one by one against the population growth rate. Variables with larger p-values obtained by univariate linear regression analyses were dropped in the pairs having an $|r| > 0.5$ [44].

We formulated a set of working hypotheses for the remaining variables (Table 2). Model I represents the effect of climatic variables on the population trends of the Greater White-Fronted Goose. Model II and III represent the effect of ecological variables and landscape metrics, respectively.

Table 2. Working hypotheses to test the effect of different variables on population trends of the Greater White-Fronted Goose (*Anser albifrons*) at the Shengjin Lake National Reserve between 1996 and 2019. The number of variables was reduced to avoid multicollinearity. H_1 indicated the predicted effect. + = positive effect, − = negative effect. For variable abbreviations, see Table 1.

Models	Variables	Types	H_1
Model I	MAT	CLIM	+
	MTCQ	CLIM	+
	MTCM	CLIM	+
	MAP	CLIM	+
	PCQ	CLIM	−
	PDQ	CLIM	−
Model II	MA	ECOL	+
	GA	ECOL	+
	EVA	ECOL	+
	NDVI	ECOL	+
Model III	MCONNECT	LAND	+
	GLPI	LAND	+
	GCONNECT	LAND	+
	EVCONNECT	LAND	+
	LACONNECT	LAND	+
	LASIMPEVE	LAND	−

Multiple linear regression models were then used to detect the effect of each variable set on population trends for the Greater White-Fronted Goose from 1996 to 2019. We ranked all possible subset models based on Akaike's information criterion for a small sample size (AICc). In addition, Akaike weights (*wi*) were also calculated to estimate the likelihood of each model [45]. Model averaging was then applied to obtain parameter estimates for each variable. The model averaging was done with a cut-off $\Delta AICc \leq 2$ [45]. The explanatory power of each variable set was calculated by averaging the adjusted-R^2 (Adj.R^2) among the best models [46].

All analyses were carried out in R 4.0.3 with the package rtrim and MuMIn.

3. Results

3.1. Changing of Population Sizes and Population Trends

Generally speaking, the population of the Greater White-Fronted Goose increased during the past two decades at the Shengjin Lake National Reserve (Figure 2). Before 2009, the population changes were relatively small, but larger population fluctuations were detected after 2009. However, population size experienced significant declines in the years of 2012 and 2013, and especially in 2019 when the population of the Greater White-Fronted Goose decreased by nearly 50% (Figure 2).

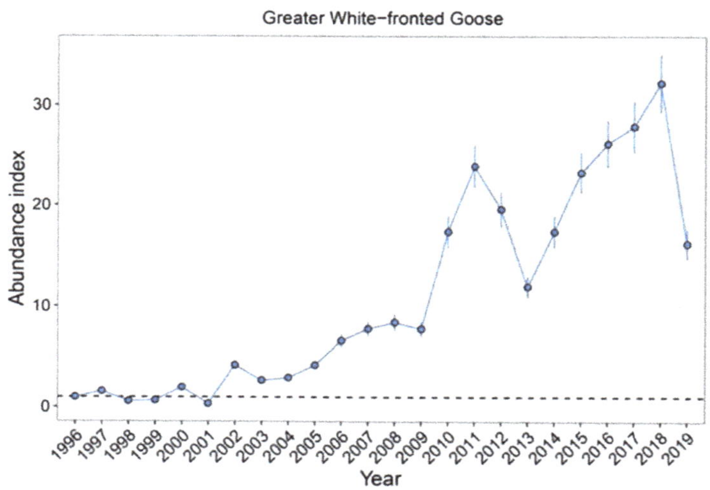

Figure 2. Changes of the population trend for the Greater White-Fronted Goose in the past two decades at Shengjin Lake National Reserve in the Yangtze River Floodplain. Error bars indicate the 95% CI.

3.2. Changing of Habitat from 1996 to 2019

The habitat of the Greater White-Fronted Goose has gradually changed during the last 20 years at the Shengjin Lake (Figure 3). The size of the grassland area increased from ca. 19 km^2 to ca. 25 km^2, and the connectivity of grasslands also improved. In contrast, Simpson's evenness index for the lake area decreased, indicating that the landscape tends to be formed by several dominating types (Figure 3).

3.3. Effect of Different Variables on Wintering Population Trends of the Greater White-Fronted Goose

The multiple linear regression models with all possible combinations of each variable category illustrated that most of the potential predictor variables were featured in the best fitting models ($\Delta AICc < 2$) affecting the population trends of the Greater White-Fronted Goose (Table 3).

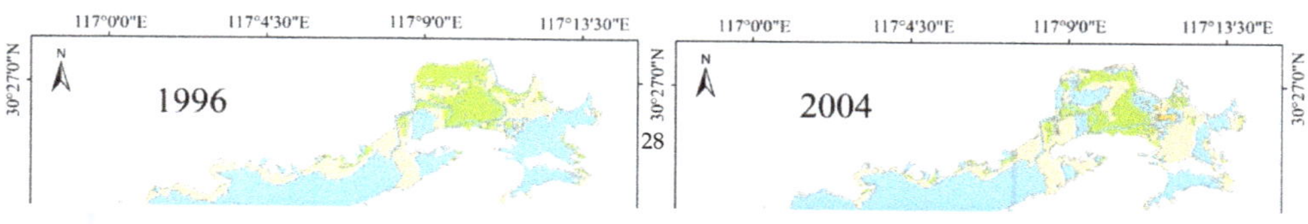

In the climate models, temperature and precipitation mostly had a positive effect on the population trend of the Greater White-Fronted Goose (Figure 4). For ecological models, grassland area (GA) and mudflat area (MA) were positively related to population trend, while emergent vegetation area (EVA) had a negative effect (Figure 4). In landscape models, connectance and dominance index of grassland areas (GCONNECT, GLPI) displayed a strong positive relationship with the population trend, but Simpson's evenness index of lake area (LASIMPEVE) showed a negative relationship with the population trend (Figure 4).

When comparing the performances of the models among each variable category, we found that landscape models scored the highest explanatory power (adj.R^2 = 0.641 ± 0.006), indicating their primary importance in explaining population trends. The ecological models also showed considerable explanatory power (adj.R^2 = 0.380 ± 0.008), which means that they also played an important role in governing the changes of population. Climate variable model had the lowest explanatory power (adj.R^2 = 0.095 ± 0.023).

Table 3. Variables of each category featured in the best models, explaining the Greater White-Fronted Goose population trend in order of increasing AICc (Akaike information criterion for small sample size), ΔAICc, and *wi* (Akaike weights) based on the multiple linear regression models. CLIM = climate factors; ECOL = ecological factors; LAND = landscape factors. For variables' abbreviations, see Table 1. df = degrees of freedom, logLik = log likelihood, adj.R^2 = adjusted R^2.

Categories	Top Models	df	logLik	AICc	ΔAICc	*wi*	adj.R^2
CLIM	MAT	3	−31.810	70.820	0	0.078	0.095
CLIM	PCQ	3	−32.326	71.852	1.032	0.047	0.055
CLIM	MAT + PCQ	4	−30.927	71.959	1.139	0.044	0.119
CLIM	MAT + MTCQ	4	−31.043	72.192	1.372	0.039	0.111
CLIM	MAP + MAT	4	−31.195	72.495	1.675	0.034	0.099
CLIM	MAT + PDQ	4	−31.356	72.817	1.996	0.029	0.087
ECOL	GA + MA	4	−26.704	63.513	0	0.533	0.381
ECOL	EVA + GA + MA	5	−26.144	65.511	1.999	0.186	0.379
LAND	GCONNECT + GLPI + LASIMPEVE	5	−19.567	52.468	0	0.474	0.641
LAND	GCONNECT + GLPI + LACONNECT + LASIMPEVE	6	−18.984	54.449	1.981	0.140	0.640

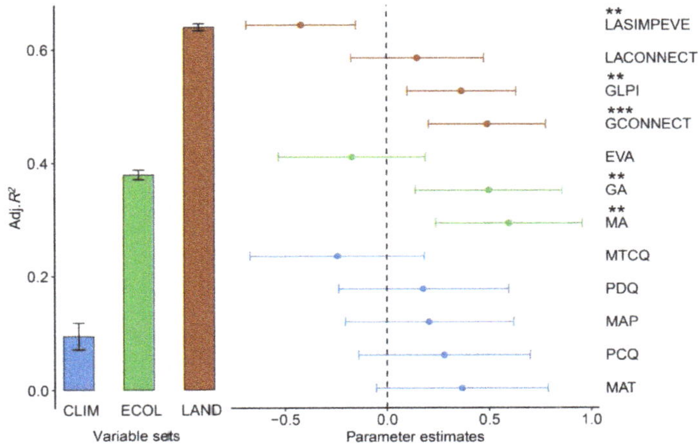

Figure 4. The overall performances of multiple regression models (histogram plot) for each variable category, explaining the population trend of the Greater White-Fronted Goose and regression coefficients of each variable (forest plot) featured in the best models. Adj.R^2 = adjusted R^2. Error bars indicate 95% CI. For variables' abbreviations, see Table 1. ** indicates $p < 0.01$; *** indicates $p < 0.001$.

4. Discussions

Since the early Anthropocene, the intensification of anthropogenic activities has significantly changed wetlands, and hence their biodiversity. Applying long time-series data to detect the effects of human activities on biodiversity is essential to offer better conservation measures [47]. In this study, using more than twenty years of wintering waterbirds monitoring data, we detected an increasing population trend in the Greater White-Fronted Goose at Shengjin Lake, although the population abundance was often reported to display a decreasing trend in some other Yangtze wetlands [48]. We then demonstrated that the availability of suitable habitat and landscape attributes are the key drivers governing population trends, while the effects of climate factors are weak. Our findings may serve as a successful conservation case to inspire practical conservation measures and better safeguard regional biodiversity.

Unexpectedly, and in opposition with most other species, we found a locally increasing population size for the Greater White-Fronted Goose over the last 24 years at Shengjin Lake. The shift in survey protocols and efforts may, however, affect our census data, and hence the population trend [49]. Only a few geese were counted during the earlier years of our study period (Figure 2) and the population abundance might be underestimated because of the lower accessibility to some habitats and the lack of skilled observers. However, the population size increase after 2007 is unlikely to be an artifact as survey protocols changed that year and were maintained consistently until 2019, and systematic surveys were conducted to cover the totality of the lake.

The increasing population trend may be related to the reproductive success in the breeding area. There are two biological flyways for Greater White-Fronted Geese in eastern Asia. Within the branch of the flyway studied here, the population size has been in critical decline since the 1990s [50,51], suggesting that the increasing population trend is hardly explained by the improvement of reproductive success. In addition, the declining population trend is often reported in some other Yangtze wetlands [48], which indicates that the improvement in wintering habitat at the Shengjin Lake may attract more birds than the other Yangtze wetlands. Our result is also in line with a former study that found that the population trend of the Greater White-Fronted Goose was stable at wetlands with the highest protection level in this region [15]. Niche theory may also be used to explain this increasing trend. In Shengjin Lake, the Swan Goose (*Anser cygnoides*) population has dramatically decreased because of the collapse of submerged macrophytes [52]. Hence, waterbird species relying on other resources, such as the Greater White-Fronted Goose, may migrate to occupy the niche, leading to an increase in population size.

Our results showed that changes in landscape attributes have the highest explanatory power for Greater White-Fronted Goose population trends. Among them, the connectance index (GCONNECT), largest patch index (GLPI) of grassland, and Simpson's evenness index of lake area (LASIMPEVE) were the most important determinants (Figure 4). Similarly, an earlier study indicated that landscape variables better explained patterns in bird richness [53]. Our results also found that connectivity and integrity of grasslands had a strong positive effect on goose population trends (Figure 4), emphasizing that habitat fragmentation caused by human activities such as aquaculture has an important effect and promotes a decrease in waterbird population sizes. Our results also showed that the population of the Greater White-Fronted Goose will probably decrease with an increase in landscape evenness (Figure 4). Habitat preference is frequently treated as one of the most important species traits associated with bird population abundance [7,54]. Wintering Greater White-Fronted Geese strictly forage on recessional grassland in this region [31], and population dynamics are thus predicted by the individual–area relationships [17], where a dominant preferred habitat may favor population maintenance. This may also be explained by a positive relationship between the largest patch index of grassland (GLPI) and the population trend of Greater White-Fronted Goose (Figure 4).

Regarding the ecological variables, the area of preferred habitat has a good explanatory power for the trend in the Greater White-Fronted Goose population. As predicted and consistent with theories, the area of grassland and mudflat positively correlated with goose population trend (Figure 4). In addition, an earlier study showed that habitat size positively affects the population abundance of herbivorous goose species [55].

The size of recessional grassland highly depends on water level fluctuation [56,57]. Hydrological changes caused by hydroelectric and water diversion projects such as the Three Gorges Dam have been frequently documented in the Yangtze wetlands, reducing river discharge and changing the vegetation structure [58,59]. However, despite an increasing population trend at the Shengjin Lake, it cannot be concluded that the Greater White-Fronted Goose may benefit from those changes. The operation of the Three Gorges Dam may accelerate downstream drainage, prolong the growing period of recessional grassland in winter [5,60], and thus negatively affect herbivorous geese foraging by resulting in a large amount of tall and old grass [18,61]. As the Shengjin Lake is located in the lower Yangtze Floodplain Plain, and the effect of hydrological changes on downstream wetlands may have a time-lag, future population changes are unlikely to be positive. In addition, we detected a rapid decline in population abundance after 2018 (Figure 2), probably triggered by the prohibition of grazing by buffaloes since 2017. Larger livestock such as buffalos may remove the old and low nutrition grass, change the vegetation structure, and thereafter facilitate grazing by smaller species such as geese [55,62]. Hence, further studies are highly suggested to verify the effects of changing grazing regimes on geese species when the acquired data are available.

Climate change is often blamed to be one of the important driving forces causing population decreases [63,64]. However, evidence is often obtained through larger scale analyses. At the local scale, other factors such as human activities and landscape changes may alter the effects of climate change [65]. In our results, climate models had the lowest explanatory power, implying the effect of climate changes on Greater White-Fronted Goose population trends is weak over the 24 years of our study.

5. Conclusions

In this paper, we demonstrated an increasing population trend in the Greater White-Fronted Goose (*Anser albifrons*) at Shengjin Lake National Reserve. However, this does not mean that the conservation status of the geese has improved in the whole region, as population decreases were generally reported in some other wetlands in the region [66]. Hence, to better protect this valuable species, further conservation measures should be considered for the Yangtze wetlands as a whole. Our results can also inspire several practical countermeasures. Firstly, as water level regimes largely determine recessional grassland availability and quality, hydrological regulation rules that allow for the preferred habitats to be gradually extended should be enacted and enforced. Secondly, aquaculture and related activities within wetlands, such as the construction of fishing nets and cofferdams, should continue to be strictly controlled as conservation profits have been preliminarily documented through an increase in grassland area and connectivity. Lastly, although it is not verified because future data are needed, the grazing prohibition policy that took effect in recent years in the region might result in changes in the vegetation such as quantity, quality, and heterogeneity, and thus possibly threaten wintering herbivorous geese. Hence, we highly recommend the adjustment of the current policy and explore an optimum grazing intensity to facilitate geese foraging.

Author Contributions: Conceptualization, Y.Z., S.C. and L.M.; methodology, Y.Z.. S.C., A.B., M.Z. and T.L.; formal analysis, S.C., A.B. and Y.Z.; investigation, S.C., Y.Z., M.Z., T.L. and H.S.; resources, L.M. and Y.Z.; data curation, S.C., Y.Z., H.S., W.X. and Y.S.; writing—original draft preparation, S.C., Y.Z. and A.B.; writing—review and editing, M.Z., T.L., H.S., B.C. and L.M.; visualization, S.C. and Y.Z.; supervision, Y.Z.; project administration, L.M.; funding acquisition, Y.Z., B.C. and L.M. All authors have read and agreed to the published version of the manuscript.

Funding: This study was supported by the Strategic Priority Research Program of the Chinese Academy of Sciences (Grant No. XDB31000000), Natural Science Foundation of Jiangsu Province (Grant No. BK20170922), and the National Natural Science Foundation of China (32071526).

Institutional Review Board Statement: Ethical review and approval were waived for this study, due to only non-experimental clinical veterinary practices were performed and no handling of animals related to this research was carried out.

Informed Consent Statement: Not applicable.

Data Availability Statement: The data presented in this study are available on request from the corresponding author.

Acknowledgments: We thank Chao Jiang, Wei Shen, and Zixi Zhao for their assistance during the surveys.

Conflicts of Interest: The authors declare no conflict of interest.

Appendix A

Table A1. The lake area during the high water period in 1989, 2002, and 2019.

Year	Area/km^2
1989	131.0022
2002	131.1035
2019	131.9908

Table A2. Landsat images used in the map processing and the dates when the images were taken. Path/row: 121/39 (Landsat Path/Row World Reference System).

Category	TM	ETM+	OLI
	5 July 1989	10 December 1999	2 August 2013
	9 April 1996	5 July 2000	24 December 2013
	25 December 1996	2 November 2000	1 May 2014
	22 August 1997	24 July 2001	5 August 2014
	7 September 1997	11 July 2002	5 June 2015
	10 November 1997	24 November 2005	30 December 2015
	8 July 1998	30 November 2007	23 June 2016
	15 December 1998	27 July 2008	16 December 2016
	29 September 1999	5 December 2009	12 July 2017
	21 November 2001	18 August 2010	19 December 2017
	8 November 2002	8 December 2010	3 August 2019
	7 August 2003	11 December 2011	23 November 2019
	13 December 2003	22 July 2012	
	9 August 2004	27 November 2012	
	15 December 2004	17 November 2014	
	12 August 2005	24 August 2018	
	30 July 2006	28 November 2018	
	21 November 2006		
	2 August 2007		
	10 December 2008		
	4 June 2009		
	28 July 2011		

References

1. Johnson, C.N.; Balmford, A.; Brook, B.W.; Buettel, J.C.; Galetti, M.; Lei, G.C.; Wilmshurst, J.M. Biodiversity losses and conservation responses in the Anthropocene. *Science* **2017**, *356*, 270–274. [CrossRef]
2. Stephens, P.A.; Mason, L.R.; Green, R.E.; Gregory, R.D.; Sauer, J.R.; Alison, J.; Aunins, A.; Brotons, L.; Butchart, S.H.M.; Campedell, T.; et al. Consistent response of bird populations to climate change on two continents. *Science* **2016**, *352*, 84–87. [CrossRef]
3. Pecl, G.T.; Araujo, M.B.; Bell, J.D.; Blanchard, J.; Bonebrake, T.C.; Chen, I.C.; Clark, T.D.; Colwell, R.K.; Danielsen, F.; Evengarc, B.; et al. Biodiversity redistribution under climate change: Impacts on ecosystems and human well-being. *Science* **2017**, *35*. [CrossRef]

4. Jia, Q.; Wang, X.; Zhang, Y.; Cao, L.; Fox, A.D. Drivers of waterbird communities and their declines on Yangtze River floodplain lakes. *Biol. Conserv.* **2018**, *218*, 240–246. [CrossRef]
5. Lehikoinen, A.; Brotons, L.; Calladine, J.; Campedelli, T.; Escandell, V.; Flousek, J.; Grueneberg, C.; Haas, F.; Harris, S.; Herrando, S.; et al. Declining population trends of European mountain birds. *Glob. Chang. Biol.* **2019**, *25*, 577–588. [CrossRef]
6. VanDerWal, J.; Murphy, H.T.; Kutt, A.S.; Perkins, G.C.; Bateman, B.L.; Perry, J.J.; Reside, A.E. Focus on poleward shifts in species' distribution underestimates the fingerprint of climate change. *Nat. Clim. Chang.* **2013**, *3*, 239–243 [CrossRef]
7. Bowler, D.E.; Heldbjerg, H.; Fox, A.D.; de Jong, M.; Bohning-Gaese, K. Long-term declines of European insectivorous bird populations and potential causes. *Conserv. Biol.* **2019**, *33*, 1120–1130. [CrossRef]
8. Keddy, P.A. *Wetland Ecology: Principles and Conservation*; Cambridge University Press: Cambridge, UK, 2010.
9. Costanza, R. Nature: Ecosystems without commodifying them. *Nature* **2006**, *443*, 749. [CrossRef]
10. Davidson, N.C. How much wetland has the world lost? Long-term and recent trends in global wetland area. *Mar. Freshw. Res.* **2014**, *65*, 934–941. [CrossRef]
11. Mammides, C. A global assessment of the human pressure on the world's lakes. *Glob. Environ. Chang.* **2020**, *63*, 102084. [CrossRef]
12. Meng, W.; He, M.; Hu, B.; Mo, X.; Li, H.; Liu, B.; Wang, Z. Status of wetlands in China: A review of extent, degradation, issues and recommendations for improvement. *Ocean Coast. Manag.* **2017**, *146*, 50–59. [CrossRef]
13. Gibbs, J.P. Wetland loss and biodiversity conservation. *Conserv. Biol.* **2000**, *14*, 314–317. [CrossRef]
14. Cao, L.; Zhang, Y.; Barter, M.; Lei, G. Anatidae in eastern China during the non-breeding season: Geographical distributions and protection status. *Biol. Conserv.* **2010**, *143*, 650–659. [CrossRef]
15. Zhang, Y.; Jia, Q.; Prins, H.H.T.; Cao, L.; de Boer, W.F. Effect of conservation efforts and ecological variables on waterbird population sizes in wetlands of the Yangtze River. *Sci. Rep.* **2015**, *5*, 17136. [CrossRef] [PubMed]
16. Connor, E.F.; Courtney, A.C.; Yoder, J.M. Individuals–area relationships: The relationship between animal population density and area. *Ecology* **2000**, *81*, 734–748.
17. Heuermann, N.; van Langevelde, F.; van Wieren, S.E.; Prins, H.H.T. Increased searching and handling effort in tall swards lead to a Type IV functional response in small grazing herbivores. *Oecologia* **2011**, *166*, 659–669. [CrossRef] [PubMed]
18. Hsu, C.-H.; Chou, J.-Y.; Fang, W.-T. Habitat selection of wintering birds in farm ponds in Taoyuan, Taiwan. *Animals* **2019**, *9*, 113. [CrossRef]
19. Durant, D.; Fritz, H.; Blais, S.; Duncan, P. The functional response in three species of herbivorous Anatidae: Effects of sward height, body mass and bill size. *J. Anim. Ecol.* **2003**, *72*, 220–231. [CrossRef]
20. Xie, C.; Huang, X.; Mu, H.Q.; Yin, W. Impacts of land-use changes on the lakes across the Yangtze floodplain in China. *Environ. Sci. Technol.* **2017**, *51*, 3669–3677. [CrossRef]
21. Fang, J.; Wang, Z.; Zhao, S.; Li, Y.; Tang, Z.; Yu, D.; Ni, L.; Liu, H.; Xie, P.; Da, L. Biodiversity changes in the lakes of the Central Yangtze. *Front. Ecol. Environ.* **2006**, *4*, 369–377. [CrossRef]
22. Madsen, J.; Cracknell, G.; Fox, T. *Goose Populations of the Western Palearctic: A Review of Status and Distribution*; 8777724372; Wetland International: Wageningen, The Netherlands, 1999.
23. Baker, J.; French, K.; Whelan, R.J. The edge effect and ecotonal species: Bird communities across a natural edge in southeastern Australia. *Ecology* **2002**, *83*, 3048–3059. [CrossRef]
24. Zhang, W.; Li, X.; Yu, L.; Si, Y. Multi-scale habitat selection by two declining East Asian waterfowl species at their core spring stopover area. *Ecol. Indic.* **2018**, *87*, 127–135. [CrossRef]
25. Wachob, D.G. The effect of thermal microclimate on foraging site selection by wintering mountain chickadees. *Condor* **1996**, *98*, 114–122. [CrossRef]
26. Brown, J.H.; Gillooly, J.F.; Allen, A.P.; Savage, V.M.; West, G.B. Toward a metabolic theory of ecology. *Ecology* **2004**, *85*, 1771–1789. [CrossRef]
27. Guo, Q.; Hu, Z.M.; Li, S.G.; Li, X.R.; Sun, X.M.; Yu, G.R. Spatial variations in aboveground net primary productivity along a climate gradient in Eurasian temperate grassland: Effects of mean annual precipitation and its seasonal distribution. *Glob. Chang. Biol.* **2012**, *18*, 3624–3631. [CrossRef]
28. Fox, A.D.; Elmberg, J.; Tombre, I.M.; Hessel, R. Agriculture and herbivorous waterfowl: A review of the scientific basis for improved management. *Biol. Rev.* **2017**, *92*, 854–877. [CrossRef] [PubMed]
29. Nolet, B.A.; Fuld, V.N.; Van Rijswijk, M.E.C. Foraging costs and accessibility as determinants of giving-up densities in a swan-pondweed system. *Oikos* **2006**, *112*, 353–362. [CrossRef]
30. Fan, Y.; Zhou, L.; Cheng, L.; Song, Y.; Xu, W. Foraging behavior of the Greater White-fronted Goose (*Anser albifrons*) wintering at Shengjin Lake: Diet shifts and habitat use. *Avian Res.* **2020**, *11*, 3. [CrossRef]
31. Zhao, M.J.; Cao, L.; Klaassen, M.; Zhang, Y.; Fox, A.D. Avoiding competition? Site use, diet and foraging behaviours in two similarly sized geese wintering in China. *Ardea* **2015**, *103*, 27–38. [CrossRef]
32. Cao, L.; Barter, M.; Zhao, M.; Meng, H.; Zhang, Y. A systematic scheme for monitoring waterbird populations at Shengjin Lake, China: Methodology and preliminary results. *Chin. Birds* **2011**, *2*, 1–17. [CrossRef]
33. Scaramuzza, P.; Micijevic, E.; Chander, G. SCL Gap-Filled Products: Phase One Methodology. Available online: https://www.usgs.gov/media/files/landsat-7-slc-gap-filled-products-phase-one-methodology (accessed on 9 May 2020).
34. Xu, H. Modification of normalised difference water index (NDWI) to enhance open water features in remotely sensed imagery. *Int. J. Remote Sens.* **2006**, *27*, 3025–3033. [CrossRef]

35. Tucker, C.J. Red and photographic infrared linear combinations for monitoring vegetation. *Remote Sens. Environ.* **1979**, *8*, 127–15 [CrossRef]
36. Meehan, T.D.; Jetz, W.; Brown, J.H. Energetic determinants of abundance in winter landbird communities. *Ecol. Lett.* **200**, 532–537. [CrossRef]
37. Marklund, O.; Sandsten, H.; Hansson, L.A.; Blindow, I. Effects of waterfowl and fish on submerged vegetation and macroinvertebrates. *Freshw. Biol.* **2002**, *47*, 2049–2059. [CrossRef]
38. Tischendorf, L. Can landscape indices predict ecological processes consistently? *Landsc. Ecol.* **2001**, *16*, 235–254. [CrossRef]
39. Li, X.; Si, Y.; Ji, L.; Gong, P. Dynamic response of East Asian Greater White-fronted Geese to changes of environment during migration: Use of multi-temporal species distribution model. *Ecol. Model.* **2017**, *360*, 70–79. [CrossRef]
40. Gao, B.; Gong, P.; Zhang, W.; Yang, J.; Si, Y. Multiscale effects of habitat and surrounding matrices on waterbird diversity in the Yangtze River Floodplain. *Landsc. Ecol.* **2020**, *36*, 179–190. [CrossRef]
41. McGarigal, K. *FRAGSTATS: Spatial Pattern Analysis Program for Quantifying Landscape Structure*; US Department of Agriculture Forest Service, Pacific Northwest Research Station: Corvallis, OR, USA, 1995; Volume 351.
42. McGarigal, K.; Cushman, S.A.; Ene, E. FRAGSTATS v4: Spatial Pattern Analysis Program for Categorical and Continuous Maps Computer Software Program Produced by the Authors at the University of Massachusetts, Amherst. 2012. Available online: http://www.umass.edu/landeco/research/fragstats/fragstats.html (accessed on 3 October 2020).
43. Pannekoek, J.; Van Strien, A. *Trim 3 Manual (TRends & Indices for Monitoring Data)*; Statistics Netherlands: Voorburg, The Netherlands, 2005.
44. Dormann, C.F.; Elith, J.; Bacher, S.; Buchmann, C.; Carl, G.; Carré, G.; Marquéz, J.R.G.; Gruber, B.; Lafourcade, B.; Leitao, P.J. Collinearity: A review of methods to deal with it and a simulation study evaluating their performance. *Ecography* **2013**, *36*, 27–4 [CrossRef]
45. Burnham, K.P.; Anderson, D.R. *Model Selection and Multimodel Inference: A Practical Information-Theoretic Approach*, 2nd ed. Springer: New York, NY, USA, 2002; p. 267.
46. Callaghan, C.T.; Major, R.E.; Lyons, M.B.; Martin, J.M.; Kingsford, R.T. The effects of local and landscape habitat attributes on bird diversity in urban greenspaces. *Ecosphere* **2018**, *9*, e02347. [CrossRef]
47. Jefferies, R.L.; Drent, R.H.; Bakker, J.P. Connecting arctic and temperate wetlands and agricultural landscapes: The dynamics of goose populations in response to global change. In *Wetlands and Natural Resource Management*; Springer: Berlin/Heidelberg, Germany, 2006; pp. 293–314.
48. Zou, Y.-A.; Zhang, P.-Y.; Zhang, S.-Q.; Chen, X.-S.; Li, F.; Deng, Z.-M.; Yang, S.; Zhang, H.; Li, F.-Y.; Xie, Y.-H. Crucial sites and environmental variables for wintering migratory waterbird population distributions in the natural wetlands in East Dongting Lake, China. *Sci. Total Environ.* **2019**, *655*, 147–157. [CrossRef]
49. Snall, T.; Kindvall, O.; Nilsson, J.; Part, T. Evaluating citizen-based presence data for bird monitoring. *Biol. Conserv.* **2011**, *144*, 804–810. [CrossRef]
50. Jia, Q.; Koyama, K.; Choi, C.-Y.; Kim, H.-J.; Cao, L.; Gao, D.; Liu, G.; Fox, A.D. Population estimates and geographical distributions of swans and geese in East Asia based on counts during the non-breeding season. *Bird Conserv. Int.* **2016**, *26*, 397–417. [CrossRef]
51. Deng, X.; Zhao, Q.; Solovyeva, D.; Lee, H.; Bysykatova-Harmey, I.; Xu, Z.; Ushiyama, K.; Shimada, T.; Koyama, K.; Park, J. Contrasting trends in two East Asian populations of the Greater White-fronted Goose *Anser albifrons*. *Wildfowl* **2020**, 181–205.
52. Fox, A.D.; Cao, L.; Zhang, Y.; Barter, M.; Zhao, M.J.; Meng, F.J.; Wang, S.L. Declines in the tuber-feeding waterbird guild at Shengjin Lake National Nature Reserve, China—A barometer of submerged macrophyte collapse. *Aquat. Conserv.* **2011**, *21* 82–9 [CrossRef]
53. Xu, C.; Huang, Z.Y.X.; Chi, T.; Chen, B.J.W.; Zhang, M.; Liu, M. Can local landscape attributes explain species richness patterns at macroecological scales? *Glob. Ecol. Biogeogr.* **2014**, *23*, 436–445. [CrossRef]
54. Iknayan, K.J.; Beissinger, S.R. Collapse of a desert bird community over the past century driven by climate change. *Proc. Natl. Acad. Sci. USA* **2018**, *115*, 8597–8602. [CrossRef]
55. Zhang, Y.; Jia, Q.; Prins, H.H.T.; Cao, L.; de Boer, W.F. Individual-area relationship best explains goose species density in wetlands. *PLoS ONE* **2015**, *10*, e0124972. [CrossRef] [PubMed]
56. Tang, X.; Li, H.; Xu, X.; Yang, G.; Liu, G.; Li, X.; Chen, D. Changing land use and its impact on the habitat suitability for wintering Anseriformes in China's Poyang Lake region. *Sci. Total Environ.* **2016**, *557*, 296–306. [CrossRef] [PubMed]
57. Zhang, Y.; Zhou, L.; Cheng, L.; Song, Y. Water level management plan based on the ecological demands of wintering waterbirds at Shengjin Lake. *Glob. Ecol. Conserv.* **2021**, *27*, e01567. [CrossRef]
58. Mei, X.F.; Dai, Z.J.; van Gelder, P.H.A.J.M.; Gao, J.J. Linking three gorges dam and downstream hydrological regimes along the Yangtze river, China. *Earth Space Sci.* **2015**, *2*, 94–106. [CrossRef]
59. Han, X.X.; Feng, L.; Hu, C.M.; Chen, X.L. Wetland changes of China's largest freshwater lake and their linkage with the Three Gorges Dam. *Remote Sens. Environ.* **2018**, *204*, 799–811. [CrossRef]
60. Lai, X.J.; Liang, Q.H.; Jiang, J.H.; Huang, Q. Impoundment effects of the three-gorges-dam on flow regimes in two China's large freshwater lakes. *Water Resour. Manag.* **2014**, *28*, 5111–5124. [CrossRef]
61. Zhang, P.Y.; Zou, Y.A.; Xie, Y.H.; Zhang, S.Q.; Chen, X.S.; Li, F.; Deng, Z.M.; Zhang, H.; Tu, W. Hydrology-driven responses of herbivorous geese in relation to changes in food quantity and quality. *Ecol. Evol.* **2020**, *10*, 5281–5292. [CrossRef]

62. Bakker, E.S.; Olff, H.; Gleichman, J.M. Contrasting effects of large herbivore grazing on smaller herbivores. *Basic Appl. Ecol.* **2009**, *10*, 141–150. [CrossRef]
63. Hole, D.G.; Willis, S.G.; Pain, D.J.; Fishpool, L.D.; Butchart, S.H.M.; Collingham, Y.C.; Rahbek, C.; Huntley, B. Projected impacts of climate change on a continent-wide protected area network. *Ecol. Lett.* **2009**, *12*, 420–431. [CrossRef] [PubMed]
64. Martay, B.; Brewer, M.J.; Elston, D.A.; Bell, J.R.; Harrington, R.; Brereton, T.M.; Barlow, K.E.; Botham, M.S.; Pearce-Higgins, J.W. Impacts of climate change on national biodiversity population trends. *Ecography* **2017**, *40*, 1139–1151. [CrossRef]
65. Amano, T.; Szekely, T.; Wauchope, H.S.; Sandel, B.; Nagy, S.; Mundkur, T.; Langendoen, T.; Blanco, D.; Michel, N.L.; Sutherland, W.J. Responses of global waterbird populations to climate change vary with latitude. *Nat. Clim. Chang.* **2020**, *10*, 959–964. [CrossRef]
66. Zhao, M.J.; Cong, P.H.; Barter, M.; Fox, A.D.; Cao, L. The changing abundance and distribution of Greater White-fronted Geese Anser albifrons in the Yangtze River floodplain: Impacts of recent hydrological changes. *Bird Conserv. Int.* **2012**, *22*, 135–143. [CrossRef]

land

Article

Linking Land Use and Plant Functional Diversity Patterns in Sabah, Borneo, through Large-Scale Spatially Continuous Sentinel-2 Inference

Leon T. Hauser [1,*], Joris Timmermans [1,2,3], Nadejda A. Soudzilovskaia [1,4] and Peter M. van Bodegom [1]

1 Department of Environmental Biology, Institute of Environmental Sciences, Leiden University, P.O. Box 9518, 2300 RA Leiden, The Netherlands; j.timmermans@uva.nl (J.T.); nadia.soudzilovskaia@uhasselt.be (N.A.S.); p.m.van.bodegom@cml.leidenuniv.nl (P.M.v.B.)

2 Lifewatch ERIC, Centre for the Virtual Laboratories and Innovations Centre, 1090 GE Amsterdam, The Netherlands

3 Biogeography & Macroecology Lab, Department Theoretical and Computational Ecology, Institute for Biodiversity and Ecosystem Dynamics, University of Amsterdam (UvA), 1090 GE Amsterdam, The Netherlands

4 Centre for Environmental Sciences, Hasselt University, Martelarenlaan 42, 3500 Hasselt, Belgium

* Correspondence: l.t.hauser@cml.leidenuniv.nl

Abstract: Global biodiversity losses erode the functioning of our vital ecosystems. Functional diversity is increasingly recognized as a critical link between biodiversity and ecosystem functioning. Satellite earth observation was proposed to address the current absence of information on large-scale continuous patterns of plant functional diversity. This study demonstrates the inference and spatial mapping of functional diversity metrics through satellite remote sensing over a large key biodiversity region (Sabah, Malaysian Borneo, ~53,000 km^2) and compares the derived estimates across a land-use gradient as an initial qualitative assessment to test the potential merits of the approach. Functional traits (leaf water content, chlorophyll-a and -b, and leaf area index) were estimated from Sentinel-2 spectral reflectance using a pre-trained neural network on radiative transfer modeling simulations. Multivariate functional diversity metrics were calculated, including functional richness, divergence, and evenness. Spatial patterns of functional diversity were related to land-use data distinguishing intact forest, logged forest, and oil palm plantations. Spatial patterns of satellite remotely sensed functional diversity are significantly related to differences in land use. Intact forests, as well as logged forests, featured consistently higher functional diversity compared to oil palm plantations. Differences were profound for functional divergence, whereas functional richness exhibited relatively large variances within land-use classes. By linking large-scale patterns of functional diversity as derived from satellite remote sensing to land-use information, this study indicated initial responsiveness to broad human disturbance gradients over large geographical and spatially contiguous extents. Despite uncertainties about the accuracy of the spatial patterns, this study provides a coherent early application of satellite-derived functional diversity toward further validation of its responsiveness across ecological gradients.

Keywords: biodiversity; land use; functional diversity; oil palm; Borneo; logging; Sentinel-2; plant diversity; satellite remote sensing; trait-based ecology

1. Introduction

The rampant decline in global biodiversity over the last few decades has become a major threat to the ecosystems on which humans depend [1]. Land-use change fuelled by agricultural expansion has been a particularly salient direct driver of biodiversity losses in terrestrial ecosystems as forests, wetlands, and grasslands have been converted [1]. Current estimates suggest that land-use-related pressures have reduced local biodiversity intactness beyond the planetary boundaries for biosphere integrity as a safe operating space for humanity [2].

In many tropical species-rich regions, the conversion of forests by land-use change has led to stark contrasts in the spatial distribution of biodiversity [3,4]. Global demands for food, biofuel, and other commodities have driven the rapid expansion of oil palm and paper and pulp industries at the expense of lowland rainforests, jeopardizing forest biota [5,6]. For instance, in Sabah, northern Borneo, records show that over 39.5% of the forest has been cleared and just 19.1% of the land surface of Sabah remained as intact primary forest in the period between 1973 and 2010 [7–9]. These trends have resulted in high rates of biodiversity loss and degradation of the forest landscape, threatening more than 1000 taxa of endemic plants present, including the iconic Dipterocarpaceae species, as well as the unique variety of fauna that are dependent on these habitats [3,4,10]. Small-scale field studies indicated that oil palm plantations support substantially fewer plant species [11,12], non-volant small mammals [13], and butterflies [14], and have lower arthropod and multi-trophic functional diversity in comparison to natural and logged forests [15,16].

The threats associated with land-use pressure are likely to further disrupt the remaining intact primary lowland rainforests in the region and decrease biodiversity [17]. One of the biggest challenges to combat these threats is that monitoring the effects of land use in the region is subject to the relative paucity of data on basic ecology [18]. In order to effectively preserve and monitor plant biodiversity, it is necessary to quantify diversity patterns and understand the behaviors and ecologies governing the distribution and abundance of species, communities, and ecosystems. In particular, the accurate characterization of plant functional diversity over large spatial and temporal scales would greatly improve our ability to track the status and resilience of key biodiversity areas, such as the entire biodiversity hotspot of Borneo, and understand the effects of human interventions [19].

Functional diversity is associated with taxonomic and phylogenetic biodiversity measures, yet tends to respond more consistently to environmental drivers [20–22]. It captures the ranges and abundances of the combined functional traits of the organisms present in an ecosystem and is strongly indicative of the way these ecosystems operate [23,24]. To maintain crucial ecosystem functions, a growing body of research highlights the importance of preserving plant functional diversity as the fundament of ecosystem functioning [25–27]. Thus, large-scale functional diversity maps, currently absent, would provide essential information for biodiversity conservation.

Mapping plant functional diversity requires extensive, consistent, and repeated data on traits over (continuous) regional and global scales, which we still lack [18]. Most field measurements of plant traits in terrestrial ecosystems were rather small-scale and limited in spatial extent due to their laboriousness and related costs [28–30]. Attempts to combine disparate in situ research activities in global traits databases tend to be temporally and geographically constrained, suffer from sampling inhomogeneities, largely exclude remote areas, and are bound by the limitations of interpolation [18,31,32]. Moreover, existing field sampling efforts were most sparse in the high-biodiversity tropics, leading to a strong underrepresentation [33].

Satellite remote sensing techniques are increasingly used to monitor and study large-scale temporal and spatial landscape changes across many parts of the world [34]. These observations are uniquely valuable because they can provide complete repeated spatial sampling, even when the measurements reveal only part of the complex reality. With ongoing technological advances, satellite remote sensing observations are poised to move beyond the monitoring of land cover change and quantify plant functional diversity across regional landscapes [33,35]. Indeed, a growing number of studies demonstrate the retrieval and analysis of individual canopy and leaf traits in discrete plots and locations through satellite remote sensing [36–42]. Yet, the step from using internally consistent trait estimates to deriving functional diversity patterns from satellite remote sensing is undertaken less often and remains limited to spatially discrete observations as opposed to the potential to facilitate spatially continuous 'wall-to-wall' inference [42–45].

However, satellite remote sensing of plant functional diversity presents a paradox: the difficulties in acquiring temporally and spatially consistent ground data over large areas highlight the clear added value of satellite remote sensing to achieve such. At the same time, the scarcity and mismatch of present data provide significant challenges to training data models for retrieving plant diversity estimates and validating the outcomes of satellite earth observation inferences [18]. Moreover, the coarse (multiple canopies aggregated), spectrally derived, timely, and pixel-wise observations from satellite remote sensing present a challenge regarding a direct coupling of traditional, static, leaf-level trait measurements done in the field [46,47]. Instead, the evaluation of functional diversity patterns across well-studied ecological gradients can provide a means to qualitatively assess the responsiveness of satellite-based plant functional diversity estimates [48,49].

In this paper, we present a spatially continuous ('wall-to-wall') inference of functional diversity estimates from satellite remote sensing retrieved over the biodiverse and heterogenous region of Sabah, Malaysian Borneo. The region's vastness, complexity, inaccessibility, and regulatory constraints (private ownership) are exemplary of the use of remote sensing to overcome the difficulties to conduct large-scale field campaigns. We apply the ESA's SNAP biophysical processor to retrieve spectral trait indicators—leaf area index, leaf water content, and leaf chlorophyll content—from the spectral bands of Sentinel-2 imagery at 20 m resolution. Through multivariate diversity metrics [50], an analysis of the functional diversity found in these combined traits is conducted and the large spatially continuous retrieval of functional diversity estimates is held against a land-use gradient with a large number of observations (N = 5626) comprising of large (40 ha) remotely sensed single land-use plots. Linking space-borne functional diversity to land use provides an initial step for the explorative assessment of the potential merits and challenges of large-scale spatially explicit inference of functional diversity metrics from satellite remote sensing.

2. Methods

2.1. Study Area

This study focused on the Malaysian province of Sabah, which is located on the northern tip of Borneo (115°12′27.317″ E–117°59′5.608″ E, 4°26′3.612″ N–7°13′51.89″ N). The region represents a crucial global biodiversity hotspot [3] with well-studied gradients of elevation [51–53] and validated maps of relevant land-use types [7,9] (Figure A1b). Sabah consists of lowlands, as well as mountainous forested territories with elevations ranging from sea level to over 4000 m at Mt. Kinabalu's peak. Sabah records an average annual rainfall of 2890 mm and a mean annual temperature of 27.8 °C [54]. Over the past few decades, widespread forest conversion for oil palm and timber/pulp production has significantly altered and threatened biodiversity, including over 1000 taxa of endemic plants, as well as the unique variety of fauna dependent on these habitats [3,8,10]. The study area is characterized by strong contrasts in human disturbance, from intact forests to intensively managed plantations, and elevational gradients that affect plant functional diversity and are expected to be reflected in satellite remote sensing estimates of these patterns [15,52].

2.2. Retrieval of Functional Traits

Sentinel-2 L1c data were acquired through ESA's Copernicus Scientific Hub over the extent of our study area. The Multi-Spectral Imager (MSI) offers observations over 13 spectral bands, the majority of which are at a 20 m spatial resolution [55]. Optical remote sensing acquisition above Sabah, Malaysia, is challenged by its year-round high average cloud cover. For this study, we acquired the Sentinel-2 observations for the 9th of July 2017 covering the study area (Figure A1a). The cloud cover on this date was one of the lowest since the launch of Sentinel-2 in 2014. Mosaicking multi-temporal images together would have introduced temporal deviances that could lead to diverse artifacts based on the inference date rather than vegetation characteristics. After the acquisition, the data were atmospherically corrected using the Sen2Cor processor [56,57]. Stringent quality flags stemming from both the atmospheric correction and the biophysical processor (see below) were applied to mask all areas affected by cloud contamination, poor atmospheric correction, poor trait retrievals (outside the physical range of variation), and shadows [57,58]. Additional cautionary buffers 100 m in radius were applied around the quality flags to further limit the influence of clouds and cloud shadows on the spectral properties of the imagery. Non-vegetated areas with a fractional cover (FC) below 30% were masked out to remove non-/marginally vegetated areas.

Estimation of canopy traits from the Sentinel-2 imagery was conducted using the biophysical processor of the ESA's Sentinel Application Platform (SNAP) Sentinel-2 toolbox [58]. SNAP uses an artificial neural network (ANN) inversion that was pre-trained on a PROSAIL simulated database that included canopy reflectance and the corresponding set of input parameters. SNAP includes an unreleased version of PROSPECT prior to PROSPECT-4 [59], coupled with the SAIL model [60,61]. The value, range, and distribution followed for each input parameter of the models are described in [58] and aim to provide general global applicability without the ingestion of ecosystem-specific ancillary data, although that comes at the cost of precision [58]. PROSAIL is bound by strong simplification of canopies that assumes a homogenous turbid medium where absorption is defined by leaf, canopy, soil, and angular properties [62]. Therefore, the interpretation needs to be done in consideration of the underlying assumptions and application of a 1D model [63]. Despite its limitations, earlier validation studies reported reasonable performance of SNAP retrievals in forested regions [39,42] and are further addressed in the discussion section.

Traits in SNAP are derived at the canopy level and include leaf area index (LAI), canopy chlorophyll (CAB*LAI), and canopy water (EWT*LAI). We reversed the multiplication by LAI to arrive at leaf level estimates: leaf chlorophyll (CAB) and leaf equivalent water thickness (EWT). LAI specifies the leaf surface area per unit ground surface area. The retrieval conducted here produces a measure of the 'effective LAI,' which does not account for clumping factors and therefore differs from a 'true LAI' measure [64,65]. LAI (m^2/m^2) determines how much light can be captured to influence primary production, but also transpiration and energy exchange [65–67]. Complementary to LAI, CAB and EWT influence processes occurring at the leaf level [68]. CAB ($\mu g/cm^2$) corresponds to a surface-based leaf content of chlorophyll-a and -b. Chlorophyll has an important role in determining the photosynthetic capacity and resource strategy of plants [69,70]. EWT (in g/cm^2) refers to the water mass stored in leaves per leaf surface area. EWT is important for the physiological plant performance and regulatory mechanisms that play a role in drought and stress tolerance [71–76]. Taken together, the traits retrievable through SNAP are ecologically meaningful.

2.3. Estimating Functional Diversity

The spatially continuous retrieved traits from the Sentinel-2 spectral reflectance were used to further estimate the functional diversity. Functional diversity is commonly partitioned into three complementary aspects of functional diversity: richness, evenness, and divergence [50,77,78]. Functional richness is a measure of the functional space occupied by a community and was calculated based on the convex hull volume [79]. Functional divergence and evenness metrics describe how trait combinations are distributed within the community's trait space, which are indicative of niche differentiation and niche space optimization, respectively [77,80–82]. Functional divergence was calculated with Euclidian distances applied to a centroid-based approach [50] that was adapted by Schneider et al. [83] for a pixel-based approach. Functional evenness was determined through branch length variation of the minimum spanning tree of a trait distance matrix signaling the regularity of the distribution across the trait space [50]. The functional diversity metrics were calculated over the 95% centermost data points, as determined by kernel density estimates, to limit the influence of extreme values, noise, and possible retrieval artifacts.

We opted for large plots, sized equally at 40 ha. (1000-pixel observations), to calculate the functional diversity metrics. This allowed for showcasing the capability of satellite remote sensing to map large spatially continuous plots to capture a large share of the variability in canopy compositions to base the functional diversity calculations on. With

1000-pixel observations, this design potentially harnesses robustness against noise in the observations. Despite the patchiness of the data due to the masks applied and the mosaic of land-use patterns, the 40 ha plots still offered a large number of data plots per land-use type (N = 5626). The plots were drawn using an algorithm based on minimal Euclidian distances to the starting pixel of the plot while solving the condition of meeting a single land-use continuous area of 40 ha [84].

2.4. Land-Use Data

Land use was derived from CIFOR's open-access and validated 'Atlas of deforestation and industrial plantations in Borneo' (https://www.cifor.org/map/atlas/ accessed on 3 July 2018) [7,9]. Their validated maps are based on longitudinal, up to 30 m spatial resolution, LandSat satellite imagery (1973–2016) with additional visual, expert-based interpretation methods and maps of oil palm and pulpwood concessions. Here, we assessed the three largest vegetated land-use types: (1) 'Intact Forest', which are old-growth forests. The overstory of these forest ecosystems is generally characterized by old closed-canopy emergent trees. Dipterocarpaceae are the dominant tree species in primary forests, accounting for 25% of the tree population and 60% of the standing volume [85]. (2) 'Logged Forest', which are intact forests that have been impacted by industrial-scale mechanized selective logging at some point since 1973. (3) 'Industrial Oil Palm Plantations', which are production systems mainly revolving around monoculture planting of Elaeis guineensis jacq. Small-holder oil palm cannot be consistently distinguished in the land-use maps. An overview of the spatial distribution of these three dominant land-use types can be found in Figure A1b. Oil palm plantations were found to only occur in the lowlands (<500 m ASL) (Table A1). The Shuttle Radar Topography Mission (30 m spatial resolution) was used to map elevation. Elevation may affect differences in functional diversity within and between land-use types.

2.5. Data Analysis

To assess the plausibility of the trait values on which we based functional diversity estimates, the performance of the inversion of SNAP's biophysical processor retrieval from spectra to traits was examined. First, we conducted a sensitivity analysis to assess the spectral layout of Sentinel-2 bands in terms of receptiveness to retrieve these traits. The analysis was based on repeated PROSAIL simulations with random variations of the trait values while mapping the spectral responses and the correlation between Sentinel-2's spectral bands and trait variations.

Second, we re-modeled the spectra based on PROSAIL in forward mode from the estimated trait values. The simulated spectra were compared against the observed Sentinel-2 spectra to assess the performance across different land-use types. This was done for 20,000 randomly drawn pixels over the study area. The search ranges for the remaining PROSAIL traits were constrained to the distribution of input variables described by [58]. We used spectral angle mapper (SAM), mean absolute error (MAE), and root mean squared error (RMSE) to assess the deviation between the simulated and actual spectra. We evaluated differences in errors across land-use types to examine whether differential performance across land-use types could have affected our results.

Third, we compared retrieved trait distributions against in situ measured traits of common species in the different studied land-use types [64,86–88]. For LAI, the study by Hadi et al. [64] provided measurements of effective LAI across sampling sites in Sabah consisting of unlogged forest, logged forest, and oil palm plantations, although the plantations sampled were relatively young (planted <10 years ago). The CABs of forests and oil palm plantations were available for top-of-canopy chlorophyll for adult oil palms (Elaeis guineensis) from [88] and those of intact forests in Danum Valley were from [86]. Relevant field measurements on EWT relevant to our study area were difficult to acquire. However, we were able to model EWT based on data on LMA available from the 'Traits of Bornean Trees Database' [89], which is part of the global TRY plant trait database [90]. For the 'Traits of Bornean Trees Database', we focussed on Dipterocarpaceae specifically, considering their dominant role in the primary forest in Sabah, accounting for 25% of the tree population and 60% of the standing volume [85]. To model EWT, we assumed a leaf water content of 63% following findings by [91] for tropical evergreen forests. We simulated the range of EWT values based on these data through the following equation:

$$EWT = \left(\frac{LWC_{\%}}{1 - LWC_{\%}}\right) * (LMA) \tag{1}$$

Differences in functional diversity metrics between land-use types were assessed using ANOVAs. The assumptions of the ANOVAs were evaluated and a log transformation of functional richness was conducted to ensure the normality of residuals in an otherwise strongly skewed distribution. The explained variance was expressed through eta^2 values which is a measure of effect size for use in ANOVA, analogous to R^2 in multiple linear regression. Significant differences between individual land-use types were further analyzed using a post hoc Tukey's HSD test.

3. Results

3.1. From Mapping Spatially Explicit Spectral Trait Indicators to Functional Diversity Estimates

The trait maps presented a high degree of patchiness due to the applied masks (including clouds, quality flags, vegetation cover, and land use) (Figure 1a). The plausibility of the spectral trait indicators was assessed through three analyses to test their validity. First, the sensitivity analysis of Sentinel-2's bands to the spectral trait indicators (LAI, CAB, and EWT) retrieved in this study showed promising responsiveness (Figure A2). Second, the reversely estimated spectra modeled using PROSAIL—simulated from the SNAP-retrieved trait estimates as input—had an RMSE of 0.012, which was a 7.6 percent mean deviation from the actually observed spectra. No profound differences in errors were observed between the individual land-use types, which suggested there was no structural bias in the inversion performance for approximation of the vegetation present in different land uses (Table A2). Finally, the comparison of the retrieved trait values to those presented in trait databases and the literature suggested that the estimated ranges of the retrieved traits were to a large degree in line with the ranges of trait values previously measured in Borneo (Figure A3).

3.2. Land-Use Patterns of Plant Functional Diversity

The spatially continuous maps of functional diversity estimates were exemplified by functional richness in Figure 1d. All functional diversity metrics appeared to differ significantly across land-use types, with all p-values well below 0.001 (Figure 2). The post hoc analysis revealed that logged forest and intact forest were statistically similar in terms of functional richness and functional evenness. For all functional diversity metrics, intact and logged forests were significantly different from oil palm plantations (Figure 2).

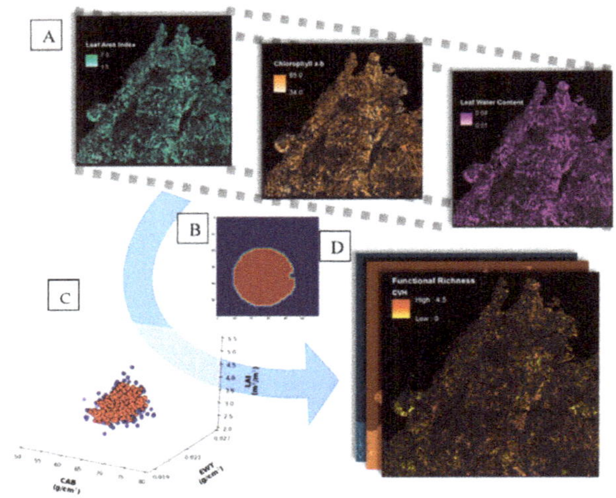

Figure 1. The data pipeline from satellite remotely sensed traits to functional diversity metrics maps. (**A**) Maps of leaf area index (m^2/m^2) (LAI), leaf chlorophyll-a and -b content ($\mu g/cm^2$) (CAB), and leaf equivalent water thickness (g/cm^2) (EWT). Masks are portrayed in dark grey. (**B**) Pixel-wise selection of 40 ha plots, i.e., 1000 pixels, based on a minimal Euclidian distance drawing new pixels relative to the starting pixel of the plot while remaining within land-use and quality masks. (**C**) Functional diversity metrics calculated over 40 ha plots, i.e., 1000 pixels, as illustrated in the 3D plot example with axes representative of different functional traits. (**D**) Maps of functional diversity over the study area of Sabah, Malaysia, exemplified here by functional richness as one of the three functional diversity metrics.

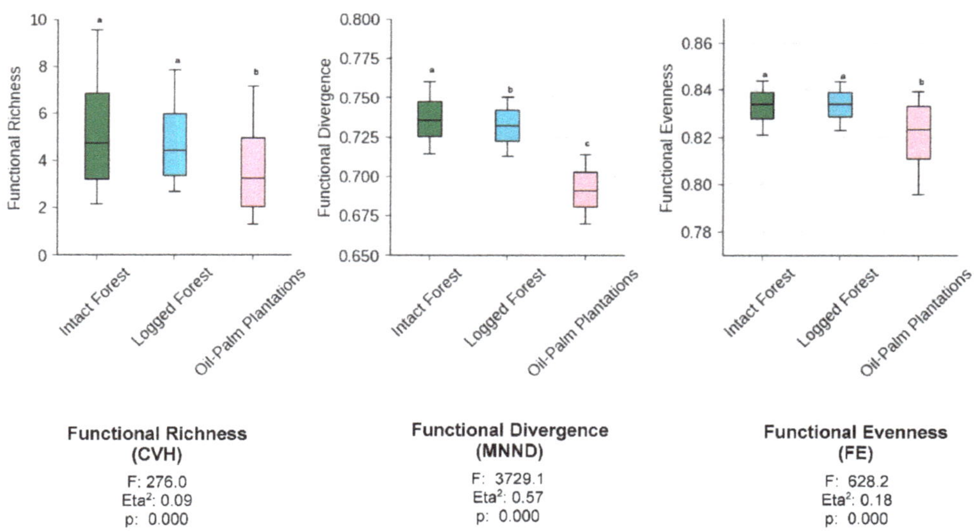

Figure 2. Differences in functional diversity metrics across land-use types are displayed in boxplots with the corresponding ANOVA results. The boxplots' boxes represent the 25th, 50th, and 75th percentiles and the whiskers depict the 10th and 90th percentiles. The annotated letters (a, b, c) indicate the significance of pairwise differences in retrieved trait estimates between land-use types according to post hoc Tukey's HSD tests.

4. Discussion

Our case study in Sabah is illustrative of a biodiversity hotspot at risk of commodity-driven land-use changes [3,8,9]. The impact of land use on functional biodiversity has been widely studied in ecology, generally through small-scale local field studies [21,22]. We have re-projected these ecological expectations to scale up considerably from traditional field assessments to the potential to study large-scale landscapes, whole regions, and terrestrial ecosystems instantaneously and with spatial continuity. Despite still being an early adoption, the results indicate initial correspondence of satellite remote-sensing-derived functional diversity metrics to broad human disturbance gradients shaped by land use. These insights may be further developed when in situ validation data and relevant ancillary ecosystem-specific information would become available. In this discussion, we address the current workflow and results in light of the limitations of the methodology applied and the value of qualitative assessments against historic trait measurements and land-use gradients.

4.1. Trait Retrieval for Functional Diversity

This study applied a workflow (Figure 1) to map the functional diversity using Sentinel-2 over a large continuous region without a heavy reliance on a priori in situ canopy trait measurements. The latter is relevant as matched field samples are hard to obtain across large geographic extents and pixel-based scales. The validity of such large-scale functional diversity estimates depends on the accuracy and representation of traits. Trait selection is constrained by the specifications of the sensor and the parameters of the radiative transfer models applied. Here, the selected traits were related to relevant carbon and water fluxes and offered three axes of functional differentiation [65,69,76,92]. The current selection of traits was pragmatic, where a larger number of traits might increase the ability to detect functional differences between observations [24]. In practice, however, model inversion becomes increasingly challenged by ill-posedness with a larger number of traits to be retrieved with the same amount of input data, especially in multi-spectral broadband inference.

For the current trait selection (LAI, CAB, and EWT), the sensitivity analysis of the Sentinel-2's bands showed promising responsiveness (Figure A2). This analysis served as a precursor of the performance of PROSAIL within SNAP when applied to Sentinel-2. Due to its ease of use, genericity, and, most importantly, functionality without hard-to-obtain ancillary data, the SNAP biophysical processor is currently a likely first port of call for many users [39], although it clearly also has limitations.

The ANN in SNAP is meant to provide generic global applicability without the ingestion of ecosystem-specific ancillary data [58]. However, it was shown that local optimization of the trait ranges under study for inversion can improve retrieval performance [42,93–95]. SNAP's biophysical processor does not facilitate a re-training of the ANN on PROSAIL simulations specific to the ecosystems studied here. Such functionality could theoretically improve the accuracy of spectral trait indicators. Additionally, the implementation of active learning heuristics can prove helpful regarding adding constraints to inversion space and facilitating intelligent sampling for the training of retrieval algorithms to overcome some of the ill-posedness and optimize a simulation subset to the ecosystem under study [40,96,97].

Similarly, it may be argued that there are radiative transfer models that are better suited for heterogeneous forest canopies as compared to PROSAIL, which is bound by 1D simplification of canopies [98]. However, alternative radiative transfer models, e.g., INFORM [99] and FRT [100], are complex and require a larger number of biophysical parameters. Without a priori information, the heavy parameterization in inversion may further induce ill-posedness and therefore, in fact, hamper retrieval performance and feasibility, especially in multi-spectral settings [39,101].

Despite its limitations, the suitability of SNAP for agricultural applications was confirmed in several studies [102–106]. For forests, Brown et al. [39], Hauser et al. [42], and Nguyen et al. [97] reported reasonable performances of the SNAP retrieval algorithm in a deciduous broadleaf forest site in Southern England, a heterogenous shrub-forest landscape in Portugal, and mangrove forests, respectively. Moreover, the performance of SNAP in a heterogeneous mixed mountain forest in Bavaria, Germany was similar to that of the input parameter heavy INFORM inversion [37]. While none of these studies applied SNAP in rainforest ecosystems relevant to our study area, these quantitative findings suggested that our pipeline is reasonable for indicative large-scale applications.

Consistent with that assessment is our finding that there was no indication of structural biases in inversion between the land-use types (Table A2) or deviation from regionally relevant trait ranges presented in literature and the TRY database (Figure A3). The most notable deviance against historic trait measurements is an overestimation of the observed LAI of oil palm plantations compared to Hadi et al.'s [64] effective LAI measurements. Differences in LAI between young plantations versus older plantations might have been responsible for this discrepancy. In our study, observed LAI values were the highest in plantations rather than tropical evergreen broadleaf forests (Figure A3). This was also reported by [65]. Intensive management regimes—through planting schemes, fertilizer, and pest management—aim to maximize production in these plantations and both resonate with relatively high chlorophyll and LAI ranges. Industrial oil palm plantations are often located on more favorable lands for high productivity in terms of elevation and slope (Table A1).

Nonetheless, the comparison against historic trait measurements should be considered carefully. First, there are temporal mismatches between the data of Sentinel-2 acquisition and the in situ trait measurements used for comparison. This likely results in inconsistencies in the comparisons made, especially given the cyclical nature and phenology of ecosystem processes [33,107]. Second, the used historic trait data were sampled at the level of individual canopies/branches/leaves, whereas the Sentinel-2 inference results in an aggregation of multiple canopies in pixels scaled in a 20 m resolution raster [47]. Indeed, the field-based estimates of EWT and CAB of individuals exhibited a larger variation as compared to satellite-observed trait indicators (Figure A3). The pixel-based aggregation of multiple species of the latter might have leveled out some of the individual variations found in the former. Third, the aggregation of canopies in Sentinel-2 pixels resulted in a dominant signal of the overstory vegetation and, in particular, the top of the canopy. Accurate correspondence of field measurements to top-of-the-canopy sunlit overstory samples is not necessarily warranted in trait databases [108]. Lastly, in contrast to species-mean trait estimates, the large-scale continuous inference through remote sensing will likely include the local representation of intra-specific variation [33].

4.2. Land-Use Gradient as Qualitative Assessment

The scale and extent of satellite remote sensing in comparison to common field observations seriously challenges quantitative validation efforts for plant diversity assessments. In light of the scarcity of relevant ground measurements, this study aimed to exploit the spatial continuity and synoptic inference of satellite remote sensing observations via qualitative testing against well-studied and large-scale ecological gradients. Airborne studies, e.g., [48,49,109], examined remotely sensed trait diversity against hypothesized linkages between functional diversity and ecosystem productivity, spatial scales, elevation, and climatic gradients. Here, we have scaled up to Sentinel-2 to compare functional diversity against different land-use types given that land use is known to strongly impact functional diversity [22,110]. The ability of satellite-derived functional diversity metrics to detect such differences at regional spatial extents offers an initial evaluation of its potential to study broad ecological phenomena from space.

The results indicated significant differences between land-use classes across functional diversity metrics. Although different spatial scales and even taxa present a challenge for quantitative comparisons, our findings corresponded qualitatively with conclusions from field studies using ground measurements; intact forests and logged forests harbor significantly more functional diversity than oil palm plantations [15,16,111,112]. The significance of differences in functional diversity did vary strongly across the metrics applied. Land use explained approximately 9% of the variance in functional richness (Figure 2). Functional divergence and evenness patterns differed more strongly across the land-use gradient, explaining 57% and 17% of the observed variance, respectively.

The low explained variance for functional richness corresponded with its high variation within the different land-use types (Table 1, Figure 2). Part of this variation could be related to a variation in landscape properties over the large geographic extent. The variation in terms of soil, elevation, slope, landscape heterogeneity (geodiversity), and microclimates may all affect functional richness [113,114]. Additional analysis (Figure A4) indeed confirmed a strong decrease in functional richness with increasing elevation for the different land-use types. By reducing the spatial extent of analysis to a smaller subset within the study area, the explanatory power of land-use types increased substantially. The findings from the subset revealed a significant increase in the explained variance of land use for functional richness (~15%) while still signaling a large variance within land-use classes. Importantly, the convex hull volume—on which the calculation of functional richness is based—is relatively sensitive to anomalous observations [115]. Our analysis was somewhat conservative by looking only at the 95% centermost observations through kernel density estimates. Nevertheless, the large landscape heterogeneity combined with noise and inconsistencies in the surface reflectance and spectral trait indicator retrieval can introduce variations that strongly affect overall functional richness.

Table 1. Descriptive statistics per functional diversity metric across land-use types in Sabah.

	N	Functional Richness		Functional Divergence		Functional Evenness	
		Mean	St. Dev.	Mean	St. Dev.	Mean	St. Dev.
Intact Forest	1192	5.474	3.354	0.737	0.020	0.833	0.010
Logged Forest	3018	4.971	2.331	0.732	0.015	0.833	0.008
Oil Palm Plantations	1416	3.974	2.876	0.692	0.018	0.820	0.018

Despite large within-group variation in functional richness, the distinction between forested and oil palm plantation land-use types remained significant and prevailed across a large spatial extent and number of plots. The highest mean functional richness was found in intact forests, closely followed by logged forests. Previous studies pointed out that functional richness can peak in sites that are exposed to moderate disturbances, creating heterogeneity in ontogenetic stages and landscape characteristics, which could play a role in the comparable functional richness between intact and logged forests [24,116,117].

Functional divergence showed more profound differences across land-use types. Intact forest recorded the highest levels of functional divergence in our analysis, followed by logged forests and oil palm plantations. The interpretation of functional divergence is tied to niche differentiation within ecological communities [77,80–82]. Niche differentiation contributes to the maintenance of functional diversity in natural tropical systems [82]. More intensively managed land-use regimes, which are characterized by monoculture, planting schemes, and pest management, diminish natural processes that allow for niche differentiation [118–120]. This resonates with the main differences found in our results, with oil palm plantations scoring significantly lower on functional divergence (Figure 2).

Functional evenness relates to how evenly trait space is filled [50,78,80]. Even distributions can be tied to a degree of optimization of trait space in response to resource availability, as well as to less functional redundancy. In man-made landscapes, such drivers are suppressed by monoculture planting, pest management, and homogenous ontogenetic stages/tree sizes. Oil palm plantations exhibited significantly lower divergence and evenness within the trait space observed (Figure 2). Both functional divergence and evenness were relatively unchanged across elevations (Figure A4). This coincided with findings by [49] in tropical forests across an Amazon-to-Andes elevation gradient.

4.3. Outlook

Beyond the shortcomings, the study provides a new platform and geographic extent (spatial scale) that indicated that oil palm plantations exhibited significantly lower functional diversity than intact or logged forests, which can contribute to the debate on oil palm's impact on tropical biodiversity and further build a case against controversial proposals to reclassify oil palm as a forest cover type [12,121,122]. Our exploration builds a case for further investment in quantitative validation through dedicated field campaigns. From this point, investing in and overcoming the challenges of representative in situ validation of pixel-based functional diversity is the next stride to further assess its accuracy and more subtle differences, as well as the robustness of these metrics against uncertainties in trait estimates. Moreover, in terms of retrieval, further maturation of the current approach toward full spatio-temporal continuity will be needed. This will include multi-temporal observations; data assimilation of multi-sensor, multi-scale remote sensing observations; and ancillary ecological datasets to achieve the following: (i) study change in functional diversity over time (important to study the impact of human disturbances; [18,123]); (ii) include a wider range of consistent traits and deal with the ill-posedness of such retrieval; (iii) allow for cross-validation with spatial, temporal, and spectral consistency across instruments; (iv) study the effects of scale and grain; and (v) overcome data gaps caused by clouds (highly relevant for humid tropical rainforests such as those found in Sabah) and other interferences [124–127].

5. Conclusions

By linking the functional diversity estimates derived from remote sensing to land-use information, this study showed the potential merits to study broad ecological patterns over large geographical extents using satellite earth observation. The study provides a new platform and geographic extent (spatial scale) that strongly indicated that oil palm plantations exhibited significantly lower functional diversity as compared to intact and logged forests, coinciding with results from earlier field studies. Specifically, we found that the observed differences in functional diversity across the land-use gradient were significant. Notably, profound differences between metrics were observed: a large variance within land-use types was observed for functional richness, while functional divergence exhibited particularly strong responsiveness to land use. With that in mind, this study acts as an exemplar for satellite-derived monitoring of functional diversity for a key biodiversity area for which traditionally little information is available. The study provides an early application toward the maturation of a spatially and temporally explicit method that hopefully fuels further validation efforts and assessment of its responsiveness across ecological gradients.

Author Contributions: Conceptualization, L.T.H., J.T., N.A.S. and P.M.v.B.; Formal analysis, L.T.H.; Funding acquisition, P.M.v.B.; Methodology, L.T.H.; Supervision, J.T., N.A.S. and P.M.v.B.; Visualization, L.T.H.; Writing—original draft, L.T.H.; Writing—review & editing, L.T.H., J.T., N.A.S. and P.M.v.B. All authors have read and agreed to the published version of the manuscript.

Funding: This project has received funding from the European Union's Horizon 2020 research and innovation program through the MULTIPLY project (http://www.multiply-h2020.eu/) under grant agreement no. 687320.

Conflicts of Interest: The authors declare no conflict of interest.

Appendix A.

Figure A1. (**A**) Map of the Sentinel-2 observation, as seen through a true visible range composite image, over the study area in Sabah, Malaysia, on the 9th of July 2017 after atmospheric corrections. (**B**) Map of the three land-use classes within the Sabah study area used for analysis.

Table A1. Descriptive statistics of elevation and slope observed between plots across land-use types. Data is derived from digital elevation models observed by the 30 m spatial resolution Shuttle Radar Topography Mission (SRTM).

	Land Use	N	Mean Elevation (m)	Mean Slope (% Change)
	Intact Forest	1192	439.90	1.88
	Logged Forest	3018	419.71	1.30
	Oil Palm Plantations	1416	63.84	0.74

Appendix A.1. Qualitative Assessment of the Validity of Spectral Trait Indicators

To assess the plausibility of the derived spectral trait indicators on which we based functional diversity estimates, the performance of the inversion of SNAP's biophysical processor retrieval from spectra to traits was examined.

First, we conducted a sensitivity analysis to assess the spectral layout of the Sentinel-2 bands in terms of receptiveness to retrieve these traits (Figure A2). The analysis was based on repeated PROSAIL simulations with random variations of the trait values while mapping the spectral responses and the correlation between Sentinel-2's spectral bands and trait variation. The findings demonstrated significant spectral sensitivity to the studied traits (LAI, EWT, and CAB) (Figure A2). These findings matched the sensitivity analyses of Sentinel-2 for retrieval of the traits under study (LAI, CAB, and EWT) that were demonstrated in previous sensitivity analyses [43,128–130].

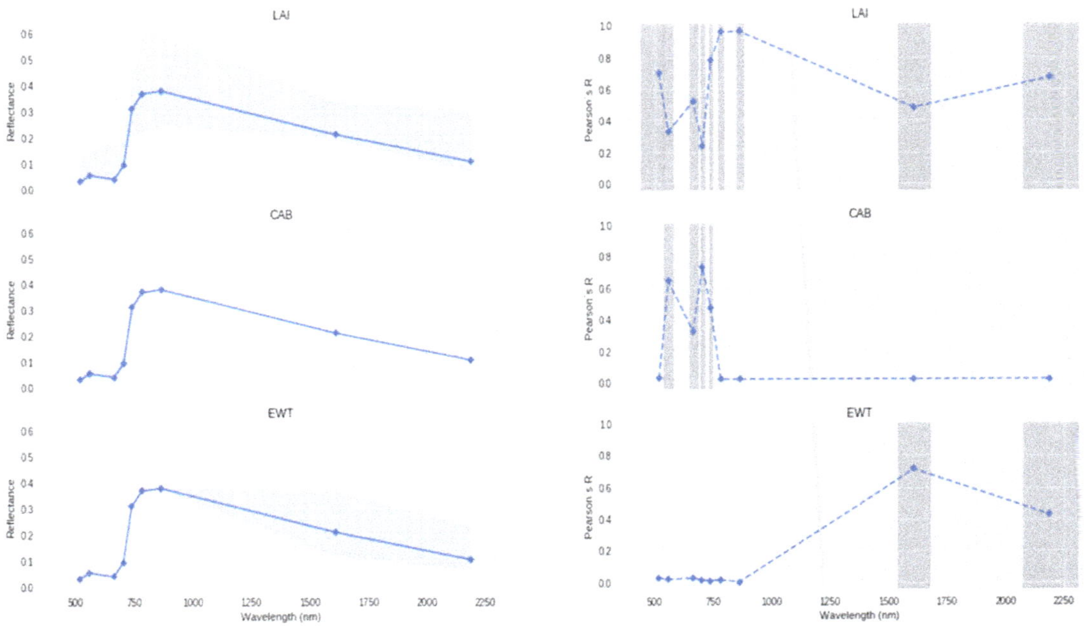

Figure A2. Sensitivity analysis of PROSAIL's LAI, CAB, and EWT parameters to the spectral layout of Sentinel-2 MSI bands used in this study. The left pane illustrates the range of variability in spectral response to changes in the parameters, while all other parameters were kept constant in the modes defined in [58]. The right pane depicts the correlation (Pearson's R) of different bands to changes in the parameters using a range of simulations defined in [58]. Significant correlations are indicated by the grey highlighted bars.

Second, the presence of biases in the derived spectral trait indicators across land-use types was examined by reversing the inversion process. From the spectrally retrieved trait estimates obtained through inversion, we re-modeled the spectra based on PROSAIL in forward mode. The simulated spectra were compared against the actual observed Sentinel-2 spectra to assess the performance (size of the error) between different land-use types/canopy types. This was done for 20,000 randomly drawn pixels over the study area. Search ranges for the remaining PROSAIL traits were constrained to the distribution of input variables described by [58]. We used spectral angle mapper (SAM), mean absolute error (MAE), and root mean squared error (RMSE) to assess the deviation between the simulated and actual spectra. We evaluated differences in the size of errors across land-use types to examine whether such biases could have affected our results. On average, across the land-use types, an RMSE of 0.012 was observed, which was a 7.6 percent mean deviation from the actually observed spectra (Table A2). No profound differences in errors were observed between the individual land-use types, which suggested that there was no structural bias in the inversion performance for approximation of the vegetation present in different land uses.

Table A2. Comparison of errors between simulated spectra with retrieved trait estimates as input and actual Sentinel-2 reflectance spectra stratified across land-use types.

	RMSE		MAE		%nRMSE	Sampled Pixels (N)
Land Use	μ	σ	μ	σ	μ	
Intact Forest	0.012	0.009	0.007	0.005	7.6	4774
Logged Forest	0.014	0.01	0.008	0.006	8.9	4274
Oil Palm Plantations	0.012	0.01	0.007	0.006	7.2	3952

Third, we compared the distributions of spectral trait indicators against in situ measured traits of common species in the different studied land-use types. Figure A3 indicates that the spectral trait indicators were to a large degree in line with the range of measurements from field studies. The most notable deviances were as follows: (1) An overestimation of LAI in oil palm plantations compared to Hadi et al.'s [130] effective LAI measurements. Hadi et al.'s [130] measurements are conducted in relatively recently planted oil palm plantations. Differences in LAI between young plantations versus older plantations might be responsible for this discrepancy. (2) A much larger variation of EWT and CAB in the species measurement conducted in the field. Notably, these measurements consisted of a variety of individual species, whereas our observations were based on the pixel-based aggregation of multiple species, possibly leveling out some of the variations.

Functional richness decreased along the elevational gradient. In line with ecological theory, this indicates a stronger functional convergence with elevation linked to stronger environmental filtering of fitness in higher altitudes [49]. A steep drop in functional richness was particularly observable above 1400 m ASL. These findings corresponded with earlier studies on elevational patterns of tree species richness on Mount Kinabalu, Borneo [51–53]. Similar to the functional richness in our study, tree species were found to decrease rapidly with elevation above 1500 m [53]. Moreover, using airborne remote sensing, studies showed a lower functional richness at a higher elevation, suggesting a smaller range of resource availability at higher altitudes, whereas, again, functional divergence remained relatively unaffected by elevation [49].

Generally, we observed a negative correlation between functional richness and elevation of Pearson's R = −0.18 for both land-use types and Pearson's R = −0.31 for intact forests specifically. For logged forests, we observed an initial increase in functional richness before following a similar decrease as seen in intact forests. Functional divergence and functional evenness, on the other hand, indicated little variation with elevation. Oil palm plantations were not found to be grown at elevations above 500 m above sea level (Table A1). Therefore, this did not allow for studying elevational effects in oil palm plantations.

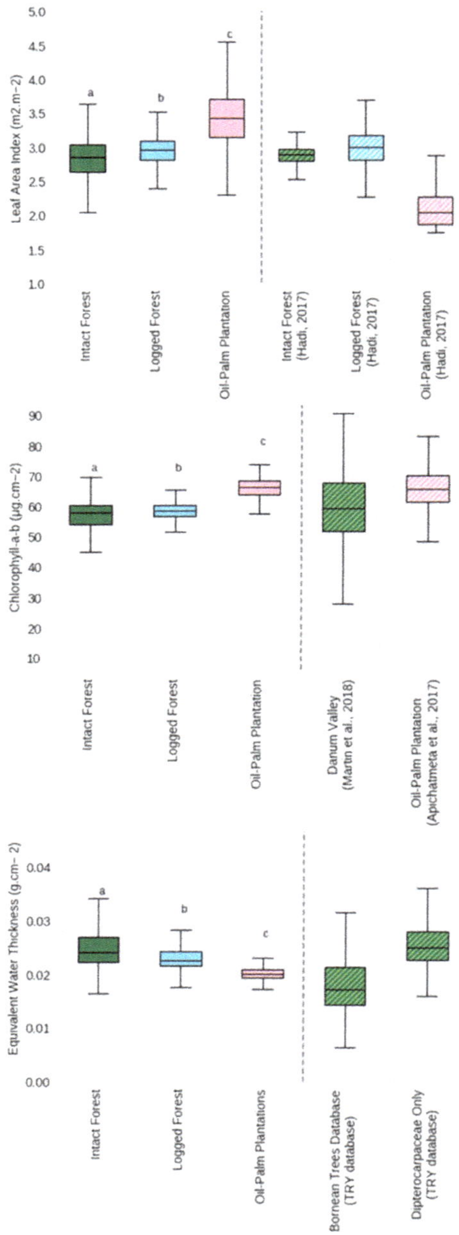

Figure A3. Comparison of the retrieval of spectral trait indicators from SNAP's biophysical processor derived from Sentinel-2 reflectance spectra to relevant in situ trait ranges in the TRY plant trait database [89,90] and other regionally relevant field campaigns [64,86,88,131]. Differences in retrieved trait estimates between land-use types according to a post hoc Tukey's HSD test are indicated with different annotated letters (a, b, c).

Appendix A.2. Functional Diversity across Elevations

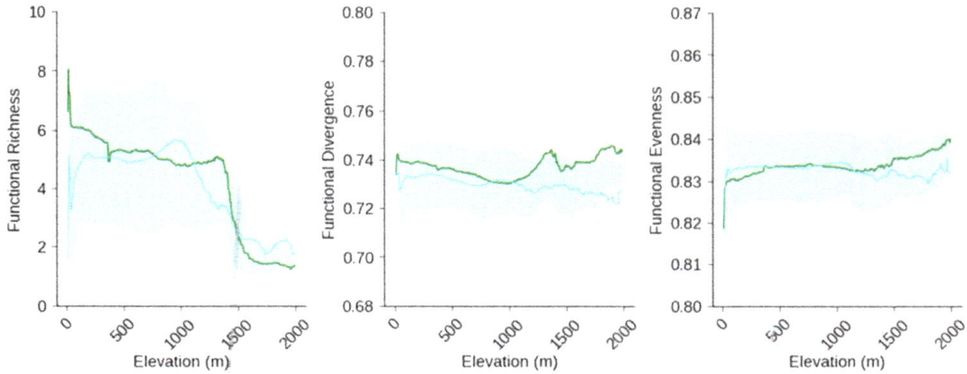

Figure A4. *Variation* of functional diversity metrics along the average elevation of the plot. Green represents intact forests and blue indicates logged forests. Oil palm plantations were only found below 500 m ASL.

References

1. IPBES. *Global Assessment Report on Biodiversity and Ecosystem Services of the Intergovernmental Science—Policy Platform on Biodiversity and Ecosystem Services*; IPBES Secretariat: Bonn, Germany, 2019.
2. Newbold, T.; Hudson, L.N.; Arnell, A.P.; Contu, S.; De Palma, A.; Ferrier, S.; Hill, S.L.L.; Hoskins, A.J.; Lysenko, I.; Phillips, H.R.P.; et al. Has land use pushed terrestrial biodiversity beyond the planetary boundary? A global assessment. *Science* **2016**, *353*, 288–291. [CrossRef]
3. UNDP. *Biodiversity Conservation in Multiple-Use Forest Landscapes in Sabah, Malaysia*; UNDP: Sabah, Malaysia, 2012.
4. Wilcove, D.S.; Giam, X.; Edwards, D.P.; Fisher, B.; Koh, L.P. Navjot's nightmare revisited: Logging, agriculture, and biodiversity in Southeast Asia. *Trends Ecol. Evol.* **2013**, *28*, 531–540. [CrossRef]
5. Sodhi, N.S.; Koh, L.P.; Brook, B.W.; Ng, P.K.L. Southeast Asian biodiversity: An impending disaster. *Trends Ecol. Evol.* **2004**, *19*, 654–660. [CrossRef]
6. Sodhi, N.S.; Koh, L.P.; Clements, R.; Wanger, T.C.; Hill, J.K.; Hamer, K.C.; Clough, Y.; Tscharntke, T.; Posa, M.R.C.; Lee, T.M. Conserving Southeast Asian forest biodiversity in human-modified landscapes. *Biol. Conserv.* **2010**, *143*, 2375–2384. [CrossRef]
7. Gaveau, D.L.A.; Sloan, S.; Molidena, E.; Yaen, H.; Sheil, D.; Abram, N.K.; Ancrenaz, M.; Nasi, R.; Quinones, M.; Wielaard, N.; et al. Four decades of forest persistence, clearance and logging on Borneo. *PLoS ONE* **2014**, *9*, e101654. [CrossRef]
8. Bryan, J.E.; Shearman, P.L.; Asner, G.P.; Knapp, D.E.; Aoro, G.; Lokes, B. Extreme Differences in Forest Degradation in Borneo: Comparing Practices in Sarawak, Sabah, and Brunei. *PLoS ONE* **2013**, *8*, e69679. [CrossRef]
9. Gaveau, D.L.A.; Sheil, D.; Salim, M.A.; Arjasakusuma, S.; Ancrenaz, M.; Pacheco, P.; Meijaard, E. Rapid conversions and avoided deforestation: Examining four decades of industrial plantation expansion in Borneo. *Sci. Rep.* **2016**, *6*, 32017. [CrossRef]
10. Maycock, C.R.; Kettle, C.J.; Khoo, E.; Pereira, J.T.; Sugau, J.B.; Nilus, R.; Ong, R.C.; Amaludin, N.A.; Newman, M.F.; Burslem, D.F.R.P. A Revised Conservation Assessment of Dipterocarps in Sabah. *Biotropica* **2012**, *44*, 649–657. [CrossRef]
11. Fitzherbert, E.B.; Struebig, M.J.; Morel, A.; Danielsen, F.; Brühl, C.A.; Donald, P.F.; Phalan, B. How will oil palm expansion affect biodiversity? *Trends Ecol. Evol.* **2008**, *23*, 538–545. [CrossRef]
12. Koh, L.P.; Wilcove, D.S. Is oil palm agriculture really destroying tropical biodiversity? *Conserv. Lett.* **2008**, *1*, 60–64. [CrossRef]
13. Bernard, H.; Fjeldså, J.; Mohamed, M. A case study on the effects of disturbance and conversion of tropical lowland rain forest on the non-volant small mammals in north Borneo: Management implications. *Mammal Study* **2009**, *34*, 85–96. [CrossRef]
14. Benedick, S.; Hill, J.K.; Mustaffa, N.; Chey, V.K.; Maryati, M.; Searle, J.B.; Schilthuizen, M.; Hamer, K.C. Impacts of rain forest fragmentation on butterflies in northern Borneo: Species richness, turnover and the value of small fragments. *J. Appl. Ecol.* **2006**, *43*, 967–977. [CrossRef]
15. Barnes, A.D.; Jochum, M.; Mumme, S.; Haneda, N.F.; Farajallah, A.; Widarto, T.H.; Brose, U. Consequences of tropical land use for multitrophic biodiversity and ecosystem functioning. *Nat. Commun.* **2014**, *5*, 6351. [CrossRef]
16. Edwards, F.a.; Edwards, D.P.; Larsen, T.H.; Hsu, W.W.; Benedick, S.; Chung, a.; Vun Khen, C.; Wilcove, D.S.; Hamer, K.C. Does logging and forest conversion to oil palm agriculture alter functional diversity in a biodiversity hotspot? *Anim. Conserv.* **2014**, *17*, 163–173. [CrossRef]
17. Neo, L.; Tan, H.T.W.; Wong, K.M. Too little, too late? Conservation exigencies for Borneo inferred from biogeographic considerations of its endemic plant genera against intense landscape modifications. *Biodivers. Conserv.* **2021**, *31*, 59–76. [CrossRef]

18. Jetz, W.; Cavender-Bares, J.; Pavlick, R.; Schimel, D.; Davis, F.W.; Asner, G.P.; Guralnick, R.; Kattge, J.; Latimer, A.M.; Moorcroft, P.; et al. Monitoring plant functional diversity from space. *Nat. Plants* **2016**, *2*, 16024. [CrossRef]
19. Hortal, J.; De Bello, F.; Diniz-Filho, J.A.F.; Lewinsohn, T.M.; Lobo, J.M.; Ladle, R.J. Seven Shortfalls that Beset Large-Scale Knowledge of Biodiversity. *Annu. Rev. Ecol. Evol. Syst.* **2015**, *46*, 523–549. [CrossRef]
20. Petchey, O.L.; Gaston, K.J. Functional diversity (FD), species richness and community composition. *Ecol. Lett.* **2002**, *5*, 402–411. [CrossRef]
21. Flynn, D.F.B.; Gogol-Prokurat, M.; Nogeire, T.; Molinari, N.; Richers, B.T.; Lin, B.B.; Simpson, N.; Mayfield, M.M.; DeClerck, F. Loss of functional diversity under land use intensification across multiple taxa. *Ecol. Lett.* **2009**, *12*, 22–33. [CrossRef]
22. de Souza, D.M.; Flynn, D.F.B.; DeClerck, F.; Rosenbaum, R.K.; de Melo Lisboa, H.; Koellner, T. Land use impacts on biodiversity in LCA: Proposal of characterization factors based on functional diversity. *Int. J. Life Cycle Assess.* **2013**, *18*, 1231–1242. [CrossRef]
23. Diaz, S.; Cabido, M. Vive la difference: Plant functional diversity matters to ecosystem processes: Plant functional diversity matters to ecosystem processes. *Trends Ecol. Evol.* **2001**, *16*, 646–655. [CrossRef]
24. Cadotte, M.W.; Carscadden, K.; Mirotchnick, N. Beyond species: Functional diversity and the maintenance of ecological processes and services. *J. Appl. Ecol.* **2011**, *48*, 1079–1087. [CrossRef]
25. Fukami, T.; Bezemer, T.M.; Mortimer, S.R.; Van Der Putten, W.H. Species divergence and trait convergence in experimental plant community assembly. *Ecol. Lett.* **2005**, *8*, 1283–1290. [CrossRef]
26. Violle, C.; Reich, P.B.; Pacala, S.W.; Enquist, B.J.; Kattge, J. The emergence and promise of functional biogeography. *Proc. Natl. Acad. Sci. USA* **2014**, *111*, 13690–13696. [CrossRef]
27. Hooper, D.U. Species diversity, functional diversity and ecosystem functioning. In *Biodiversity and Ecosystem Functioning: Synthesis and Perspectives*; Oxford University Press on Demand: Oxford, UK, 2002.
28. Asner, G.P.; Martin, R.E.; Anderson, C.B.; Knapp, D.E. Quantifying forest canopy traits: Imaging spectroscopy versus field survey. *Remote Sens. Environ.* **2015**, *158*, 15–27. [CrossRef]
29. Götzenberger, L.; de Bello, F.; Bråthen, K.A.; Davison, J.; Dubuis, A.; Guisan, A.; Lepš, J.; Lindborg, R.; Moora, M.; Pärtel, M.; et al. Ecological assembly rules in plant communities-approaches, patterns and prospects. *Biol. Rev.* **2012**, *87*, 111–127. [CrossRef]
30. Chiarucci, A.; Bacaro, G.; Scheiner, S.M. Old and new challenges in using species diversity for assessing biodiversity. *Philos. Trans. R. Soc. B Biol. Sci.* **2011**, *366*, 2426–2437. [CrossRef]
31. Granger, V.; Bez, N.; Fromentin, J.; Meynard, C.; Jadaud, A.; Mérigot, B. Mapping diversity indices: Not a trivial issue. *Methods Ecol. Evol.* **2015**, *6*, 688–696. [CrossRef]
32. Takashina, N.; Economo, E.P. Developing generalized sampling schemes with known error properties: The case of a moving observer. *Ecography* **2021**, *44*, 293–306. [CrossRef]
33. Schimel, D.; Pavlick, R.; Fisher, J.B.; Asner, G.P.; Saatchi, S.; Townsend, P.; Miller, C.; Frankenberg, C.; Hibbard, K.; Cox, P. Observing terrestrial ecosystems and the carbon cycle from space. *Glob. Chang. Biol.* **2015**, *21*, 1762–1776. [CrossRef]
34. Jensen, J.R. *Remote Sensing of the Environment: Pearson New International Edition: An Earth Resource Perspective*; Pearson Education Limited: London, UK, 2013; ISBN 9781292034935.
35. Butler, D. Earth observation enters next phase. *Nature* **2014**, *508*, 160–161. [CrossRef]
36. Serbin, S.P.; Wu, J.; Ely, K.S.; Kruger, E.L.; Townsend, P.A.; Meng, R.; Wolfe, B.T.; Chlus, A.; Wang, Z.; Rogers, A. From the Arctic to the tropics: Multibiome prediction of leaf mass per area using leaf reflectance. *New Phytol.* **2019**, *224*, 1557–1568. [CrossRef]
37. Ali, A.M.; Darvishzadeh, R.; Skidmore, A.; Gara, T.W.; O'Connor, B.; Roeoesli, C.; Heurich, M.; Paganini, M. Comparing methods for mapping canopy chlorophyll content in a mixed mountain forest using Sentinel-2 data. *Int. J. Appl. Earth Obs. Geoinf.* **2020**, *87*, 102037. [CrossRef]
38. Ali, A.M.; Darvishzadeh, R.; Skidmore, A.; Heurich, M.; Paganini, M.; Heiden, U.; Mücher, S. Evaluating prediction models for mapping canopy chlorophyll content across biomes. *Remote Sens.* **2020**, *12*, 1788. [CrossRef]
39. Brown, L.A.; Ogutu, B.O.; Dash, J. Estimating Forest Leaf Area Index and Canopy Chlorophyll Content with Sentinel-2: An Evaluation of Two Hybrid Retrieval Algorithms. *Remote Sens.* **2019**, *11*, 1752. [CrossRef]
40. Verrelst, J.; Rivera-Caicedo, J.P.; Reyes-Muñoz, P.; Morata, M.; Amin, E.; Tagliabue, G.; Panigada, C.; Hank, T.; Berger, K. Mapping landscape canopy nitrogen content from space using PRISMA data. *ISPRS J. Photogramm. Remote Sens.* **2021**, *178*, 382–395. [CrossRef]
41. Aguirre-gutiérrez, J.; Rifai, S.; Shenkin, A.; Oliveras, I.; Patrick, L.; Svátek, M.; Girardin, C.A.J.; Both, S.; Riutta, T.; Berenguer, E.; et al. Pantropical modelling of canopy functional traits using Sentinel-2 remote sensing data. *Remote Sens. Environ.* **2021**, *252*, 112122. [CrossRef]
42. Hauser, L.T.; Féret, J.-B.; An Binh, N.; van der Windt, N.; Sil, Â.F.; Timmermans, J.; Soudzilovskaia, N.A.; van Bodegom, P.M. Towards scalable estimation of plant functional diversity from Sentinel-2: In-situ validation in a heterogeneous (semi-)natural landscape. *Remote Sens. Environ.* **2021**, *262*, 112505. [CrossRef]
43. Rossi, C.; Kneubühler, M.; Schütz, M.; Schaepman, M.E.; Haller, R.M.; Risch, A.C. From local to regional: Functional diversity in differently managed alpine grasslands. *Remote Sens. Environ.* **2020**, *236*, 111415. [CrossRef]
44. Ma, X.; Mahecha, M.D.; Migliavacca, M.; van der Plas, F.; Benavides, R.; Ratcliffe, S.; Kattge, J.; Richter, R.; Musavi, T.; Baeten, L.; et al. Inferring plant functional diversity from space: The potential of Sentinel-2. *Remote Sens. Environ.* **2019**, *233*, 111368. [CrossRef]

5. Hauser, L.T.; Timmermans, J.; van der Windt, N.; Sil, Â.F.; César de Sá, N.; Soudzilovskaia, N.A.; van Bodegom, P.M. Explaining discrepancies between spectral and in-situ plant diversity in multispectral satellite earth observation. *Remote Sens. Environ.* **2021**, *265*, 112684. [CrossRef]
6. Anderson, C.B. Biodiversity monitoring, earth observations and the ecology of scale. *Ecol. Lett.* **2018**, *21*, 1572–1585. [CrossRef] [PubMed]
7. Abelleira Martínez, O.J.; Fremier, A.K.; Günter, S.; Ramos Bendaña, Z.; Vierling, L.; Galbraith, S.M.; Bosque-Pérez, N.a.; Ordoñez, J.C. Scaling up functional traits for ecosystem services with remote sensing: Concepts and methods. *Ecol. Evol.* **2016**, *6*, 4359–4371. [CrossRef] [PubMed]
8. Chaurasia, A.N.; Dave, M.G.; Parmar, R.M.; Bhattacharya, B.; Marpu, P.R.; Singh, A.; Krishnayya, N.S.R. Inferring species diversity and variability over climatic gradient with spectral diversity metrics. *Remote Sens.* **2020**, *12*, 2130. [CrossRef]
9. Durán, S.M.; Martin, R.E.; Díaz, S.; Maitner, B.S.; Malhi, Y.; Salinas, N.; Shenkin, A.; Silman, M.R.; Wieczynski, D.J.; Asner, G.P.; et al. Informing trait-based ecology by assessing remotely sensed functional diversity across a broad tropical temperature gradient. *Sci. Adv.* **2019**, *5*, aaw8114. [CrossRef]
0. Villéger, S.; Mason, N.W.H.; Mouillot, D. New multidimensional functional diversity indices for a multifaceted framework in functional ecology. *Ecology* **2008**, *89*, 2290–2301. [CrossRef]
1. Kitayama, K. An altitudinal transect study of the vegetation on Mount Kinabalu, Borneo. *Vegetatio* **1992**, *102*, 149–171. [CrossRef]
2. Aiba, S.; Kitayama, K. Structure, Composition and Species Diversity in an Altitude-Substrate Matrix of Rain Forest Tree Communities on Mount Kinabalu, Borneo. *Plant Ecol.* **2010**, *140*, 139–157. [CrossRef]
3. Grytnes, J.A.; Beaman, J.H. Elevational species richness patterns for vascular plants on Mount Kinabalu, Borneo. *J. Biogeogr.* **2006**, *33*, 1838–1849. [CrossRef]
4. Döbert, T.F.; Webber, B.L.; Sugau, J.B.; Dickinson, K.J.M.; Didham, R.K. Can leaf area index and biomass be estimated from Braun-Blanquet cover scores in tropical forests? *J. Veg. Sci.* **2015**, *26*, 1043–1053. [CrossRef]
5. ESA. *Sentinel-2 User Handbook*; ESA Standard Document, Date 24/07/2015, Issue 1, Rev 2; ESA Paris, France, 2015; p. 64.
6. Gascon, F.; Cadau, E.; Colin, O.; Hoersch, B.; Isola, C.; Fernández, B.L.; Martimort, P. Copernicus Sentinel-2 Mission: Products, Algorithms and Cal/Val. In *Proceedings of the Earth Observing Systems XIX*; Butler, J.J., Xiong, X., Gu, X., Eds.; International Society for Optics and Photonics: Bellingham, WA, USA, 2014; Volume 9218, pp. 455–463.
7. Louis, J.; Debaecker, V.; Pflug, B.; Main-Knorn, M.; Bieniarz, J.; Mueller-Wilm, U.; Cadau, E.; Gascon, F. Sentinel-2 Sen2Cor: L2A Processor for Users. In Proceedings of the Living Planet Symposium 2016, Prague, Czech Republic, 9–13 May 2016; pp. 1–8.
8. Weiss, M.; Baret, F. *Sentinel-2 ToolBox Level 2 Biophysical Product Algorithms*; Version 1.1; ESA: Paris, France, 2016.
9. Feret, J.B.; François, C.; Asner, G.P.; Gitelson, A.a.; Martin, R.E.; Bidel, L.P.R.; Ustin, S.L.; le Maire, G.; Jacquemoud, S. PROSPECT-4 and 5: Advances in the leaf optical properties model separating photosynthetic pigments. *Remote Sens. Environ.* **2008**, *112*, 3030–3043. [CrossRef]
0. Fourty, T.; Baret, F. *Amelioration de la Precision des Coefficients D'absorption Specifique de la Matière Seche et des Pigments Photosynthetiques*; INRA: Avignon, France, 1997.
1. Verhoef, W. Light scattering by leaf layers with application to canopy reflectance modeling: The SAIL model. *Remote Sens. Environ.* **1984**, *16*, 125–141. [CrossRef]
2. Jacquemoud, S.; Verhoef, W.; Baret, F.; Zarco-Tejada, P.J.; Asner, G.P.; François, C.; Ustin, S.L. PROSPECT + SAIL: 15 Years of Use for Land Surface Characterization. In Proceedings of the 2006 IEEE International Symposium on Geoscience and Remote Sensing, Denver, CO, USA, 31 July–4 August 2006; Volume 516, pp. 1992–1995. [CrossRef]
3. Jacquemoud, S.; Verhoef, W.; Baret, F.; Bacour, C.; Zarco-Tejada, P.J.; Asner, G.P.; François, C.; Ustin, S.L. PROSPECT + SAIL models: A review of use for vegetation characterization. *Remote Sens. Environ.* **2009**, *113*, S56–S66. [CrossRef]
4. Pfeifer, M.; Korhonen, L.; Wheeler, C.; Rautiainen, M. Remote Sensing of Environment Forest canopy structure and reflectance in humid tropical Borneo: A physically-based interpretation using spectral invariants. *Remote Sens. Environ.* **2017**, *201*, 314–330. [CrossRef]
5. Asner, G.P.; Scurlock, J.M.O.; Hicke, J.A. Global synthesis of leaf area index observations: Implications for ecological and remote sensing studies. *Glob. Ecol. Biogeogr.* **2003**, *12*, 191–205. [CrossRef]
6. Castillo, J.; Apan, A.A.; Maraseni, T.N.; Salmo, S.G. Estimation and mapping of above-ground biomass of mangrove forests and their replacement land uses in the Philippines using Sentinel imagery. *ISPRS J. Photogramm. Remote Sens.* **2017**, *134*, 70–85. [CrossRef]
7. Zheng, G.; Moskal, L.M. Retrieving Leaf Area Index (LAI) Using Remote Sensing: Theories, Methods and Sensors. *Sensors* **2009**, *9*, 2719–2745. [CrossRef]
8. Poorter, L.; Bongers, F. Leaf traits are good predictors of plant performance across 53 rain forest species. *Ecology* **2006**, *87*, 1733–1743. [CrossRef]
9. Croft, H.; Chen, J.M.; Luo, X.; Bartlett, P.; Chen, B.; Staebler, R.M. Leaf chlorophyll content as a proxy for leaf photosynthetic capacity. *Glob. Chang. Biol.* **2017**, *23*, 3513–3524. [CrossRef]
0. Cao, K. Leaf anatomy and chlorophyll content of 12 woody species in contrasting light conditions in a Bornean heath forest. *Can. J. Bot.* **2000**, *78*, 1245–1253.
1. Weiher, E.; van der Werf, A.; Thompson, K.; Roderick, M.; Garnier, E.; Eriksson, O. Challenging Theophrastus: A common core list of plant traits for functional ecology. *J. Veg. Sci.* **1999**, *10*, 609–620. [CrossRef]

72. Saura-Mas, S.; Lloret, F. Leaf and Shoot Water Content and Leaf Dry Matter Content of Mediterranean Woody Species with Different Post-fire Regenerative Strategies. *Ann. Bot.* **2017**, *99*, 545–554. [CrossRef] [PubMed]
73. Van Nieuwstadt, M.G.L.; Sheil, D. Drought, fire and tree survival in a Borneo rain forest, East Kalimantan, Indonesia. *J. Eco* **2005**, *93*, 191–201. [CrossRef]
74. Lawlor, D.W.; Cornic, G. Photosynthetic carbon assimilation and associated metabolism in relation to water deficits in higher plants. *Plant Cell Environ.* **2002**, *25*, 275–294. [CrossRef] [PubMed]
75. Asbjornsen, H.; Goldsmith, G.R.; Alvarado-Barrientos, M.S.; Rebel, K.; Van Osch, F.P.; Rietkerk, M.; Chen, J.; Gotsch, S.; Tobón, C Geissert, D.R.; et al. Ecohydrological advances and applications in plant–water relations research: A review. *J. Plant Ecol.* **2011**, 3–22. [CrossRef]
76. Damm, A.; Paul-Limoges, E.; Haghighi, E.; Simmer, C.; Morsdorf, F.; Schneider, F.D.; van der Tol, C.; Migliavacca, M.; Rascher, U. Remote sensing of plant-water relations: An overview and future perspectives. *J. Plant Physiol.* **2018**, *227*, 3–19. [CrossRef]
77. Mason, N.W.H.; Mouillot, D.; Lee, W.G.; Wilson, J.B. Functional richness, functional evenness and functional divergence: Th primary components of functional diversity. *Oikos* **2005**, *111*, 112–118. [CrossRef]
78. Mouillot, D.; Mason, W.H.N.; Dumay, O.; Wilson, J.B. Functional regularity: A neglected aspect of functional diversity. *Oecologi* **2005**, *142*, 353–359. [CrossRef]
79. Cornwell, W.K.; Schwilk, D.W.; Ackerly, D.D. A Trait-Based Test for Habitat Filtering: Convex Hull Volume. *Ecology* **2006**, *87* 1465–1471. [CrossRef]
80. Mouchet, M.a.; Villéger, S.; Mason, N.W.H.; Mouillot, D. Functional diversity measures: An overview of their redundancy and their ability to discriminate community assembly rules. *Funct. Ecol.* **2010**, *24*, 867–876. [CrossRef]
81. Aiba, M.; Katabuchi, M.; Takafumi, H.; Matsuzaki, S.I.S.; Sasaki, T.; Hiura, T. Robustness of trait distribution metrics fo community assembly studies under the uncertainties of assembly processes. *Ecology* **2013**, *94*, 2873–2885. [CrossRef] [PubMed]
82. Kraft, N.J.B.; Valencia, R.; Ackerly, D.D. Functional Traits and Niche-Based Tree Community Assembly in an Amazonian Forest *Science* **2008**, *322*, 580–582. [CrossRef]
83. Schneider, F.D.; Morsdorf, F.; Schmid, B.; Petchey, O.L.; Hueni, A.; Schimel, D.S.; Schaepman, M.E. Mapping functional diversity from remotely sensed morphological and physiological forest traits. *Nat. Commun.* **2017**, *8*, s41467–s42017. [CrossRef] [PubMed]
84. Fornberg, B.; Flyer, N. *A Primer on Radial Basis Functions with Applications to the Geosciences*; SIAM, Society for Industrial and Applied Mathematics: Philadelphia, PA, USA, 2015.
85. Sist, P.; Saridan, A. Stand structure and floristic composition of a primary lowland dipterocarp forest in East Kalimantan. *J. Trop For. Sci.* **1999**, *11*, 704–722.
86. Martin, R.; Chadwick, K.; Brodrick, P.; Carranza-Jimenez, L.; Vaughn, N.; Asner, G. An Approach for Foliar Trait Retrieval from Airborne Imaging Spectroscopy of Tropical Forests. *Remote Sens.* **2018**, *10*, 199. [CrossRef]
87. Kattge, J.; Bönisch, G.; Díaz, S.; Lavorel, S.; Prentice, I.C.; Leadley, P.; Tautenhahn, S.; Werner, G.D.A.; Aakala, T.; Abedi, M.; et a TRY plant trait database—Enhanced coverage and open access. *Glob. Chang. Biol.* **2020**, *26*, 119–188. [CrossRef]
88. Apichatmeta, K.; Sudsiri, C.J.; Ritchie, R.J. Photosynthesis of oil palm (Elaeis guineensis). *Sci. Hortic.* **2017**, *214*, 34–40. [CrossRef]
89. Kurokawa, H.; Nakashizuka, T. Leaf herbivory and decomposability in a Malaysian tropical rain forest. *Ecology* **2008** 2645–2656. [CrossRef]
90. Kattge, J.; Díaz, S.; Lavorel, S.; Prentice, I.C.; Leadley, P.; Bönisch, G.; Garnier, E.; Westoby, M.; Reich, P.B.; Wright, I.J.; et al. TRY—A global database of plant traits. *Glob. Chang. Biol.* **2011**, *17*, 2905–2935. [CrossRef]
91. Wu, J.; Chavana-Bryant, C.; Prohaska, N.; Serbin, S.P.; Guan, K.; Albert, L.P.; Yang, X.; van Leeuwen, W.J.D.; Garnello, A.J Martins, G.; et al. Convergence in relationships between leaf traits, spectra and age across diverse canopy environments and two contrasting tropical forests. *New Phytol.* **2017**, *214*, 1033–1048. [CrossRef]
92. de Bello, F.; Lavorel, S.; Díaz, S.; Harrington, R.; Cornelissen, J.H.C.; Bardgett, R.D.; Berg, M.P.; Cipriotti, P.; Feld, C.K Hering, D.; et al. Towards an assessment of multiple ecosystem processes and services via functional traits. *Biodivers. Conserv* **2010**, *19*, 2873–2893. [CrossRef]
93. Combal, B.; Baret, F.; Weiss, M. Improving canopy variables estimation from remote sensing data by exploiting ancillary information. Case study on sugar beet canopies. *Agronomie* **2002**, *22*, 205–215. [CrossRef]
94. Verrelst, J.; Malenovský, Z.; Van der Tol, C.; Camps-Valls, G.; Gastellu-Etchegorry, J.P.; Lewis, P.; North, P.; Moreno, J. Quantifying Vegetation Biophysical Variables from Imaging Spectroscopy Data: A Review on Retrieval Methods. *Surv. Geophys.* **2019**, 589–629. [CrossRef]
95. Dorigo, W.; Richter, R.; Baret, F.; Bamler, R.; Wagner, W. Enhanced automated canopy characterization from hyperspectral data b a novel two step radiative transfer model inversion approach. *Remote Sens.* **2009**, *1*, 1139–1170. [CrossRef]
96. Berger, K.; Pablo, J.; Caicedo, R.; Martino, L.; Wocher, M.; Hank, T. A Survey of Active Learning for Quantifying Vegetation Trait from Terrestrial Earth Observation Data. *Remote Sens.* **2021**, *13*, 287. [CrossRef]
97. Nguyen, B.A.; Hauser, L.T.; Hoa, P.V.; Thi, G.; Thao, P. Quantifying mangrove leaf area index from Sentinel-2 imagery using hybrid models and active learning. *Int. J. Remote Sens.* **2022**, 1–22. [CrossRef]
98. Berger, K.; Atzberger, C.; Danner, M.; D'Urso, G.; Mauser, W.; Vuolo, F.; Hank, T. Evaluation of the PROSAIL Model Capabilitie for Future Hyperspectral Model Environments: A Review Study. *Remote Sens.* **2018**, *10*, 85. [CrossRef]
99. Schlerf, M.; Atzberger, C. Inversion of a forest reflectance model to estimate structural canopy variables from hyperspectral remote sensing data. *Remote Sens. Environ.* **2006**, *100*, 281–294. [CrossRef]

100. Kuusk, A.; Nilson, T. A directional multispectral forest reflectance model. *Remote Sens. Environ.* **2000**, *72*, 244–252. [CrossRef]
101. Huang, J.; Zeng, Y.; Kuusk, A.; Wu, B.; Dong, L.; Mao, K.; Chen, J. Inverting a forest canopy reflectance model to retrieve the overstorey and understorey leaf area index for forest stands. *Int. J. Remote Sens.* **2011**, *32*, 7591–7611. [CrossRef]
102. Pasqualotto, N.; D'Urso, G.; Bolognesi, S.F.; Belfiore, O.R.; Van Wittenberghe, S.; Delegido, J.; Pezzola, A.; Winschel, C.; Moreno, J. Retrieval of evapotranspiration from sentinel-2: Comparison of vegetation indices, semi-empirical models and SNAP biophysical processor approach. *Agronomy* **2019**, *9*, 663. [CrossRef]
103. Vinué, D.; Camacho, F.; Fuster, B. Validation of Sentinel-2 LAI and FAPAR products derived from SNAP toolbox over a cropland site in Barrax and over an agroforested site in Liria. In Proceedings of the Fifth Recent Advances in Quantitative Remote Sensing, Torrent, Spain, 18–22 September 2018; p. 248.
104. Xie, Q.; Dash, J.; Huete, A.; Jiang, A.; Yin, G.; Ding, Y.; Peng, D.; Hall, C.C.; Brown, L.; Shi, Y.; et al. Retrieval of crop biophysical parameters from Sentinel-2 remote sensing imagery. *Int. J. Appl. Earth Obs. Geoinf.* **2019**, *80*, 187–195. [CrossRef]
105. Danner, M.; Berger, K.; Wocher, M.; Mauser, W.; Hank, T. Efficient RTM-based training of machine learning regression algorithms to quantify biophysical & biochemical traits of agricultural crops. *ISPRS J. Photogramm. Remote Sens.* **2021**, *173*, 278–296. [CrossRef]
106. Vanino, S.; Nino, P.; De Michele, C.; Falanga Bolognesi, S.; D'Urso, G.; Di Bene, C.; Pennelli, B.; Vuolo, F.; Farina, R.; Pulighe, G.; et al. Capability of Sentinel-2 data for estimating maximum evapotranspiration and irrigation requirements for tomato crop in Central Italy. *Remote Sens. Environ.* **2018**, *215*, 452–470. [CrossRef]
107. Noormets, A. *Phenology of Ecosystem Processes*; Springer Science & Business Media: New York, NY, USA, 2013; Volume 53, ISBN 978-1-107-05799-1.
108. Gara, T.W.; Darvishzadeh, R.; Skidmore, A.K.; Wang, T.; Heurich, M. Accurate modelling of canopy traits from seasonal Sentinel-2 imagery based on the vertical distribution of leaf traits. *ISPRS J. Photogramm. Remote Sens.* **2019**, *157*, 108–123. [CrossRef]
109. Asner, G.P.; Martin, R.E.; Anderson, C.B.; Kryston, K.; Vaughn, N.; Knapp, D.E.; Bentley, L.P.; Shenkin, A.; Salinas, N.; Sinca, F.; et al. Scale dependence of canopy trait distributions along a tropical forest elevation gradient. *New Phytol.* **2017**, *214*, 973–988. [CrossRef]
110. Duraiappah, A.K.; Naeem, S.; Agardy, T.; Ash, N.J.; Cooper, H.D.; Diaz, S.; Faith, D.P.; Mace, G. McNeely, J.A.; Mooney, H.A.; et al. *Ecosystems and Human Well-Being: Biodiversity Synthesis: A Report of the Millennium Ecosystem Assessment*; IPBES Secretariat: Bonn, Germany, 2005.
111. Lin, B.B.; Flynn, D.F.B.; Bunker, D.E.; Uriarte, M.; Naeem, S. The effect of agricultural diversity and crop choice on functional capacity change in grassland conversions. *J. Appl. Ecol.* **2011**, *48*, 609–618. [CrossRef]
112. Maeshiro, R.; Kusumoto, B.; Fujii, S.; Shiono, T.; Kubota, Y. Using tree functional diversity to evaluate management impacts in a subtropical forest. *Ecosphere* **2013**, *4*, 6. [CrossRef]
113. Wang, Z.; Rahbek, C.; Fang, J. Effects of geographical extent on the determinants of woody plant diversity. *Ecography* **2012**, *35*, 1160–1167. [CrossRef]
114. Zarnetske, P.L.; Read, Q.D.; Record, S.; Gaddis, K.; Pau, S.; Hobi, M.; Malone, S.; Costanza, J.; Dahlin, K.M.; Latimer, A.M.; et al. Towards connecting biodiversity and geodiversity across scales with satellite remote sensing. *Glob. Ecol. Biogeogr.* **2019**, *28*, 548–556. [CrossRef]
115. Blonder, B.; Lamanna, C.; Violle, C.; Enquist, B.J. The n-dimensional hypervolume. *Glob. Ecol. Biogeogr.* **2014**, *23*, 595–609. [CrossRef]
116. Bongers, F.; Poorter, L.; Hawthorne, W.D.; Sheil, D. The intermediate disturbance hypothesis applies to tropical forests, but disturbance contributes little to tree diversity. *Ecol. Lett.* **2009**, *12*, 798–805. [CrossRef] [PubMed]
117. Biswas, S.R.; Mallik, A.U. Disturbance effects on species diversity and functional diversity in riparian and upland plant communities. *Ecology* **2010**, *91*, 28–35. [CrossRef] [PubMed]
118. Suárez-Castro, A.F.; Raymundo, M.; Bimler, M.; Mayfield, M.M. Using multi-scale spatially explicit frameworks to understand the relationship between functional diversity and species richness. *Ecography* **2022**, 1–18. [CrossRef]
119. Shipley, B. Community assembly, natural selection and maximum entropy models. *Oikos* **2010**, *119*, 604–609. [CrossRef]
120. Huston, M.A.; Huston, M.A. *Biological Diversity: The Coexistence of Species*; Cambridge University Press: Cambridge, UK, 1994.
121. Jong, H.N. Proposal could redefine palm oil-driven deforestation as reforestation in Indonesia. *Mongabay*, 7 January 2022.
122. Hinkes, C. Adding (bio)fuel to the fire: Discourses on palm oil sustainability in the context of European policy development. *Environ. Dev. Sustain.* **2020**, *22*, 7661–7682. [CrossRef]
123. Pakeman, R.J. Functional diversity indices reveal the impacts of land use intensification on plant community assembly. *J. Ecol.* **2011**, *99*, 1143–1151. [CrossRef]
124. Hauser, L.T.; Binh, N.A.; Hoa, P.V.; Quan, N.H.; Timmermans, J. Gap-free monitoring of annual mangrove forest dynamics in ca mau province, vietnamese mekong delta, using the landsat-7-8 archives and post-classification temporal optimization. *Remote Sens.* **2020**, *12*, 3729. [CrossRef]
125. Lahoz, W.A.; Schneider, P. Data assimilation: Making sense of Earth Observation. *Front. Environ. Sci.* **2014**, *2*, 16. [CrossRef]
126. Lewis, P.; Gomez-Dans, J.; Kaminski, T.; Settle, J.; Quaife, T.; Gobron, N.; Styles, J.; Berger, M. An Earth Observation Land Data Assimilation System (EO-LDAS). *Remote Sens. Environ.* **2012**, *120*, 219–235. [CrossRef]
127. Belda, S.; Pipia, L.; Morcillo-Pallarés, P.; Rivera-Caicedo, J.P.; Amin, E.; De Grave, C.; Verrelst, J. DATimeS: A machine learning time series GUI toolbox for gap-filling and vegetation phenology trends detection. *Environ. Model. Softw.* **2020**, *127*, 104666. [CrossRef]

128. Gu, C.; Du, H.; Mao, F.; Han, N.; Zhou, G.; Xu, X.; Sun, S.; Gao, G. Global sensitivity analysis of PROSAIL model parameters when simulating Moso bamboo forest canopy reflectance. *Int. J. Remote Sens.* **2016**, *37*, 5270–5286. [CrossRef]
129. Verrelst, J.; Vicent, J.; Rivera-Caicedo, J.P.; Lumbierres, M.; Morcillo-Pallarés, P.; Moreno, J. Global sensitivity analysis of leaf-canopy-atmosphere RTMs: Implications for biophysical variables retrieval from top-of-atmosphere radiance data. *Remote Sens.* **2019**, *11*, 1923. [CrossRef]
130. de Sá, N.C.; Baratchi, M.; Hauser, L.T.; van Bodegom, P.M. Exploring the Impact of Noise on Hybrid Inversion of PROSAIL RTM on Sentinel-2 Data. *Remote Sens.* **2021**, *13*, 648. [CrossRef]
131. Azzeme, A.M.; Abdullah, S.N.A.; Aziz, M.A.; Wahab, P.E.M. Oil palm leaves and roots differ in physiological response, antioxidant enzyme activities and expression of stress-responsive genes upon exposure to drought stress. *Acta Physiol. Plant.* **2016**, *38*, 52. [CrossRef]

land

Article

The Spatial and Temporal Evolution and Drivers of Habitat Quality in the Hung River Valley

Xin Fan [1,2,†], Xinchen Gu [3,4,†], Haoran Yu [5,†], Aihua Long [4,*], Damien Sinonmatohou Tiando [1], Shengya Ou [6], Jiangfeng Li [1], Yuejing Rong [7,8], Guiling Tang [9], Yanjun Zheng [10], Mingjie Shi [11], Mengwen Wang [12,13], Xiong Wang [14] and Chunbo Huang [14]

1 School of Public Administration, China University of Geosciences, Wuhan 430074, China; fanfx8@edu.cug.cn (X.F.); intlstudents@cug.edu.cn (D.S.T.); jfli@cug.edu.cn (J.L.)
2 State Key Laboratory of Earth Surface Processes and Resource Ecology, Beijing Normal University, Beijing 100875, China
3 State Key Laboratory of Hydraulic Engineering Simulation and Safety, School of Civil Engineering, Tianjin University, Tianjin 300072, China; gxc@tju.edu.cn
4 State Key Laboratory of Simulation and Regulation of Water Cycle in River Basin, China Institute of Water Resources and Hydropower Research, Beijing 100044, China
5 Anhui Urbanization Development Research Center, Hefei 230022, China; yuhaoran2021@ahjzu.edu.cn
6 School of Geography and Tourism, Shaanxi Normal University, Xi'an 710119, China; osy201705598@snnu.edu.cn
7 State Key Laboratory of Urban and Regional Ecology, Research Center for Eco-Environmental Sciences, Chinese Academy of Sciences, Beijing 100085, China; yjrong_st@rcees.ac.cn
8 University of Chinese Academy of Sciences, Beijing 100049, China
9 School of International Education, China University of Geosciences, Wuhan 430078, China; tangguiling@cug.edu.cn
10 Student Innovation & Entrepreneurship Guidance Centre, China University of Geosciences, Wuhan 430074, China; Zhengyanjun@cug.edu.cn
11 College of Resource and Environment, Xinjiang Agricultural University, Urumqi 830052, China; 320192307@xjau.edu.cn
12 College of Resources and Environment, Shandong Agricultural University, Taishan 271018, China; tsuls_wmw@tsu.edu.cn
13 School of Economics and Management, Shandong Agricultural University, Taishan 271018, China
14 Research Center for Spatial Planning and Human-Environment System Simulation, School of Geography and Information Engineering, China University of Geosciences, Wuhan 430074, China wxrwx@cug.edu.cn (X.W.); huangchunbo@cug.edu.cn (C.H.)
* Correspondence: ahlong@iwhr.com
† These authors contributed equally to this work and should be considered co-first authors.

Abstract: The survival and sustainability of regional species is constrained by habitat quality. In recent decades, the intensification of human activities on a global scale has had a profound impact on regional ecosystems and poses a serious threat to regional sustainable development. Scientific measurement of the drivers of habitat quality can provide important support for the development of effective biodiversity conservation and sustainable land-use policies. Taking the Hung River Valley as an example, the InVEST model was used to assess the habitat quality of the study area in 2000, 2005, 2010, 2015, and 2020 and to explore its spatial and temporal variation and distribution characteristics in combination with the spatial autocorrelation model, and the geographically weighted regression (GWR) model was used to explore the drivers of habitat quality change. The results show the following: (1) The overall habitat quality shows an increasing trend during 2000–2020, but the expansion of construction land in the central region plays a dominant role in the degradation of regional habitat quality. (2) The "Guide-Ledu" line is the dividing line of habitat quality in the Hung River Valley, with a general distribution of "south is good, north is bad" and "south is hot, north is cold". (3) Natural factors such as slope and elevation basically shape the overall distribution pattern of habitat quality, while urbanisation factors such as population density, gross domestic product, and the night-time lighting index are generally negatively correlated with habitat quality. The results of the study can reveal the linkage between ecosystems and land-use change in the context of urbanisation.

Keywords: geographically weighted regression (GWR); habitat quality; Hung River Valley; landscape pattern; urbanisation

1. Introduction

Habitat quality is an important indicator of an ecosystem's ability to provide suitable conditions for the growth, development, and distribution of species, based on the availability of subsistence resources [1,2]. Since the beginning of industrial society, human activities have caused a series of ecological problems, such as habitat fragmentation and a loss of species diversity, which have led to serious threats to the overall ecological security of the region. In this context, the spatial and temporal distribution characteristics and evolutionary mechanisms of habitat quality have gradually become a hot topic in related research fields [3–6].

At present, the research on habitat quality in China and abroad has achieved advanced results from macroscopic to microscopic scales. Whether at the local, watershed, or regional scales, scholars have conducted in-depth studies on the spatial and temporal evolution [7], distribution characteristics [8], influencing factors [9–11], driving mechanisms [12–15], and pathways to enhance habitat quality [16–18]. Although more research results have been achieved, the current research on the drivers of habitat quality needs to be further explored. There are many factors that influence habitat quality, and the degree of influence of the same influencing factor on habitat quality in different spaces can be spatially heterogeneous [11,19,20]. Simply explaining which influencing factors lead to changes in habitat quality tends to ignore the processes and relationships between subjective and objective elements, which in turn affects the accurate mining of habitat quality influencing factors in the future [21]. Habitat quality is primarily a matter of selecting an evaluation model, and most habitat quality evaluations are based on landscape pattern-based indicator systems and model-based approaches, compared to the more scientific role of models in predicting future habitat distribution and siting protected areas [22,23]. The Integrated Valuation of Environmental Services and Tradeoffs model (InVEST), developed by Stanford University, the World Wildlife Fund, and the Nature Conservancy, has a high data demand. The InVEST (Integrated Valuation of Environmental Services and Tradeoffs) model, developed jointly by the Nature Conservancy and the Nature Fund, has been gradually applied to related studies because of its relatively small data requirements and the high visibility of the results [24–26]. In revealing the drivers of habitat quality, ordinary least square (OLS) and geographically weighted regression (GWR) models are good at detecting subtle changes in the process mechanism of habitat quality over time and space, and are an important research method for exploring the drivers of objective objects [27].

The Hung River Valley is in the transition zone between the Qinghai–Tibet Plateau and the Loess Plateau. Its ecological environment is fragile and regarded as a "sensitive area", and the current literature on the ecology of the region is relatively small [22–24]. In the context of the comprehensive pilot work of new urbanisation, the Xining–Haidong metropolitan area will be established in Qinghai Province, and the Hung River Valley area, represented by Xining and Haidong, will usher in a new round of rapid development and become the core growth pole, leading the development of Qinghai and even the northwest region. The development potential brings greater ecological risks. Based on the above research status and regional background, the main research of this paper includes (1) quantitatively assessing the spatial and temporal evolution of the landscape type, landscape pattern, and habitat quality in the Hung River Valley with the help of a land transfer matrix, a landscape pattern analysis method, and the InVEST model; (2) exploring the spatial and temporal coupling relationship between habitat quality change and urbanisation in the Hung River Valley based on the GWR model; (3) finally determining, through the above research, the habitat quality of the Hung River Valley over a 20-year period and the drivers of habitat quality, providing a scientific reference for biodiversity conservation and regional ecological development in the eastern Tibetan Plateau, providing decision support for land use, ecological red line delineation, and coordinated and sustainable economic and social development, and providing new ideas for habitat quality assessments in ecologically sensitive areas.

2. Materials and Methods

2.1. Study Site

The geographical location of the Hung River Valley is 100°51′~103°04′ E, 35°01′~38° N (Figure 1), with a total area of about 35,273.77 km^2, covering Xining City, Haidong City, Huangnan Tibetan Autonomous Prefecture, Hainan Tibetan Autonomous Prefecture, and Haibei Tibetan Autonomous Prefecture, which is the political, economic, and cultural centre of Qinghai Province. It is the political, economic, and cultural centre of Qinghai Province. The Hung River Valley is located in the Yellow River and the Huangshui River basin triangle and is the transition area between the Loess Plateau and the Tibetan Plateau, whose elevation is 1659~5149 m, from the north to the south distribution of the Datong River, the Huangshui River, the Yellow River, and Qilian Mountain, a block of two parallel ridge valleys that has created a unique "three mountains between two valleys" landform. It is in the eastern monsoon area of Qinghai Province and at the end of the eastern monsoon zone in Qinghai Province, the intersection of three natural zones: the arid zone of Northwest China, the eastern monsoon zone, and the Tibetan Plateau zone. The climate is mild, the water is abundant, and the sunshine is long, making the region a natural environment for biological reproduction, with plants such as Qilian cypress, *Pinellia pinnata*, and Bashan fir and wild animals such as Sumen antelope, rock sheep, and plateau partridge.

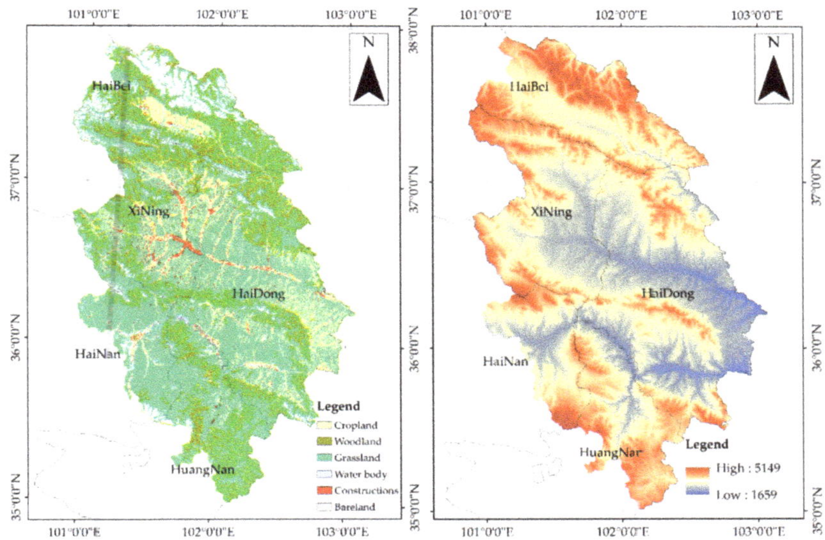

Figure 1. Location of the Hung River Valley area.

2.2. Data Source

The data used in this study include the following: (1) 5 periods of land-use data (precision: 100 × 100 m): 2000, 2005, 2010, 2015, and 2020, and the land-use types are divided into 6 primary categories: grassland, arable land, forest land, construction land, water, and bare land. The data come from the Resource and Environment Science Data Center of the Chinese Academy of Sciences (http://www.resdc.cn (accessed on 3 October

2021)), the accuracy of which meets the needs of the study [28]. (2) Basic geographic data, mainly including the carrier data of the borders of cities in Qinghai Province and four types of highways: national highways, provincial highways, highways, and railways. Road data are used to calculate habitat quality, and they were provided by the National Geographic Information Resource Directory Service System National Basic Geographic Database (http://www.webmap.cn (accessed on 3 October 2021)). (3) A digital elevation model (DEM) from the Geospatial Data Cloud (http://www.gscloud.cn (accessed on 3 October 2021)) with a spatial resolution of 30 m. (4) Corrected DMSP/OLS night-time lighting data (with an accuracy of 500 × 500 m) from the China Research Data Service Platform (CNRDS) and the National Basic Geographic Information Centre (GIC); National Polar-Orbiting Operational Environmental Satellite System Preparatory Project-Visible Infrared Imaging Radiometer (NPP-VIIRS) night-time light data (accuracy: 500 × 500 m) for 2015 and 2020; DMSP/OLS night-time light data with NPP-VIIRS night-time light data from the Earth Observation Group (EOG) website. The data resolution will affect the accuracy of the research results and facilitate spatial calculation and analysis. Therefore, the land-use data are used as the standard, and other data are sampled as 100 × 100 m. The unified coordinate system of all data is WGS_1984_UTM_Zone_50N.

2.3. Research Methodology

2.3.1. InVEST Model

InVEST is a model used for ecosystem service function assessments, in which the habitat quality evaluation module is based on the linkage between land cover and habitat threat sources. It calculates the threat intensity of threat sources by considering the radius of stress, spatial weights, and spatial attenuation types and combines the habitat adaptation of other land types and the sensitivity to threat sources to obtain the habitat quality of the area with the following equation:

$$Q_{xj} = H_j\left[1 - \left(\frac{D_{xj}^z}{D_{xj}^z + k^z}\right)\right] \tag{1}$$

where Q_{xj} represents the habitat quality index of raster x in landscape type j within the Hung River Valley; the value range of H_j is [0,1], representing the habitat suitability score of landscape type j; D_{xj} is the habitat degradation degree of grid unit x in land category j; k is the half-saturation constant; because all the resolutions in this study are 100 m, k is 50; z is the scale constant, generally taken as 2.5.

In this study, the Habitat Quality module parameter tables (Tables 1 and 2) were set based on the InVEST model manual and related studies [25].

2.3.2. Spatial Autocorrelation Analysis

The global clustering test is used for global spatial distribution patterns of habitat quality, i.e., high value aggregation or low-value aggregation [26,27,29], and is expressed as:

$$G(d) = \frac{\sum_{i=1j=1}^{n}\sum_{ij}^{n} w_{ij}(d)x_i x_j}{\sum_{i=1}^{n}\sum_{j=1}^{n} x_i x_j} \tag{2}$$

$$Z(G) = [G(d) - E(G)]/\sqrt{\text{var}(G)} \tag{3}$$

where w_{ij} is the spatial weight defined by the distance rule; x_i denotes the value of the variable in region i; x_j denotes the value of the variable in region j; $E(G)$ denotes the expected value of $G(d)$; $var(G)$ denotes the variance of $G(d)$. Based on the value of $Z(G)$, whether $G(d)$ meets the significance level and whether there is a positive or negative spatial correlation can be determined. When $G(d)$ is positive and $Z(G)$ is statistically significant, there is a high value cluster of habitat quality in the region; when $G(d)$ is negative and $Z(G)$ is statistically significant, there is a low-value cluster of agricultural habitat quality in the region.

Table 1. The maximum impact distance, weight, and the attenuation type of the threat sources.

Threat Source	Maximum Stress Distance/km	Weight	Attenuation Type
Paddy field	0.5	0.5	Linear
Non-irrigated arable land	0.5	0.5	Linear
Urban land	10.0	1.0	Exponential
Rural settlement	2.0	0.7	Exponential
Industrial and mining land	1.0	0.5	Exponential
Traffic land	3.0	1.0	Linear

Table 2. Habitat suitability of different types of land and its sensitivity to threats.

Type	Habitat Type	Habitat Suitability	Sensitivity					
			Paddy Field	Non-Irrigated Arable Land	Rural Settlement	Urban Land	Industrial and Mining Land	Traffic Land
Cropland	Paddy field	0.4	0	0.3	0.35	0.5	0.1	0.1
	dry land	0.4	0.3	0	0.35	0.5	0.1	0.1
Woodland	Forestland	1.0	0.8	0.5	0.2	0.5	0.8	0.8
	Irrigate forestland	1.0	0.8	0.9	0.7	1.0	0.5	0.5
	Sparse forestland	0.7	0.7	0.8	0.8	0.9	0.6	0.6
	Others	0.7	0.7	0.8	0.8	0.8	0.6	0.6
Grassland	High coverage	0.8	0.5	0.5	0.5	0.6	0.3	0.3
	Medium coverage	0.7	0.5	0.5	0.6	0.6	0.4	0.4
	Low coverage	0.6	0.5	0.5	0.5	0.6	0.3	0.3
Water areas	Canal	0.8	0.3	0.2	0.3	0.3	0.2	0.2
	lake	0.8	0.3	0.2	0.3	0.3	0.2	0.2
	Reservoir pond	0.7	0.2	0.2	0.3	0.3	0.1	0.1
	Snowfield	0.5	0.2	0.2	0.2	0.7	0.1	0.1
	Beach	0.5	0.2	0.2	0.2	0.7	0.1	0.1
Construction	Urban land	0	0	0	0	0	0	0
	Rural settlement	0	0	0	0	0	0	0
	Others	0	0	0	0.6	0	0	0
Bare land	Sandy land	0.2	0.1	0.5	0.6	0.9	0.6	0.6
	Gobi	0.2	0.1	0.5	0.6	0.9	0.6	0.6
	Marsh land	0.6	0.8	0.8	0.8	1.0	0.6	0.6
	Bare land	0.2	0.1	0.5	0.6	0.9	0.6	0.6
	Bare rock land	0.2	0.1	0.5	0.6	0.9	0.6	0.6
	Alpine desert	0.2	0.1	0.5	0.6	0.9	0.6	0.6

2.3.3. Hotspot Analysis

Spatial hotspot detection analysis is a test for the presence of significant high and low values in an area and can be used to reveal "hotspots" and "coldspots" in a spatial visual representation. The main study here is on habitat quality differentiation [27,29,30]. It is calculated by the formula:

$$G_i^*(d) = \frac{\sum_{j=1}^{n} w_{ij}(d)x_j}{\sum_{j=1}^{n} x_j} \tag{4}$$

where $G_i^*(d)$ is normalised in the same way as in Equation (3) to obtain $Z(G_i^*)$. If $Z(G_i^*)$ is positive and statistically significant, the value around i is higher and belongs to the "hot spot zone"; otherwise, it belongs to the "cold spot zone".

2.3.4. Land-Use Change Transfer Matrix and Landscape Pattern Analysis

With the help of ArcGIS vectorisation calculations, the process of land-use change can be analysed quantitatively. The land-use transfer matrix can clearly reflect information on the dynamic process of inter-transformation between the area of each category at the beginning and end of a period of time in the study area [21–23], with the following expression:

$$S_{ij} = \begin{bmatrix} S_{11} & S_{12} & \dots & S_{1n} \\ S_{21} & S_{22} & \dots & S_{2n} \\ \vdots & \vdots & \ddots & \vdots \\ S_{n1} & S_{n2} & \dots & S_{nn} \end{bmatrix} \tag{5}$$

where S_{ij} denotes the area of the land type, n denotes the number of types of land use, and i and j denote the serial numbers of land-use types at the beginning and end of the study period, respectively.

According to the landscape pattern characteristics of the Hung River Valley region and the purpose of the study, this paper selected the number of patches (NP), patch density (PD), maximum patch index (LPI), average patch area (AREA_MN), landscape separation index (DIVISION), landscape edge density (ED), landscape shape index (LSI), sprawl index (CONTAG), Shannon Diversity Index (SHDI), and Shannon Evenness Index (SHEI), 10 indices calculated with the help of Fragstats 4.2 software, to analyse the degree of fragmentation, shape complexity, and diversity at the landscape level. NP, PD, LPI, and AREA_MN are used to describe the scale and quantity of various types of land and reflect the spatial pattern characteristics of the land; DIVISION, ED, LSI, CONTAG, SHDI, and SHEI are used to describe the connection degree and patch shape of various types of land, diversity, etc., reflecting the spatial structure characteristics of the land. The specific indices and calculations are detailed in the methodology of references [22–24].

2.3.5. Geographically Weighted Regression (GWR) Model

The GWR model calculates regression coefficients for each location, which accurately characterise the spatial characteristics of relationships by constructing local regression equations on each grid of the study area. The GWR model reflects the differences in the influence of different regional influences on the dependent variable due to the presence of spatial autocorrelation and spatial heterogeneity [31–34]. The formula is calculated as follows:

$$Y_i = \beta_0(u_i, v_i) + \sum_{k=1}^{p} \beta k(u_i, v_i) x_{ik} + \varepsilon_i \tag{6}$$

where β_0 is the model constant; (u_i, v_i) is the coordinate of the ith sample point; β_k is the k regression parameter of the i sample point; ε_i is the residual of the i sample point. The difference from the general linear regression is that β is a function of the geographical coordinates (u_i, v_i).

3. Results

3.1. Land-Use Change Characteristics

The land-use types in the Hung River Valley region are diverse and structurally complex. Dividing the land-use types in the study area into cropland, woodland, grassland, water, construction land, and bareland, the five-phase landscape type distribution map shows (Figure 2) that the land-use types in the study area are mainly woodland, grassland, and arable land, of which woodland and grassland both account for more than 32% of the total area of the study area, followed by the area of arable land, which accounts for about 20% of the total area of the region. Overall, the woodland, grassland, and arable land cover 90% of the total area of the study area and have a greater impact on the overall landscape, while the proportion of construction land, water, and bare land is smaller, accounting for about 10% of the total area of the study area. From 2000 to 2020, the area of arable land decreased, and the area of water bodies continued to increase; from 2000 to 2020, the area of arable land transferred out was the largest, and the proportion of construction land increased the most.

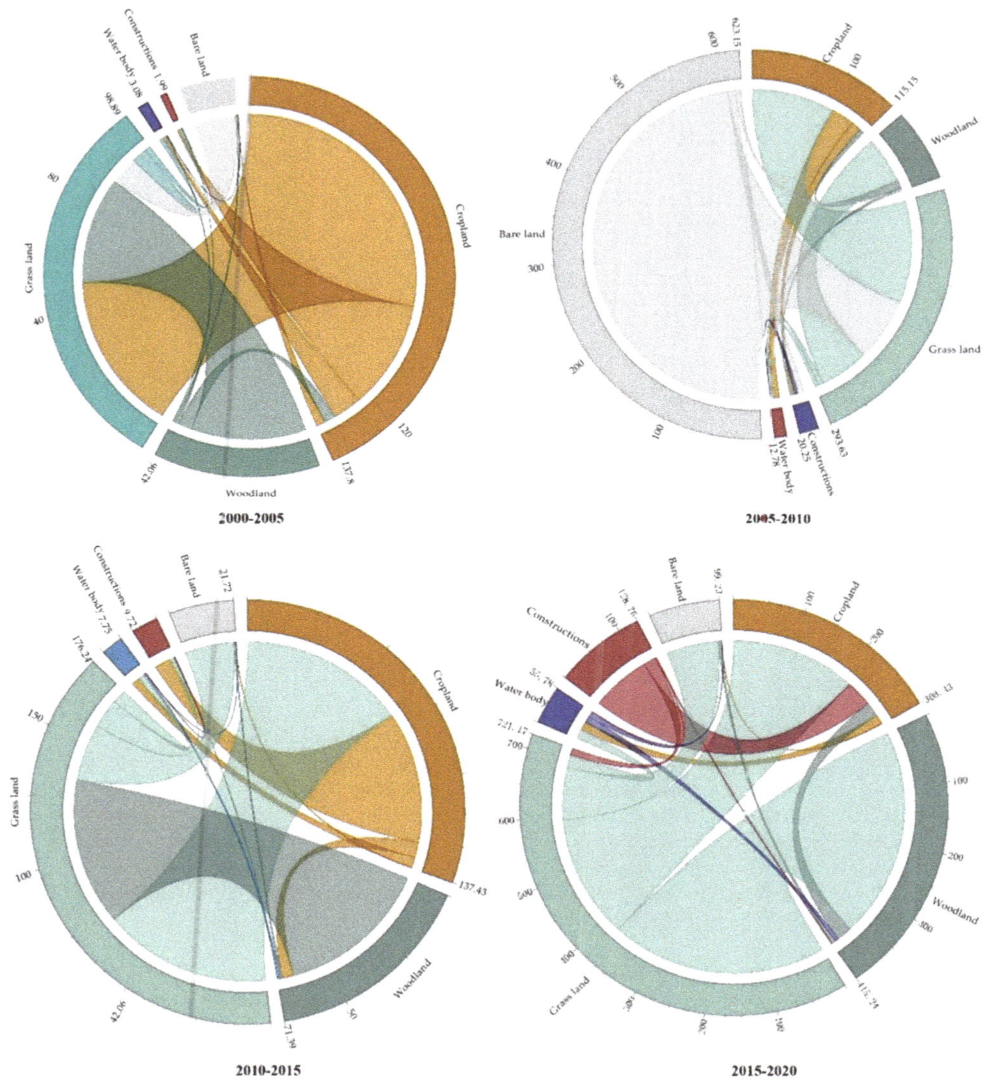

Figure 2. Sankey Diagram of Land-Use Transfer.

To fully understand the structural characteristics of land-use changes in the Hung River Valley during this period, a land-use transfer matrix was constructed to calculate the number of mutual land-use transfers from 2000 to 2020 (Table 3). As shown in the table, from 2000 to 2005, land-use shifts mainly occurred between arable land, grassland, and forest land, with a larger amount of arable land shifting to grassland and construction land. During the period 2010–2015, the trend of the previous period continued, with arable land always being the source of inflow of construction land and grassland, while the inflow and outflow of grassland and forest land were basically the same. From 2015 to 2020, the interconversion of various land uses made the inflow and outflow basically the same.

Table 3. Land-use transfer matrix for the Hung River Valley (km^2).

Years	Type	Cultivated Land	Forestland	Grassland	Water Areas	Construction	Bare Land
2000–2005	Cultivated land	—	4.29	49.7	0.9	0.96	0.25
	Forestland	3.44	—	33.78	0.37	0.1	0.63
	Grassland	88.55	35.57	—	1.49	0.8	13.32
	Water areas	8.91	1.04	4.85	—	0.1	0.15
	Construction	33.93	0.48	2	0.09	—	0.01
	Bare land	2.97	0.68	8.56	0.23	0.03	—
2005–2010	Cultivated land	—	10.52	135.32	3.61	7.5	2.26
	Forestland	6.61	—	63.14	0.85	0.21	5.6
	Grassland	105.6	61.06	—	2.55	4.56	604.67
	Water areas	13.08	2.14	6.19	—	0.51	10.38
	Construction	29.51	0.52	5.75	4.22	—	0.24
	Bare land	0.35	1.54	80.23	9.02		—
2010–2015	Cultivated land	—	4.87	70.84	3.63	7	0.36
	Forestland	6.52	—	62.74	1.06	0.21	0.69
	Grassland	70.02	64.34	—	2.45	1.83	20.45
	Water areas	4.41	0.85	6.07	—	0.23	0.22
	Construction	56.04	0.46	12.62	0.2	—	0
	Bare land	0.44	0.87	23.97	0.23	0	—
2015–2020	Cultivated land	—	26.22	219.07	15.84	104.28	1.43
	Forestland	25.19	—	376.38	11.26	1.89	4.88
	Grassland	218.72	374.21	—	17.65	21.28	91.87
	Water areas	16.17	7.09	21.97	—	1.26	1
	Construction	46.66	1.49	11.83	0.91	—	0.05
	Bare land	1.69	6.23	91.92	10.12	0.05	—

The land-use changes in the Hung River Valley in the past 20 years are mainly influenced by policies and urban expansion. In ecological protection policies, agricultural and livestock production and urbanisation, industrialisation, and other factors under the comprehensive effect of the Hung River Valley grassland, arable land, and construction land change dramatically. Each landscape flows between each other. Since 2002, Qinghai Province has fully implemented the policy of returning farmland to forest and grass, coupled with the establishment of many nature reserves. Arable land is in a net outflow situation, manifested in the expansion of grassland scale. At the same time, the Hung River Valley is an important axis of economic development in Qinghai Province, with intense human activity, high levels of urbanisation and industrialisation, and a continuous increase in the demand for construction land, the main source of which is the occupation of arable land.

3.2. Analysis of the Evolution of the Landscape Pattern Features

Since the magnitudes of different indicators are different, to facilitate comparison, the sampling of Z is standardized, as are all landscape pattern index values. As can be seen in Figure 3, NP and PD showed an increasing trend between 2000 and 2005, which is related to the partial conversion of arable land in the study area into forest land and grassland, and a large amount of arable land outflow, causing arable land fragmentation. After 2005, PD and NP gradually decreased, combined with the actual situation of the Hung River Valley. It can be seen that urban construction land encroachment on arable land, where there are many construction enclaves, merged with the original construction land, the main reason for the decline in PD and NP, in addition to the demolition and merging of rural settlements. LPI and AREA_MN increased slightly between 2000 and 2005 and fluctuated little after 2005, while DIVISION decreased slightly, indicating that the overall landscape pattern of the Hung River Valley showed a clustering trend after 2005.

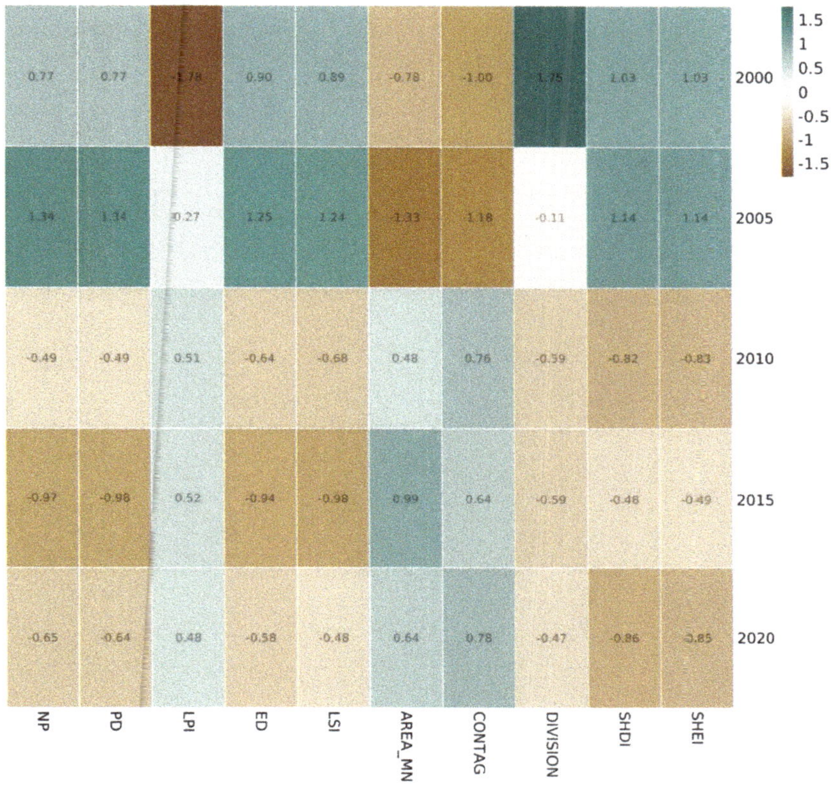

Figure 3. Hung River Valley overall landscape pattern trend evolution map, 2000–2020.

The ED and LSI decreased slightly from 2000 to 2020, indicating that the overall landscape shape of the study area tends to be clustered, with the shape changing from complex to simple. conTAG increased relatively over the 20-year period, indicating that the overall landscape connectivity in the study area has increased. The Shannon Diversity Index (SHDI) decreased, and landscape heterogeneity diminished; the Shannon Evenness Index (SHEI) shows a decreasing trend, and the landscape type dominant over the overall landscape in the study area increased.

Changes in habitat quality are an indirect reflection of changes in different land types. To reveal in more detail the relationship between landscape changes in the study area and their impact on habitat quality, the characteristics of changes in different land types between 2000 and 2020 were calculated using Fragstats 4.2 (Figure 4). Grassland and woodland are important landscapes regarding the quality of habitat, with a decrease in the number of patches (NP), a decrease in density (PD), an increase in the mean patch area (AREA_MN), and a decrease in ED, LSI, and DIVISION from 2005–2020, indicating an overall clustering of landscape patterns and an increase in disturbance resistance in grassland. It is particularly important to note that the patches of built-up land in the study area decreased, and density decreased over the 20-year period, but ED increased slightly, and LSI and DIVISION remained largely unchanged, indicating that the landscape pattern in some areas became dispersed, which may be related to the extensive expansion of towns in Xining.

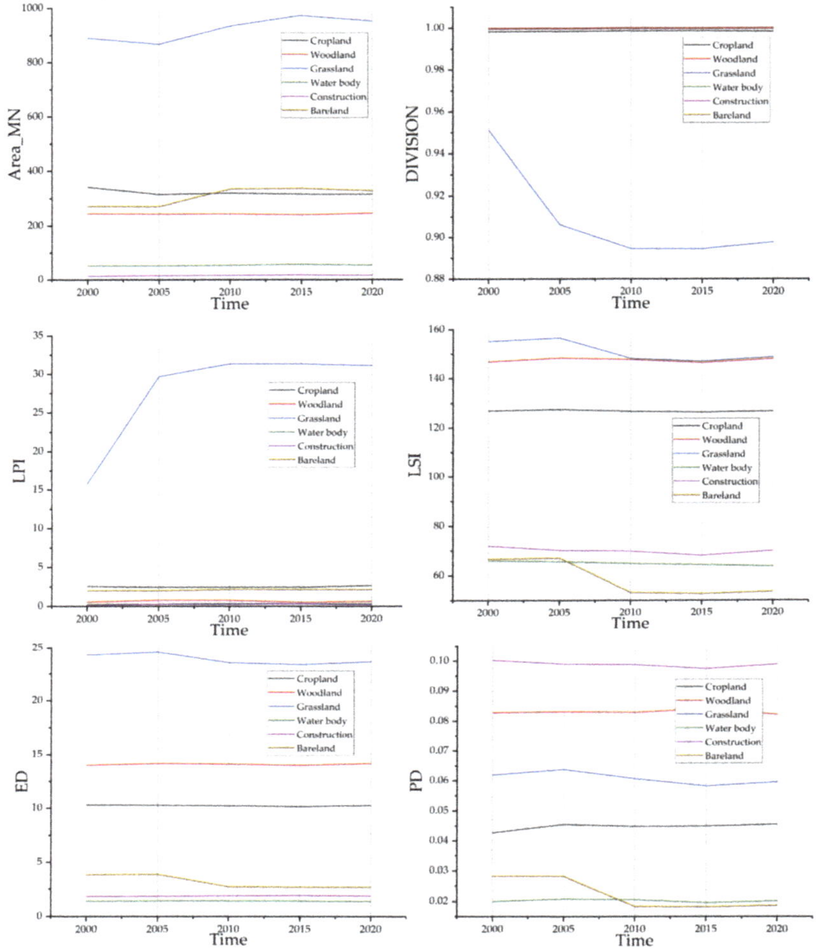

Figure 4. Evolution of landscape pattern indices for different land types in the Hung River Valley, 2000–2020.

3.3. Spatial and Temporal Variation of Habitat Quality

The Habitat Quality Index (HQI) reflects the fragmentation of habitat patches in the study area, on the one hand, and the ability of habitat patches to resist the threat of habitat degradation brought about by human activities. On the other hand, its value is a continuous value between 0 and 1: the closer to 1, the better the habitat quality, indicating that biodiversity is better maintained. The raster areas of different classes and their percentages were counted (Table 4). To more accurately portray the evolution of habitat quality, using the Re-classify tool of the ArcGIS 10.6 software platform, habitat quality was classified into five levels: very low (0–0.2), low (0.2–0.4), medium (0.4–0.6), high (0.6–0.8), and very high (0.8–1), based on the actual situation in the Hung River Valley and with reference to existing studies [13,18].

Table 4. Habitat quality percentage statistics for the Huangshan Valley, 2000–2020.

Level	2000		2005		2010		2015		2020	
	Area/km^2	Percentage/%	Area/km^2	Percentage/%	Area/km^2	Percentage/%	Area/km^2	Percentage/%	Area/km^2	Percentage/%
Very low	2939.82	8.3	2972.32	8.4	2458.76	6.9	2554.65	7.3	2485.86	7.0
Low	3803.14	10.8	3760.62	10.7	3760.60	10.7	3714.35	10.5	3738.7	10.6
Medium	5726.48	16.3	5754.36	16.3	6633.74	18.8	6595.7	18.7	6632.13	18.8
High	16,038.43	45.5	16,012.46	45.4	15,606.80	44.3	15,583.67	44.2	15,573	44.0
Very high	6727.01	19	6735.09	19	6773.13	19.2	6786.52	19.3	6788.54	19.3

From 2000 to 2020, the overall habitat quality showed an upward trend, and the global average habitat indexes were 0.656, 0.657, 0.661, 0.661, and 0.662, respectively. The main reason for this is that around 2000, Qinghai Province, based on its development orientation, successively carried out a series of ecological protection and restoration work, e.g., returning farmland to forest, returning farmland to grassland, and constructing nature reserves, which promoted the transformation of some medium- and high-grade habitat patches to higher-grade habitat patches.

On a spatial scale (Figure 5), the quality of habitats in the concentrated areas of woodland, grassland, and watersheds is high, while the quality of habitats on arable land, construction land, and bare land is low. The whole area is dominated by habitat patches of excellent grade. Habitat quality in the north, east, and west districts of Xining is low compared to other areas, due to the deteriorating ecological conditions in the central areas of Xining as a result of increasingly intense human activities. Among all areas, the habitat quality indexes of Datong County, Guide County, Mutual Aid County, Menyuan County, and Zunhua County all exceed 0.7, with Guide County being the highest, maintaining a level of 0.9; Hualong County, Jianzha County, Ping'an County, and Minhe County all show an increasing trend, with Hualong County being the highest and Minhe County being relatively low; the socio-economic development of Huangyuan County and Huanzhong County is strongly affected by the radiation of Xining City, and the social and economic development has a greater impact on the ecological environment.

3.4. Hotspot Analysis of the Spatial Distribution of Habitat Quality

3.4.1. Overall Clustering Characteristics

To explore the spatial differentiation characteristics of habitat quality in the Hung River Valley in more detail, the study area was divided into 2328 4 × 4 km grids using grid analysis. The mean values of habitat quality in 2000, 2005, 2010, 2015, and 2020 were extracted from 18 counties and cities in the Hung River Valley based on grid scale, and the ArcGIS 10.6 platform was used to calculate the spatial clustering of habitat quality in the study area from 2000 to 2020.

The Moran's I calculations for the five periods of 2000, 2005, 2010, 2015, and 2020 showed that the Z scores of the five periods were all above 75 and much higher than 2.58, and the *p*-values passed the 1% significance test, indicating that the spatial distribution of habitat quality values in the Hung River Valley was not random at a 99.9% confidence level and that there was significant spatial correlation (Table 5).

The Moran's I index for all five periods was greater than 0.65, showing a significant pattern of aggregation, i.e., high values of habitat quality clustered in space, and low values tended to be adjacent to each other. Since 2000, the aggregation effect of habitat quality in the Hung River Valley has been increasing on the whole; however, from 2015 to 2020, the aggregation effect has been on the decline, mainly because the areas with high habitat quality are affected by the expansion of urban land, eroding the original woodland, grassland, and other ecological landscapes and causing habitat fragmentation. The development of a large amount of urban land has led to an increasingly widespread distribution of areas with a low habitat quality. This is shown in Table 5.

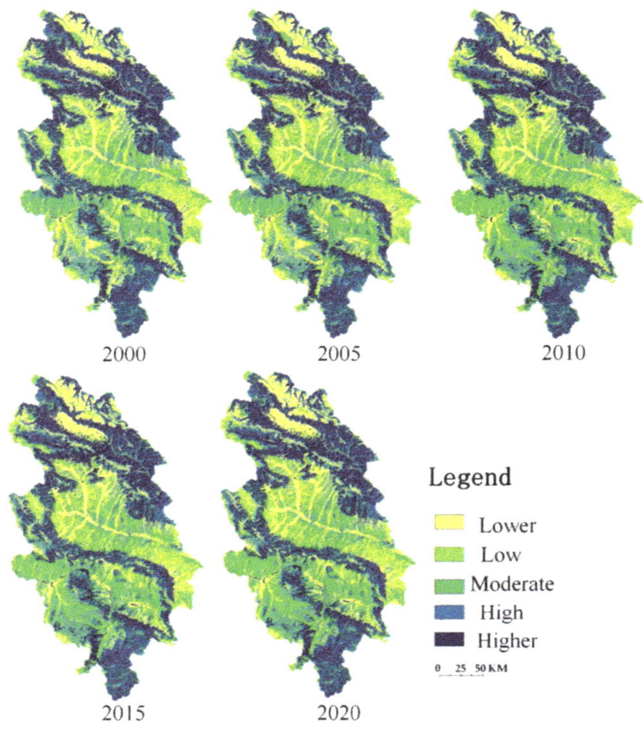

Figure 5. Spatial Distribution of habitat quality in the Hung River Valley, 2000–2020.

Table 5. Global Moran's I Index table.

Year	Global Autocorrelation Index Statistics			Result
	Moran's I	Z-Score	p-Value	
2000	0.699	46.929	0.0000	Gather
2005	0.698	46.886	0.0000	Gather
2010	0.705	47.344	0.0000	Gather
2015	0.777	51.755	0.0000	Gather
2020	0.706	47.427	0.0000	Gather

3.4.2. Local Agglomeration Characteristics

The global spatial autocorrelation can only reflect whether there are agglomerative features in the study area as a whole and cannot clarify the location distribution of agglomerative features. Based on the ArcGIS 10.6 platform, a hotspot analysis was conducted based on a grid, and cold spots and hotspots with a confidence level above 90% were selected to reflect the distribution of high- and low-value habitat quality clusters in the Hung River Valley (Figure 6).

During 2000–2020, the Hung River Valley habitat quality changes show obvious regional differences "Guide-Ledu", south of the line of the habitat quality index, including Ping'an County, Ledu District, Tongren County, and Guide County, generally improved. This is mainly due to the implementation of ecological protection policies, e.g., returning farmland to forest and grass, and establishing establishment many nature reserves, scenic spots, forest parks, and geoparks, such as the Mengda Nature Reserve and the Sanjiangyuan Nature Reserve. The overall habitat quality north of the "Guide-Ledu" linkage has a significant spatial aggregation effect, but the change effect is not obvious, and there are cold spots in Xining City and surrounding counties. The cold spot area is concentrated in Xining City, which is the political and economic centre of Qinghai Province and the gathering area of arable land, and the intense production and construction and agricultural activities have interfered with the ecological environment, resulting in the clustering and distribution of low habitats in the area. Mutual Aid County and Ping'an County are adjacent to Xining. Mutual Aid County has the largest population and the most intense human activities in Haidong, thus showing a secondary cold point concentration distribution, and the habitat cold point rose and then declined between 2000 and 2020, dropping to the lowest point in 2015 with the worst habitat quality.

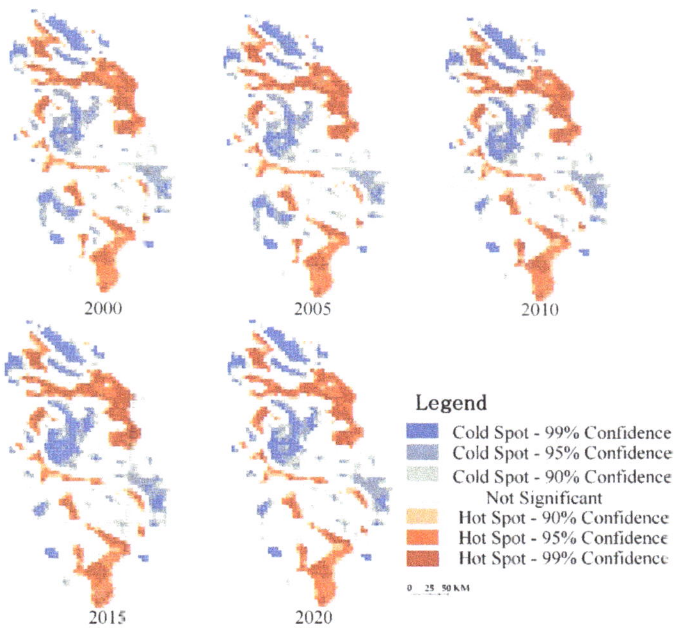

Figure 6. Habitat quality "hotspots" analysis in the Hung River Valley, 2000–2020.

3.5. Spatial Response of Habitat Quality to Urbanisation

The spatial pattern of habitat reflects that human activities and natural elements are important influencing factors of spatial differentiation of habitat quality. Among these factors, human activity status has gradually become the main independent variable of habitat quality. Natural factors, such as slope, average annual rainfall, average annual temperature, and elevation, and urbanisation process elements such as population density, gross domestic product, and the night light index, are selected as independent variables. The optimal model was selected by a comparative analysis of GWR and ordinary least squares (OLS). The coefficient of variance expansion is a measure of the severity of multiple (multiple) collinearities in a multiple linear regression model. It represents the ratio of the variance of the regression coefficient estimator to the variance, assuming that the independent variables are not linearly correlated. Usually, 10 is used as the judgment boundary. When VIF < 10, there is no multicollinearity; when $10 \leq$ VIF < 100, there is strong multicollinearity; when VIF ≥ 100, there is severe multicollinearity.

The results show that the VIF of the natural and socio-economic factors in the OLS model is less than 10, and there is no covariance between the variables, which satisfies the requirements of the explanatory variables. The explanatory power of the

OLS model for habitat quality was less than 50%, and the goodness of fit of the GWR model was significantly higher than that of the OLS model in all five time sections, with its explanatory power reaching more than 90%. Meanwhile, the Sigma and AICc in the GWR model were lower than those of the OLS model, indicating that the GWR model had better explanatory power for the factors influencing habitat quality and its model accuracy was better (Tables 6 and 7).

Table 6. GWR model goodness of fit.

Model Parameters	2000	2005	2010	2015	2020
Bandwidth	55	55	55	55	55
Residual Squares	230.377	238.716	233.701	228.572	228.182
Effective Number	536.451	534.240	532.123	529.740	529.882
Sigma	0.353	0.359	0.355	0.351	0.351
AICC	6338.648	2661.698	2604.091	2543.320	2539.722
R^2	0.903	0.900	0.902	0.904	0.904
Adjusted R^2	0.875	0.871	0.874	0.877	0.877

Table 7. GWR global regression coefficients.

Global Regression Coefficients	2000	2005	2010	2015	2020
DEM (Digital elevation model)	0.119	0.127	0.247	0.236	0.246
SLOPE	0.324	0.352	0.344	0.343	0.342
RAIN	0.187	0.190	0.175	0.171	0.175
TEM	0.217	0.234	0.348	0.339	0.349
GDP	−0.067	−0.012	0.036	0.017	0.036
LIGHT	−0.030	−0.089	−0.177	−0.187	−0.176
POP	−0.049	−0.060	−0.031	−0.015	−0.021

Slope and elevation are important natural factors influencing habitat quality. However, due to the interaction of human activities, the correlation between habitat quality and natural factors such as slope and elevation is complicated. As can be seen in Figure 7, slope, elevation, and habitat quality generally show a positive relationship, with the positive and negative effects of slope on habitat being relatively complexly distributed within a geographical area. The positive correlation between slope and habitat is mainly concentrated in mountainous and hilly areas and areas with continuous construction land, and the positive correlation between height and slope is distributed in discontinuous bands. The central region is flat. Human activities are relatively frequent, and the natural ecological space is encroached upon by construction land, resulting in serious habitat degradation. Some of the hilly areas are affected by human development and construction activities, and the habitats are damaged to a certain extent, while the green areas and water areas in the plains are better protected by ecological protection, and the overall level of habitats is higher; under the influence of human activities, the slope of the area is negatively correlated with habitat quality. The areas with a high positive correlation between elevation and habitat quality are mainly concentrated in the hilly areas, with a cluster and circle pattern of distribution. In general, natural factors such as slope and elevation play an important role in the overall pattern of habitat distribution. With higher slope and elevation, socio-economic activities are generally less frequent in the area, and the disturbance factors to the ecosystem are smaller, so the impact on habitat quality is relatively small. In plain areas, where human activities are frequent, socio-economic factors have a more prominent impact on habitat quality than do natural factors.

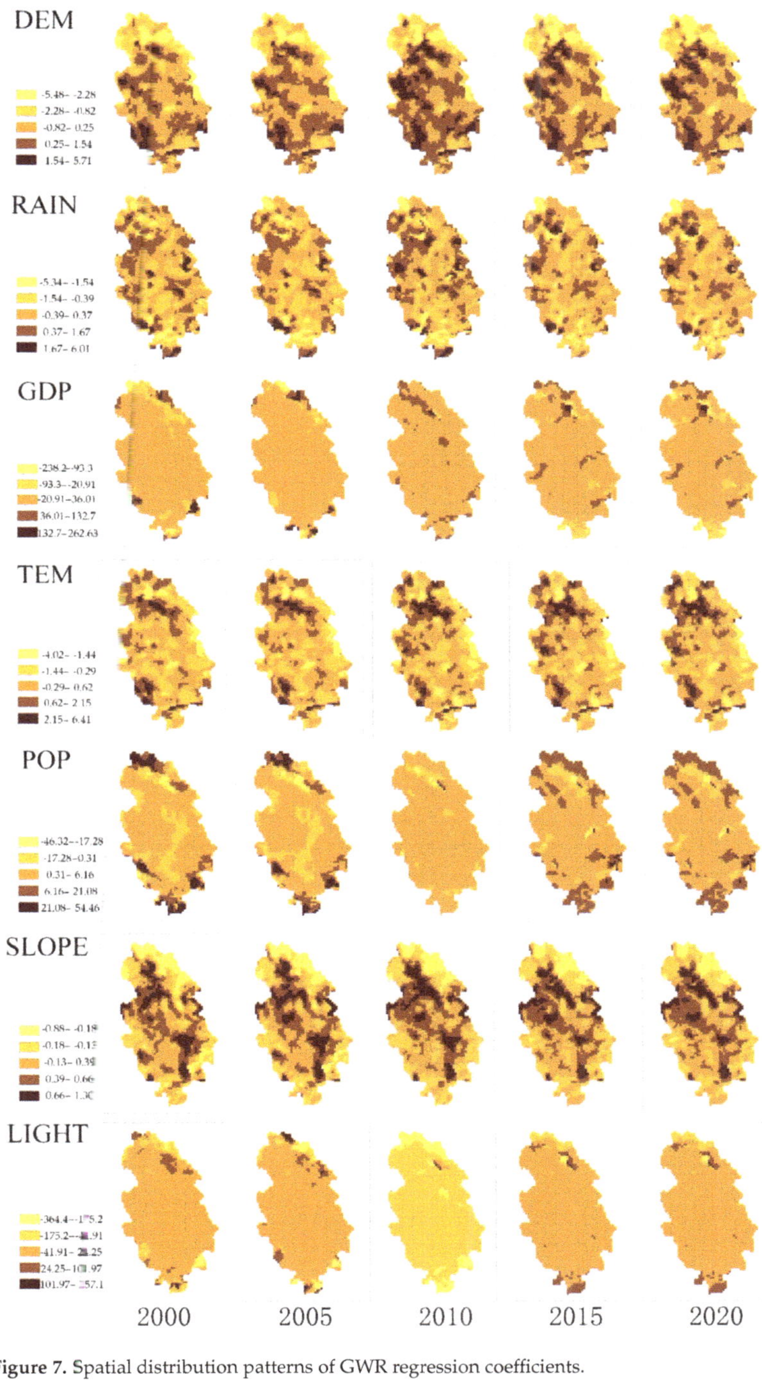

Figure 7. Spatial distribution patterns of GWR regression coefficients.

Urbanisation factors such as population density, gross domestic product, and the night-time light index have a more significant negative correlation with habitat quality (Figure 6). The negative correlation between population density and habitat quality is most significant and widely distributed: During the study period, the negative influence of the northern and central zones was further expanded, because the central zone was influenced by the economic radiation of Xining, and the towns were developed significantly, with a relatively obvious population growth, which increased the pressure on the ecological carrying capacity of the surrounding areas. The high density of economic activities also contributed to the fragmentation of the landscape pattern and the encroachment of arable land and construction land on ecological land.

The regression coefficients of the economic impact on habitat show that the negative impact in the study area is less intense. The reason for this is that the study area is affected by the policy of "returning farmland to forest" and "returning farmland to grass" in Qinghai Province, and the GDP output value of mainly arable land is low, so the negative impact on habitat is weak. The northern and southern parts of the city have seen rapid economic development, and the ecological environment has been significantly disturbed by economic activities. The night-time light index characterises the indirect disturbance effect of urban socio-economic development and high intensity human activities on the ecosystem: In 2000, the positively correlated areas were mainly scattered in clusters in the plains, while the negatively correlated areas were widely distributed, with the most intense negative impact in the mountainous hills. Between 2000 and 2010, the areas with intense negative correlation and the positively correlated areas both decreased. However, the negative correlation dominated the region, and the trend from the mountains to the plains was stronger and weaker. Between 2010 and 2020, the areas with strong negative correlations expanded again, which was related more to urban development. Overall, the agglomeration of factors in the urbanisation process is an important driver of regional habitat quality change, and the spatial heterogeneity of the impact of socio-economic factors on habitat quality is more significant as the rate of urbanisation accelerates.

4. Discussion

4.1. Scaling Effects of Land Characteristics and Habitat Quality

The coupling of landscape patterns and ecological processes reflected in land use is a central theme in landscape ecology research [35–37]. The resolution of the spatial characteristics of land-use has long been the focus and difficulty of scholars studying landscape ecology [38–40]. Because the relationship between patterns and processes is often non-linear and shows multi-factor interaction and time-lagged effects [41]. Differences in scale and accuracy can cause the type, number, and spatial distribution characteristics and configurations of land-use units to reflect different patterns and process couplings [42–44].

The use of the InVEST model to evaluate spatial and temporal changes in regional habitat quality is mainly based on land-use data for model input. In the process of revealing the drivers of habitat quality change, the drivers of the same evaluation unit will reflect different intensities of influence on different scales of habitat quality units [41,42]. For human activities, the study of land-use change processes and their ecological effects is also a process impact on patterns, but such impacts often involve time scales of years or decades, so the impact of "fast" ecological processes on landscape patterns has a lag in time scale [44]. This and numerous studies have expressed the mismatch in time scales between landscape patterns and ecological processes, leading to a certain lack of understanding of feedback mechanisms and systems as a whole [44–47].

Therefore, research on the effects of land characteristics and habitat quality scales needs to be further investigated, with a focus on the integration of natural, socio-economic, and human factors to resolve the complexity of the landscape in breadth and on the coupling of macroscopic patterns and microscopic processes [44,48]. Providing a reliable basis for macro-pattern characterisation and management strategy formulation, along with macro-pattern planning and management, will in turn strengthen the practical significance of micro-research [49]. The ultimate goal is to scientifically, rationally, and accurately reveal the important influencing factors of habitat quality change due to land-use change [50,51].

4.2. Research Shortcomings

The InVEST model is relatively mature and outperforms traditional methods in terms of spatial expression and dynamic research, but there is a certain degree of subjectivity in the parameter settings in the calculations, and the validation of the parameters and the assessment of their rationality are worth exploring in depth [10,24,26]. This study explores the effects of factor agglomeration on habitat in the urbanisation process and obtains some insights that are beneficial to ensuring ecological safety in the urbanisation process. As a complex social-ecological system, the impact of urbanisation on habitat quality includes not only economic and demographic factors, but also hidden factors such as culture and policies [42,43], and there is a complex relationship between factors in the urbanisation process, so it is necessary to introduce more social factors and clarify the interrelationships between them for a comprehensive study. This is a direction for further research.

4.3. Policy Recommendations

The Hung River Valley is a concentrated area of arable land in Qinghai Province, which requires the continuous optimisation of agricultural production methods and improvement of agricultural production efficiency. At the same time, in the context of the work on comprehensive land improvement and ecological protection and restoration, the region's landscape type characteristics and current ecological problems should be combined to promote agricultural production and ecological protection and restoration in an integrated manner using development, preparation, reclamation, and restoration. Compared to areas such as Datong and GuiDe counties, there is more room to improve the quality of habitats in the cities of Xining and Haidong, and there is a need to further promote the construction of large-scale forest farms and accelerate the implementation of ecological construction projects such as the greening of key areas in central cities, major towns and parks, and urban wetland parks, in order to systematically improve the quality of regional habitats. A fragmented and complex landscape distribution is not conducive to the protection and restoration of ecosystems; thus, when formulating and implementing relevant measures, it is necessary to adhere to the principle of integrated management of "mountains, water, forests, fields, lakes, and grasses", reduce the risk of excessive intervention caused by the management of single elements, and maintain natural and semi-natural landscapes.

5. Conclusions

(1) From 2000 to 2020, the area of grassland, construction land, and watersheds in the Hung River Valley increased year by year. The area of cropland kept decreasing. The woodland and bareland fluctuated and changed, but was basically stable. Among them, the main source of growth in construction land is the occupation of cropland, which is especially obvious in Xining and Haidong; woodland, grassland, water, and other high habitat landscape types increased steadily, thanks to the implementation of policies returning farmland to forest and grass and establishing nature reserves at all levels.

(2) From 2000 to 2020, the Hung River Valley Habitat Quality Index was stable, at around 0.66, with a slight increase. The regional habitat quality changed, and hot and cold states on both sides of the "Guide-Ledu" were differently distributed. The habitat level of Xining showed low-quality characteristics, including decreasing and gathering cold spots, while the habitat quality index of Datong County, Guide County, Mutual County, Menyuan County, and Zunhua County was higher. The ecological protection pressure is relatively small.

(3) The natural elements shaped the overall habitat distribution pattern in the Hung River Valley, with slope, elevation, and habitat quality generally showing a positive relationship, and the effect of slope on habitat was relatively complex. The effects of disturbance on the ecosystem were strong, biodiversity was destroyed, ecosystem imbalance occurred, and habitat quality was significantly degraded.

Author Contributions: Conceptualization, X.F.; methodology, X.G.; software, H.Y.; validation, A.L.; formal analysis, D.S.T.; investigation, S.O.; resources, J.L.; data curation, Y.R.; writing—original draft preparation, X.F.; writing—review and editing, X.G.; visualization, G.T.; supervision, Y.Z.; project administration, M.S., M.W. and X.W.; funding acquisition, A.L., C.H. and X.F. All authors have read and agreed to the published version of the manuscript.

Funding: This research was funded by the National Key Research and Development Program of China (Project Numbers: 2017YFC0404301), the Natural Science Foundation of China (Project Numbers: 42001218), and the State Key Laboratory of Earth Surface Processes and Resource Ecology, Beijing Normal University, China (Project Numbers: 2021-KF-03).

Institutional Review Board Statement: Not applicable.

Informed Consent Statement: Not applicable.

Conflicts of Interest: The authors declare that there is no conflict of interest.

References

1. Costanza, R.; D'Arge, R.; De Groot, R.; Farber, S.; Grasso, M.; Hannon, B.; Limburg, K.; Naeem, S.; O'Neill, R.V.; Paruelo, J.; et a. The value of the world's ecosystem services and natural capital. *Nature* **1997**, *387*, 253–260. [CrossRef]
2. Ellis, E.C.; Pascual, U.; Mertz, O. Ecosystem services and nature's contribution to people: Negotiating diverse values an trade-offs in land systems. *Curr. Opin. Environ. Sustain.* **2019**, *38*, 86–94. [CrossRef]
3. Fellman, J.B.; Hood, E.; Dryer, W.; Pyare, S. Stream physical characteristics impact habitat quality for pacific salmon in tw temperate coastal watersheds. *PLoS ONE* **2015**, *10*, e0132652. [CrossRef]
4. Gao, Y.; Ma, L.; Liu, J.; Zhuang, Z.; Huang, Q.; Li, M. Constructing ecological networks based on habitat quality assessment: A case study of Chang-zhou, China. *Sci. Rep.* **2017**, *7*, 46073. [CrossRef] [PubMed]
5. Sharp, R.; Tallis, H.T.; Ricketts, T.; Guerry, D.A.; Wood, A.S.; Chaplin-Kramer, R.; Nelson, E.; Ennaanay, D.; Wolny, S.; Olwer N.; et al. *InVEST 3.2.0 User's Guide;* The Natural Capital Project; Stanford University, University of Minnesota, The Natur Conservancy, World Wildlife Fund: Minneapolis, MN, USA, 2018.
6. Hillard, E.M.; Nielsen, C.K.; Groninger, J.W. Swamp rabbits as indicators of wildlife habitat quality in bottomland hardwo forest ecosystems. *Ecol. Indic.* **2017**, *79*, 47–53. [CrossRef]
7. Peng, J.; Pan, Y.; Liu, Y.; Zhao, H.; Wang, Y. Linking ecological degradation risk to identify ecological security patterns in a rapidl urbanizing landscape. *Habitat Int.* **2018**, *71*, 110–124. [CrossRef]
8. Barbara, R.; Stefan, L. A spatially explicit patch model of habitat quality, integrating spatio-structural indicators. *Ecol. Indic.* **201** *94*, 8–14.
9. Janus, J.; Bozek, P. Land abandonment in Poland after the collapse of socialism: Over a quarter of a century of increasing tre cover on agricultural land. *Ecol. Eng.* **2019**, *138*, 106–117. [CrossRef]
10. Liu, Y.; Huang, X.; Yang, H.; Zhong, T. Environmental effects of land use/cover change caused by urbanization and policies i South-west China Karstarea: A case study of Guiyang. *Habitat Int.* **2014**, *44*, 339–348. [CrossRef]
11. Terrado, M.; Sabater, S.; Chaplin-Kramer, B.; Mandle, L.; Ziv, G.; Acuña, V. Model development for the assessment of terrestri and aquatic habitat quality in conservation planning. *Sci. Total Environ.* **2016**, *540*, 63–70. [CrossRef]
12. Sala, O.E.; Chapin, F.S.; Armesto, J.J.; Berlow, E.L.; Bloomfield, J.; Dirzo, R.; Hubersanwald, E.; Huenneke, L.F.; Jackson, R.B Kinzig, A.P.J.S. Global biodiversity scenarios for the year 2100. *Science* **2000**, *287*, 1770–1774. [CrossRef] [PubMed]
13. Liu, C.H.; Lun, Y. Progress, achievements and prospects of biodiversity protection in Yunnan Province. *Biodivers. Sci.* **2020**, *29* 200. [CrossRef]
14. Liu, Y.; Zhou, Y.; Du, Y. Study on the Spatio-Temporal Patterns of Habitat Quality and Its Terrain Gradient Effects of the Middl of the Yangtze River Economic Belt Based on InVEST Model. Resour. *Environ. Yangtze Basin.* **2019**, *28*, 2429–2440.
15. Lei, J.C.; Liu, J.X.; Yong, F.; Liu, H.M.; Wu, J.; Ding, H.; Wang, J.M.; Wu, S.Q.; Cheng, S.; Cui, P. Multi- Scenario Ecosystem Servic Assessment of Wuma River Valley Based on CLUE- S and InVEST Models. *J. Ecol. Rural Environ.* **2017**, *33*, 1084–1093.
16. Liang, X.; Liu, X.P.; Li, X.; Chen, Y.M.; Tian, H.; Yao, Y. Delineating multi-scenario urban growth boundaries with a CA-base FLUS model and morphological method. *Landsc. Urban Plan.* **2018**, *177*, 47–63. [CrossRef]
17. Koull, N.; Chehma, A. Soil characteristics and plant distribution in saline wetlands of Oued Righ, northeastern Algeria. *J. Ari Land.* **2016**, *8*, 948–959. [CrossRef]

18. Wei, H.; Zhao, W.W.; Zhang, X.; Wang, X.Z. Regional ecosystem service value evaluation based on land use changes: A case study in Dezhou, Shandong Province, China. *Acta Ecol. Sin.* **2017**, *37*, 3830–3839.
19. Kim, T.; Song, C.; Lee, W.K.; Kim, M.; Lim, C.H.; Jeon, S.W.; Kim, J. Habitat quality valuation using InVEST model in Jeju Island. *J. Korea Soc. Environ. Restor. Technol.* **2015**, *18*, 1–11. [CrossRef]
20. Qamer, F.M.; Shehzad, K.; Abbas, S.; Murthy, M.; Xi, C.; Gilani, H.; Bajracharya, B. Mapping Deforestation and Forest Degradation Patterns in Western Himalaya, Pakistan. *Remote Sens.* **2016**, *8*, 385. [CrossRef]
21. Shalaby, A.; Tateishi, R. Remote sensing and GIS for mapping and monitoring land cover and land-use changes in the Northwestern coastal zone of Egypt. *Appl. Geogr.* **2007**, *27*, 28–41. [CrossRef]
22. Zhang, H.Y.; Wang, L.; Zhao R.; Liang, Y. Integrating landscape pattern with economic indices to conduct land use two-dimensional division: Taking Qinghai Province as an example. *Bull. Surv. Mapp.* **2013**, *3*, 82–85.
23. Zhao, D.Y.; Zeng, Y.N. Land use changes and ecological risk assessment in eastern Qinghai Plateau: A case study in Haidong, Qinghai, China. *J. Desert Res.* **2016**, *36*, 1190–1197.
24. Wu, B.W.; Yang, S.T.; Shao, N.F.; Peng, R.W.; Guan, Y.B. Effects of land use change on ecosystem service value in fragile ecological area of the Loess Plateau: A case study of Fenhe River Basin. *Res. Soil Water Conserv.* **2019**, *26*. 340–345.
25. Hou, L.W. InVEST habitat quality model parameter assignment discussion. *Environ. Dev.* **2017**, *29*, 80–81.
26. Newbold, T.; Hudson, L.N.; Hill, S.L.L.; Contu, S.; Lysenko, I.; Senior, R.A.; Börger, L.; Bennett, D.J.; Choimes, A.; Collen, B.; et al. Global effects of land use on local terrestrial biodiversity. *Nature* **2015**, *520*, 45–50. [CrossRef] [PubMed]
27. Hennig, E.I.; Schwick, C.; Soukup, T.; Orlitová, E.; Kienast, F.; Jaeger, J.A.G. Multi-scale analysis of urban sprawl in Europe: Towards a European de-sprawling strategy. *Land Use Policy* **2015**, *49*, 483–498. [CrossRef]
28. Xinliang, X.; Jiyuan, L.; Shuwen, Z.; Rendong, L.; Changzhen, Y.; Shixin, W. *China Multi-Period Land Use and Land Cover Remote Sensing Monitoring Dataset (CNLUCC)*; Data Registration and Publishing System of the Resource and Environmental Science Data Center of the Chinese: Shanghai, China, 2020. [CrossRef]
29. McDonald, R.I.; Weber, K.; Padowski, J.; Flörke, M.; Schneider, C.; Green, P.A.; Gleeson, T.; Eckman, S.; Lehner, B.; Balk, D.; et al. Water on an urban planet: Urbanization and the reach of urban water infrastructure. *Glob. Environ. Chang.* **2014**, *27*, 96–105. [CrossRef]
30. Deng, Y.J.; Yao, S.B.; Hou, M.Y.; Zhang, T.Y.; Lu, Y.N.; Gong, Z.W.; Wang, Y.F. Assessing the effects of the Green for Grain Program on ecosystem carbon storage service by linking the InVEST and FLUS models: A case study of Zichang county in hilly and gully region of Loess Plateau. *J. Nat. Resour. Energy* **2020**, *35*, 826–844.
31. Wu, Y.Z.; Shan, L.P.; Guo, Z.; Peng, Y. Cultivated land protection policies in China facing 2030: Dynamic balance system versus basic farmland zoning. *Habitat Int.* **2017**, *69*, 126–138. [CrossRef]
32. Wang, C.D.; Li, X.; Yu, H.J.; Wang, Y.T. Tracing the spatial variation and value change of ecosystem services in Yellow River Delta, China. *Ecol. Indic.* **2019**, *96*, 270–277. [CrossRef]
33. Shi, H.; Li, X.; Yang, Z.Z.; Li, T.H.; Ren, Y.; Liu, T.T.; Yang, N.D.; Zhang, H.; Chen, G.Z.; Liang, X. Tourism land use simulation for regional tourism planning using POIs and cellular automata. *Trans. GIS* **2020**, *24*, 1119–1138. [CrossRef]
34. Sharma, R.; Nehren, U.; Rahman, S.A.; Meyer, M.; Rimal, B.; Seta, G.A.; Baral, H. Modeling Land Use and Land Cover Changesn and Their Effects on Biodiversity in Central Kalimantan, Indonesia. *Land* **2018**, *7*, 57. [CrossRef]
35. Gustafson E, J. Quantifying landscape spatial pattern: What is the state of the art? *Ecosystems* **1998**, *1*, 143–156. [CrossRef]
36. Van der Biest, K.; Meire, P.; Schellekens, T.; D'hondt, B.; Bonte, D.; Vanagt, T.; Ysebaert, T. Aligning biodiversity conservation and ecosystem services in spatial planning: Focus on ecosystem processes. *Sci. Total Environ.* **2020**, *712*, 136350. [CrossRef] [PubMed]
37. Lin, W.B.; Sun, Y.M.; Nijhuis, S.; Wang, Z.L. Scenario-based flood risk assessment for urbanizing deltas using future land-use simulation (FLUS): Guangzhou Metropolitan Area as a case study. *Sci. Total Environ.* **2020**, *739*, 139899. [CrossRef]
38. Liu, X.J.; Li, X.; Liang, X.; Shi, H.; Ou, J.P. Simulating the Change of Terrestrial Carbon Storage in China Based on the FLUS-InVEST Model. *Trop. Geogr.* **2019**, *39*, 397–409.
39. Di Febbraro, M.; Sallustio, L.; Vizzarri, M.; De Rosa, D.; De Lisio, L.; Loy, A.; Eichelberger, B.; Marchetti, M. Expert-based and correlative models to map habitat quality: Which gives better support to conservation planning? *Glob. Ecol. Conserv.* **2018**, *16*, e00513. [CrossRef]
40. Zhu, J.; Ding, N.; Li, D.; Sun, W.; Xie, Y.; Wang, X. Spatiotemporal Analysis of the Nonlinear Negative Relationship between Urbanization and Habitat Quality in Metropolitan Areas. *Sustainability* **2020**, *12*, 669. [CrossRef]
41. Sun, L.; Yang, L.; Chen, L.D.; Zhao, F.K.; Li, S.J. Hydraulic redistribution and its contribution to water retention during short-term drought in the summer rainy season in a humid area. *J. Hydrol.* **2018**, *566*, 377–385. [CrossRef]
42. Priyadarshini, K.V.R.; Prins, H.H.T.; De Bie, S.; Heitknig, I.M.A.; Woodborne, S.; Gort, G.; Kirkman, K.; Ludwig, F.; Dawson, T.E.; De Kroon, H. Seasonality of hydraulic redistribution by trees to grasses and changes in their water-source use that change tree-grass interactions. *Ecohydrology* **2016**, *9*, 218–228. [CrossRef]
43. Tang, Z.; Ma, J.; Peng, H.; Wang, S.; Wei, J. Spatiotemporal changes of vegetation and their responses to temperature and precipitation in upper Shiyang river basin. *Adv. Space Res.* **2017**, *60*, 969–979. [CrossRef]
44. Hu, Y.N.; Mei, L.; Wei, J.G. Spatial differentiation and dynamic mechanism of regional travel agency efficiency in China based on GWR model. *Sci. Geogr. Sin.* **2018**, *38*, 107–113.

45. Zhai, T.; Wang, J.; Fang, Y.; Qin, Y.; Huang, L.; Chen, Y. Assessing ecological risks caused by human activities in rapid urbanization coastal areas: Towards an integrated approach to determining key areas of terrestrial-oceanic ecosystems preservation and restoration. *Sci. Total Environ.* **2020**, *708*, 135153. [CrossRef] [PubMed]
46. Li, F.; Wang, L.; Chen, Z.; Clarke, K.C.; Li, M.; Jiang, P. Extending the SLEUTH model to integrate habitat quality into urban growth simulation. *J. Environ. Manag.* **2018**, *217*, 486–498. [CrossRef]
47. Gu, X.; Long, A.; Liu, G.; Yu, J.; Wang, H.; Yang, Y.; Zhang, P. Changes in Ecosystem Service Value in the 1 km Lakeshore Zone of Poyang Lake from 1980 to 2020. *Land* **2021**, *10*, 951. [CrossRef]
48. Nematollahi, S.; Fakheran, S.; Kienast, F.; Jafari, A. Application of InVEST habitat quality module in spatially vulnerability assessment of natural habitats (case study: Chaharmahal and Bakhtiari province, Iran). *Environ. Monit. Assess.* **2020**, *192*, 1–17. [CrossRef] [PubMed]
49. Anchang, J.Y.; Ananga, E.O.; Pu, R. An efficient unsupervised index based approach for mapping urban vegetation from IKONOS imagery. *Int. J. Appl. Earth Obs. Geoinform.* **2016**, *50*, 211–220. [CrossRef]
50. Gong, P.; Li, X.; Wang, J.; Bai, Y.; Chen, B.; Hu, T.; Liu, X.; Xu, B.; Yang, J.; Zhang, W.; et al. Annual maps of global artificial impervious area (GAIA) between 1985 and 2018. *Remote Sens. Environ.* **2020**, *236*, 111510. [CrossRef]
51. Ma, Y.; Zhang, S.; Yang, K.; Li, M. Influence of spatiotemporal pattern changes of impervious surface of urban megaregion on thermal environment: A case study of the Guangdong—Hong Kong—Macao Greater Bay Area of China. *Ecol. Indic.* **2021**, *121*, 107106. [CrossRef]

land

MDPI

Article

Mechanisms of Change in Urban Green Infrastructure—Evidence from Romania and Poland

Alexandru-Ionuț Petrișor [1,2,3,*], Lidia Mierzejewska [4] and Andrei Mitrea [5]

1 Doctoral School of Urban Planning, Ion Mincu University of Architecture and Urbanism, 010014 Bucharest, Romania
2 National Institute for Research and Development in Tourism, 50741 Bucharest, Romania
3 National Institute for Research and Development in Constructions, Urbanism and Sustainable Spatial Development URBAN-INCERC, 021652 Bucharest, Romania
4 Faculty of Human Geography and Planning, Adam Mickiewicz University in Poznań, 61-680 Poznan, Poland; mierzeja@amu.edu.pl
5 Department of Urban Planning and Territorial Development, Faculty of Urban Planning, Ion Mincu University of Architecture and Urbanism, 010014 Bucharest, Romania; andrei.mitrea@uauim.ro
* Correspondence: alexandru_petrisor@yahoo.com; Tel.: +40-213-077-191

Abstract: The extent and continuity of green infrastructure can be adjusted by planning. Depending on the sense of the adjustment, the process can lead to a vicious cycle, resulting in poorer urban quality, or to a virtuous planning, thus leading to psychological wellbeing and sustainability. However, socioeconomic circumstances also play an important role in managing green infrastructure. Starting from these premises, the current study aims to take an in-depth look at the mechanisms of change in urban green infrastructure and provide concrete planning recommendations for dealing with the green infrastructure. It is based on a complex approach, combining an ecological design, including geo-statistical analyses of the structure and dynamics of different categories of green infrastructure in all Romanian and Polish cities covered by the Urban Atlas data during 2006–2018, with selected case studies for analyzing the deeper mechanisms and drivers of change in green infrastructure, and focusing on the role of different planning actors. The results indicate that green infrastructure was lost in all the cities analyzed, regardless of the different planning systems of the two countries. Based on this, specific recommendations can be phrased for all stakeholders of the planning process, including planners, local administrations, policy makers, and scientists.

Keywords: transitional dynamic; post-socialist countries; urban sprawl; derogatory planning; urban greenery

1. Introduction

The problems of urban green infrastructure (GI), in particular those related to its loss and fragmentation, have been widely discussed in the literature, thereby giving us the opportunity to review numerous titles covering the past 30 years. In summary, the review emphasized that the previous studies:

1. Justify the importance of any green infrastructure by the ecosystem services provided to the human population [1–3];
2. Show that fragmentation of the green infrastructure reduces the level of ecosystem services [4–6];
3. Indicate a significant influence of urban sprawl on fragmentation [3];
4. Reduce the planning problem to the choice between compact and dispersed cities [3,7].

However, despite the large number of studies, planning implications have not been addressed too much, and there seems to be a large gap between scientific approaches and planning practice. Where do we stand at present, from a scientific viewpoint? A quick glance at the literature reveals the following:

1. The literature kicks off sometime during the late 1990s and early 2000s, when it tries to establish an academically respectable position. Early papers touch upon the urban growth process, thereby linking to an older research strand, with an already established tradition. Once integrated, the research strand on GI substantiates its role within the urban growth tradition via the ecosystem services (ES) approach. The ecosystem approach GI research seems to gain initial momentum within the late 2000s, with a substantial increase in written output over the past few years [8].
2. We can therefore infer that GI is a rewarding research topic, especially in recent times. We are therefore entitled to conduct an appraisal of the agreed body of knowledge. The rationale behind such an assessment is simple: planners need actionable information within their planning practices.
3. Unfortunately, after nearly two decades of a relatively intensive research effort, the agreed body of knowledge looks surprisingly thin. One is bound to ask why the situation presents itself bleakly. The answers seem to be manifold, and we shall turn to them throughout this article.

The main focus of this study is the relationship between urbanization, planning, and GI, because planning can sometimes turn the vicious circle around, thereby preventing the loss of GI and consequences derived from losing its ES on urban welfare and sustainability [3,9]. The detailed mechanisms of urbanization dynamics are still debated by the existing literature. Hence, Puşcaşu [10] describes a cycle consisting of: (1) expansion of housing (second/vacation house) following the abandonment of agriculture; (2) tourism; (3) changes in large cities due to socio-economic drivers; and (4) loss of traditional rural space and consequent dissolution of borders between urban and rural areas. Díaz-Palacios-Sisternes et al. [11] present a simpler process: (1) "urbanization": transformation of agricultural land into urban land, (2) "agrarianization": development of agriculture and transformation of urban land into agricultural land, and (3) "renaturalization": colonization of urban and agricultural land by nature. A similar mechanism is presented by Grădinaru et al. [12]: (1) expansion of cities over agricultural areas; (2) abandonment of agricultural land which becomes unprofitable due to size, shape, and accessibility; and (3) transformation of agricultural land into built-up areas. Accounting for the change of political regime in Romania, Ianoş et al. [13] distinguish two drivers: (1) tourism, commercial, and residential amenities during 1990–2000; and (2) metropolitan urban sprawl after. Finally, in Eastern Europe, urbanization occurs around large cities, which add natural or agricultural land to their territory and turn it into urbanized land [14–16]. A different mechanism is underlined by other studies [17,18], namely the "derogatory planning". Hence, if the previous mechanisms accounted mostly for an "urbanization by sprawl" [19,20], "derogatory planning" is responsible for urbanization within city limits, including fragmentation and transformation of GI. In summary, the mechanism consists of real estate development occurring through a series of exemptions from planning provisions (either national or municipal), favored by local authorities for different interests (usually economic). As a result, urban development causes a densification of built-up area at the expense of nature, and in most cases in worsening the living conditions of inhabitants [21,22]. Such mechanisms can generate concrete recommendations: Our previous study [3] partially filled in the gap, but it was based on general analyses that were unable to look at the changes of particular types of GI. We therefore indicated the study of these changes as a possible continuation of research.

Therefore, our aim originates in the need for correlating "ecological studies" in their epidemiological meaning [23], i.e., carried out at the level of population units (in this case, different cities or urban areas considered each one as a whole), with an in-depth analysis of individual cases, in order to discern the mechanisms affecting the urban GI. Consequently, the study plans to combine analyses looking at land cover and use changes with analyses of individual cities, to derive some concrete planning recommendations.

2. Materials and Methods

The study covers the period of 2006–2018, but focuses on the interval 2006–2012, for the following reasons: First, the start and end years were dictated by the availability of data, collected by the Urban Atlas this way. Second, the period was selected to reflect two important socio-economic trends: the real estate boom peaking in 2008, and subsequent economic decline and partial recovery. This choice correlates with the finding that socio-economic drivers are those affecting the environment [24,25].

To overcome the effects of "ecological fallacy" that usually accompany an ecological design, i.e., meaning that conclusions obtained at population level do not necessarily hold true at the individual one [26], we used two type of methods: quantitative, applied for looking at trends of transitional dynamics affecting the GI in all analyzed cities, and qualitative, for in-depth analyses of drivers determining these dynamics in selected cities. An outline of the methodology is presented in Figure 1.

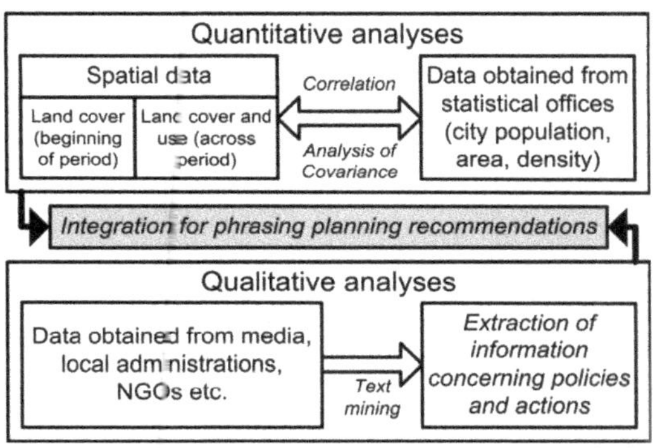

Figure 1. Outline of the study methodology, integrating quantitative and qualitative approaches to phrase planning recommendations.

The quantitative methods rely on using Urban Atlas data, appropriate for understanding phenomena specific to urban areas [27,28]. The data are freely provided in a shape file format projected to the ETRS 1989 Lambert Azimuthal Equal Area L52 M10 system [27] by the Copernicus Land Monitoring Service (https://land.copernicus.eu/ (accessed on 14 December 2020). We used the three datasets available for 2006, 2012, and 2018 land cover and use, with a resolution of a minimum mapping unit of 25 hectares and a minimum 100 m width of linear elements [29], as well as two datasets on the changes (2006–2012 and 2012–2018), with a 0.25 ha resolution for artificial surfaces and 1 ha for others [30].

The study used only Romanian and Polish cities present in the Urban Atlas datasets in 2006, 2012, and 2018, thus totaling 32 in Poland and 14 in Romania (Figure 2). As the image shows, the spatial distribution is relatively balanced from a spatial perspective in both countries. Although the Urban Atlas dataset was designed to include cities over 100,000 inhabitants, it included smaller cities, and excluded larger ones in both countries. However, our dataset was limited by the Urban Atlas selection criteria and availability of data. Moreover, since Urban Atlas data cover cities and their functional urban areas, we limited our analyses to the administrative limits of cities, such that the GI analyzed corresponds to the urban GI.

Figure 2. Cities included in the study. The sizes of the cities were enlarged for a better visualization of their position.

The methodology consisted of analyses performed using ArcView GIS 3.X and its spatial analysis extensions. First, data were converted to the Stereo 70 format and were clipped to the boundaries of administrative units. Data were recoded matching Urban Atlas categories to those of the GI [3,9,25], more specifically: 14,100—green urban areas, 14,200—sports and leisure facilities, 20,000—agricultural areas, and 30,000—natural and (semi-)natural areas. The changes for each category were coded as "loss", if land parcels labeled as GI in the beginning year had a different use in the end one, "gain" if land parcels with a different use in the beginning year became GI in the end one, or "transformation" if parcels preserved their GI label but changed their land use during the period. In addition, we reclassified the newer data, which include a more detailed classification of agricultural areas, and natural and (semi-)natural areas, to match the previous ones, in order to permit comparisons. Recoded polygons were dissolved for each city, computing the total area affected by each process, using the X-Tools extension of ArcView GIS 3. To overcome the differences due to city size, the area occupied by each category or the changes affecting it were expressed as a share of the GI or as a share of the area affected by changes of the GI, respectively.

Additional data from the statistical yearbooks were collected for each city, especially concerning its population. The administrative boundaries were used to compute the city area using the X-Tools extension of ArcView GIS 3. The additional data, including the city population, area, and density (computed based on the others) were used in analysis of co-variance (ANCOVA) looking at the dependence of the status parameters (area of each category of GI) and process ones (gain or loss of each category and transformation) on the additional variables. In addition, we tested whether the loss for each category depended on the total area covered by that category in the first year of each period. The choice of ANCOVA over multiple linear regression was justified by the fact that in addition to the numerical variables, differences were also sought by country and period, which are categorical variables. In addition to ANCOVA, correlation analyses were performed for all variables.

All statistical analyses were run at two levels of significance: 0.05, characteristic to all disciplines, and 0.1, used in environmental sciences and occasionally in other fields. The latest corresponds to an "uncertainty area" due to a reduced sample size. Hence, if significant results are detected at this level, it is likely that additional analyses, performed on larger samples, would yield results significant at the 0.05 threshold.

Provided that the statistical analyses revealed significant differences between the two countries, separate analyses were run for each country apart and overall.

Qualitative analyses compared representative case studies selected from each country, relying on information from official sources (public administrations, official statistics, and scientific literature) and mass media, in terms of their performance. Media sources included both "official" versions (i.e., pure media—newspapers, radio, or television, or official media releases of local administrations) and sources reflecting the perspective of different civil society stakeholders, especially NGOs. All sources are cited as references. We have selected two case studies from Poland (Konin and Lublin) and four from Romania (Bucureşti, Cluj Napoca, Giurgiu, and Oradea). The reason is that Polish cities were found to be on average twice as large than the Romanian ones [3], and Polish planning conditions are identical for all cities. Thus, we felt the need to include smaller cities, which were found only in Romania. Moreover, we attempted to select "well doers" and "poor performers" from among Romanian cities.

3. Results

3.1. Quantitative Analysis

Tables 1 and 2 and Figure 3 depict analyses of the structure of GI, and Table 3 shows its dynamics. The analyses describe the influence of potential drivers, based on the results of the analysis of co-variance. The Annex found in the Supplementary Materials includes two tables showing the correlation of all variables, i.e., Tables S1 and S2. In addition to the statistical analyses. In addition to the statistical analyses, Table 4 displays an overall comparison of the loss of different categories of the GI across the two periods and counties.

Table 1. Variables influencing the structure of GI categories in Romanian and Polish cities covered by Urban Atlas data. The table displays the p value associated with each relationship, using the following notations: **Bold**—significant, $p \leq 0.05$; *Italic*—significant, $p \leq 0.1$; Regular—not significant, $p > 0.1$.

Independent Variables	Dependent Variable			
	Green Urban Areas	**Sports and Leisure**	**Agricultural Areas**	**Natural and (Semi-)Natural Areas**
1. Overall				
Country	**<0.0001**	**<0.0001**	**<0.0001**	**0.0164**
Year	0.6389	0.7335	0.5020	0.5959
Area	**0.0301**	**0.0493**	*0.0813*	0.1875
Population	0.1617	0.3566	0.4706	0.4542
Density	**<0.0001**	**<0.0001**	0.2877	0.7733
2. Poland				
Year	0.9049	0.7772	0.6193	0.7028
Area	**0.0306**	*0.0564*	**0.0003**	**0.0039**
Population	0.7377	0.1531	**0.0169**	**0.0298**
Density	**<0.0001**	**<0.0001**	**0.0475**	0.3868
3. Romania				
Year	0.1316	0.8289	0.7811	0.8095
Area	0.4695	0.1739	**0.0005**	**0.0007**
Population	**0.0055**	0.3646	**0.0027**	**0.0059**
Density	**<0.0001**	**0.0141**	**0.0196**	**0.0018**

Table 2. Distribution of the GI across the different categories (14,100—green urban areas, 14,200—sports and leisure, 20,000—agricultural areas, and 30,000—natural and (semi-)natural areas) in Romanian and Polish cities covered by Urban Atlas data. The values represent the share of the area of each category from the total GI area per city.

	Green Urban Areas	Sports and Leisure	Agricultural Areas	Natural and (Semi-)Natural Areas
Poland				
Białystok	9.84	10.46	40 55	39.15
Bydgoszcz	3.64	5.50	34.78	56.09

Table 2. *Cont.*

	Green Urban Areas	Sports and Leisure	Agricultural Areas	Natural and (Semi-)Natural Areas
Poland				
Częstochowa	4.05	3.83	79.08	13.05
Gdańsk	5.84	7.62	51.61	34.92
Gorzów Wielkopolski	7.03	7.58	69.00	16.39
Jastrzębie-Zdrój	4.58	2.66	77.23	15.53
Jelenia Góra	1.65	3.54	46.24	48.57
Kalisz	2.96	3.48	85.89	7.67
Katowice	5.81	3.69	15.11	75.39
Kielce	3.81	5.80	51.38	39.01
Konin	3.28	4.22	79.02	13.48
Koszalin	3.74	4.52	42.16	49.58
Kraków	9.65	4.63	73.42	12.30
Łódź	7.19	6.53	62.56	23.72
Lublin	8.36	7.24	62.11	22.29
Nowy Sącz	1.74	3.87	62.92	31.47
Olsztyn	7.62	6.02	35.67	50.69
Opole	2.58	2.80	77.39	17.22
Ostrów Wielkopolski	3.44	8.93	63.92	23.71
Pabianice	4.77	4.49	73.46	17.28
Płock	4.30	6.25	64.59	24.86
Poznań	12.46	7.74	49.28	30.52
Radom	4.00	4.38	76.76	14.87
Rybnik	1.22	1.93	37.97	58.88
Rzeszów	3.51	4.88	80.74	10.87
Stargard	3.14	9.04	71.03	16.79
Suwałki	1.21	2.74	75.01	21.04
Szczecin	6.74	10.91	37.20	45.15
Toruń	8.03	7.32	30.45	54.20
Warszawa	12.50	9.11	38.51	39.88
Wrocław	7.08	12.28	63.98	16.66
Zielona Góra	0.86	2.17	28.20	68.78
All cities (average)	5.21	5.82	57.41	31.56
Romania				
Alba Iulia	0.78	0.24	69.33	29.65
Arad	1.19	0.34	91.98	6.49
Bacău	6.92	3.35	74.76	14.96
Brăila	9.36	2.34	78.14	10.16
Bucureşti	19.82	5.62	63.66	10.90
Călăraşi	0.48	0.22	89.85	9.45
Cluj Napoca	1.43	0.61	74.71	23.24
Craiova	6.45	2.33	80.44	10.79
Giurgiu	1.62	0.46	76.96	20.95
Oradea	2.53	0.62	91.89	4.96
Piatra Neamţ	0.43	0.63	34.99	63.95
Sibiu	1.41	0.97	68.47	29.15
Târgu Mureş	2.49	5.57	50.19	41.76
Timişoara	3.30	1.96	86.45	8.29
All cities (average)	4.16	1.80	73.70	20.34
All cities and countries (average)	4.89	4.60	62.37	28.15

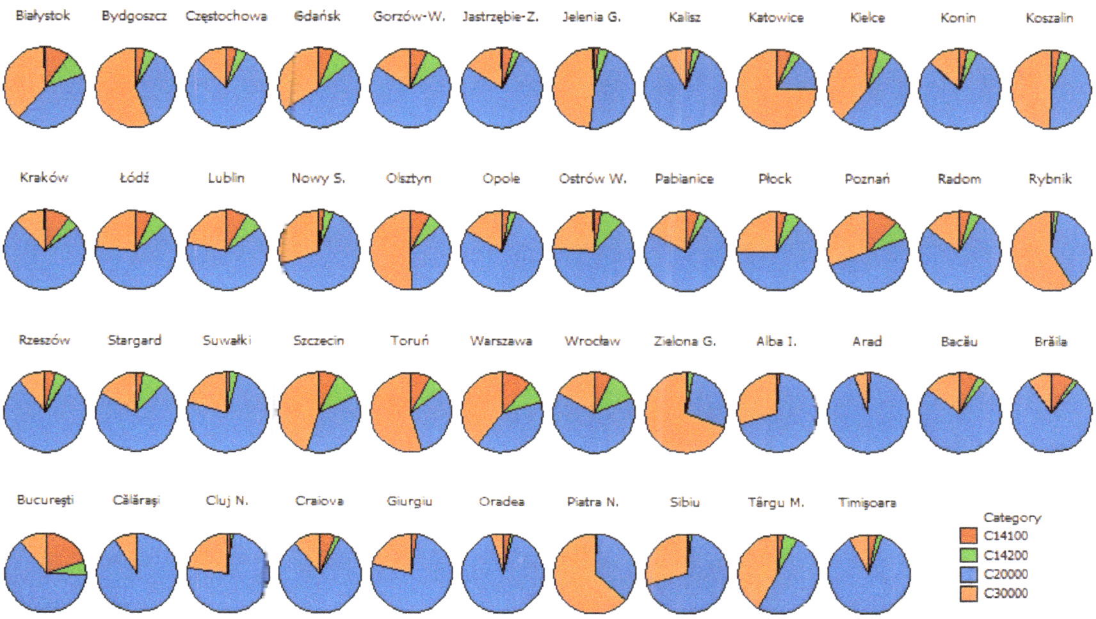

Figure 3. Distribution of the GI across the different categories (14100—green urban areas, 14200—sports and leisure, 20000—agricultural areas, and 30000—natural and (semi-)natural areas) in Romanian and Polish cities covered by Urban Atlas data. The graph is built based on the average values in 2006, 2012, and 2018, respectively.

3.2. *Qualitative Analysis*

To explore the drivers of change in more depth, the analysis continued with several representative case studies from each country (Konin and Lublin in Poland, and Bucureşti, Cluj Napoca, Giurgiu, and Oradea in Romania), based on data from the media, official statistics, or scientific literature, which can explain the processes affecting local GI. Several characteristics of the cities are displayed in Table 5.

3.2.1. Romania

We attempted to select "well doers" and "poor performers" from among Romanian cities. Hence, we have chosen two typical examples of Romanian small cities, Giurgiu and Oradea, and two large ones, Bucharest and Cluj-Napoca. Nevertheless, the structure and dynamics of their GI differ, as well as other features. Giurgiu stands out as the only city where the GI did not suffer any transformations during 2006–2012; Bucharest has the smallest share of GI (dominated by green urban areas—highest share among all case studies), and Cluj-Napoca the largest, even compared to all case studies, including the Polish ones. Bucharest also has the highest density. Finally, the GI of Oradea and Giurgiu is dominated by agricultural areas.

Table 3. Variables influencing the gain and loss of GI categories in Romanian and Polish cities covered by Urban Atlas data. The table displays the *p* value associated with each relationship, using the following notations: **Bold**—significant, $p \leq 0.05$; *Italic*—significant, $p \leq 0.1$; Regular—not significant, $p > 0.1$.

1. Overall	**Dependent Variable**										
Independent Variables	**Green Urban Areas**		**Sports and Leisure**		**Agricultural Areas**		**Natural and (Semi-) Natural Areas**		**Transformation**	**Total Gain**	**Total Loss**
	Gain	**Loss**	**Gain**	**Loss**	**Gain**	**Loss**	**Gain**	**Loss**			
Country	0.8414	0.6825	*0.0628*	0.2250	0.4820	**0.0254**	0.3140	0.7181	*0.0941*	0.6837	0.1687
Year	0.6532	**0.0011**	0.2102	0.8518	0.4377	**<0.0001**	**<0.0001**	0.4206	0.1441	**<0.0001**	0.5066
Area	0.9995	0.8110	0.5498	0.9647	0.9741	0.3471	0.7064	0.4986	0.2371	*0.0543*	**<0.0001**
Population	0.7500	0.2653	0.2379	0.6296	0.4571	0.2506	0.4186	0.4788	0.9103	0.6661	0.9756
Density	0.4096	0.2726	0.1575	0.3812	0.8754	*0.0532*	0.3519	0.4828	0.4384	0.5339	0.4182
Initial total area	–	**0.0004**	–	**0.0001**	–	**0.0015**	–	**0.0350**	–	–	–
2. Poland	**Dependent Variable**										
Independent Variables	**Green urban Areas**		**Sports And Leisure**		**Agricultural Areas**		**Natural And (Semi-) Natural Areas**		**Transformation**	**Total Gain**	**Total Loss**
	Gain	**Loss**	**Gain**	**Loss**	**Gain**	**Loss**	**Gain**	**Loss**			
Year	0.5330	**0.0005**	0.3687	0.8864	0.4829	**<0.0001**	**0.0003**	0.5784	0.1637	**0.0005**	0.9834
Area	0.6921	0.8668	0.8849	0.7543	0.6828	0.3874	0.9952	0.3337	**0.0301**	*0.0934*	**0.0008**
Population	0.7451	0.5526	0.8068	0.5360	0.4936	0.2347	0.7810	0.3865	0.1490	0.7630	0.4698
Density	0.7001	0.6654	0.9734	0.3235	0.9975	0.2695	0.5118	0.7713	0.6297	0.3974	0.0537
Initial total area	–	**0.0044**	–	**0.0009**	–	**0.0050**	–	**0.0187**	–	–	–
3. Romania	**Dependent Variable**										
Independent Variables	**Green Urban Areas**		**Sports And Leisure**		**Agricultural Areas**		**Natural And (Semi-) Natural Areas**		**Transformation**	**Total Gain**	**Total Loss**
	Gain	**Loss**	**Gain**	**Loss**	**Gain**	**Loss**	**Gain**	**Loss**			
Year	0.9662	0.7782	0.3254	0.4380	0.7763	**<0.0001**	0.0397	*0.0606*	0.6268	**0.0138**	0.2157
Area	0.8202	0.9313	0.8749	0.8025	0.4396	0.2342	0.8730	0.3735	0.6867	0.1531	**0.0367**
Population	0.6677	0.3218	0.2686	0.2093	0.9208	*0.0912*	0.5502	0.5248	**0.0176**	0.8931	0.8408
Density	0.4138	0.5720	0.1279	0.7637	0.7300	**0.0303**	0.5380	0.6001	0.5778	0.5764	0.9159
Initial total area	–	0.1917	–	0.8992	–	0.7659	–	0.5430	–	–	–

Table 4. Overall loss of different categories of GI categories across the two periods and counties in Romanian and Polish cities covered by Urban Atlas data. The table displays the average share of each category in the total loss of GI.

Period	Category	Poland	Romania	Overall
2006–2012	Green urban areas	3.9474	1.5187	3.2083
	Sports and leisure	1.1086	0.1366	0.8128
	Agricultural areas	88.9037	97.4752	91.5124
	Natural and (semi-) natural areas	6.0402	0.8695	4.4665
2012–2018	Green urban areas	1.0691	1.6006	1.2308
	Sports and leisure	1.3190	0.2404	0.9908
	Agricultural areas	6.2796	19.4738	10.2952
	Natural and (semi-) natural areas	5.1993	9.8599	6.6177
2006–2018	Green urban areas	2.5082	1.5597	2.2196
	Sports and leisure	1.2138	0.1885	0.9018
	Agricultural areas	47.5917	58.4745	50.9038
	Natural and (semi-) natural areas	5.6197	5.3647	5.5421

Table 5. Data for the individual case studies included in the qualitative analyses. The table displays physical and demographical characteristics and the dynamics of GI.

City	Konin	Lublin	Bucureşti	Cluj Napoca	Giurgiu	Oradea
Country	Poland	Poland	Romania	Romania	Romania	Romania
Area (km^2) in 2006	82	147	162	88	22	77
Area (km^2) in 2012	82	147	238	93	30	79
Area (km^2) in 2018	82	293	240	105	30	82
Population in 2006	80,471	353,483	1,931,236	305,620	69,479	205,956
Population in 2012	77,847	347,678	1,883,425	324,576	61,353	196,367
Population in 2018	74,151	342,039	2,121,794	324,267	67,402	221,398
Density (per km^2) in 2006	981	2405	11,958	3467	3226	2668
Density (per km^2) in 2012	949	2365	7918	3483	2022	2483
Density (per km^2) in 2018	904	2339	8840	3091	2221	2706
Share of GI (%) in 2006	57	60	29	75	57	62
Share of GI (%) in 2012	56	54	27	73	56	58
Share of GI (%) in 2018	55	52	27	70	56	56
Fragmentation of GI during 2006–2012	No	No	Yes	Yes	Yes	Yes
Fragmentation of GI during 2012–2018	Yes	No	Yes	Yes	Yes	Yes
Gain of GI during 2006–2012	Yes	No	Yes	Yes	Yes	Yes
Gain of GI during 2012–2018	Yes	Yes	Yes	Yes	Yes	Yes
Loss of GI during 2006–2012	Yes	Yes	Yes	Yes	Yes	Yes
Loss of GI during 2012–2018	Yes	Yes	Yes	Yes	Yes	Yes
Balance (gain/loss) of GI during 2006–2012	Loss	Loss	Loss	Loss	Loss	Loss
Balance (gain/loss) of GI during 2012–2018	Loss	Loss	Loss	Loss	Loss	Loss
Transformation of GI during 2006–2012	Yes	Yes	Yes	Yes	No	Yes
Transformation of GI during 2012–2018	Yes	No	Yes	Yes	Yes	Yes
Average share of green urban areas (%)	3	8	20	1	2	3
Average share of sports and leisure (%)	4	7	6	1	0	1
Average share of agricultural areas (%)	79	62	64	75	77	92
Average share of natural and (semi-) natural areas (%)	13	22	11	23	21	5

Giurgiu and Oradea were former industrial cities, which experienced a marked decline in industrial activity between 2006 and 2012, and, along with it, a decrease in population, which was more dramatic in Giurgiu (from 69,479 to 61,353—12%), but also present in Oradea (from 205,956 to 196,367—5%). During the next period, both cities witnessed an increase in population (to 67,402—10% in Giurgiu, and 221,398—13% in Oradea). However, the fate of industry differs in each city. In Oradea, the first industrial parks were created in 2008; they are among the few ones created in Romania thus far, and according to the definition of GI, their creation corresponds to adding new GI to the existing one. This is not something created formally; the inner regulations of industrial parks (Eurobusiness II used as an example—[31]) has clear provisions on the creation and maintenance of green spaces covering 25% of the park area. In addition to the industrial park, the public administration of Oradea has taken concrete actions, including and without limiting to, planting 23,207 trees during 2009–2019, to compensate for the 7475 cut off during the same period [32], and by the creation of 20 new landscaped green spaces in 2011 [33]. The trend continued with the creation of 8 hectares of new green spaces in 2015 [34] and another 10 in 2018, through conversion of degraded land, using European funding [35]. Giurgiu also witnessed some incentives of the local administration that consisted of creating five green spaces totaling 8000 m^2 in 2009 [36], but most likely, the changes are due to the abandonment of its industry and colonization of abandoned spaces by vegetation, as well as by landscaping the exit route to Bulgaria, which is situated on the administrative territory of Giurgiu. Apart from this, Giurgiu is referred to by the media as a city with poor green spaces [37,38].

On the opposite side, the case of Bucharest can serve as a good example of a city where planning does not seem to account for the GI. The latest Master Plan of Bucharest was approved in 2000, and its validity was extended to 2010 by the City Hall. The new Master Plan is still in the preparatory phase, although each city must renew its plan every 10 years, at most, according to the Romanian legislation. The provisions of the 2000 Master Plan included the creation of a "yellow-green belt", consisting of existing forests and agricultural areas of neighboring rural settlements. The green belt was never created; instead, small businesses built up their facilities around the belt line, and neighboring rural settlements lost their agriculture and rural functions, turning into residential areas for the citizens of Bucharest who, due to social and economic reasons, switched from living in large housing estates built in the socialist period to individual housing, bought and modified or built up from scratch. At the same time, the restitution of properties, a process formalized in 2005 by the creation of a specialized agency, resulted in accelerating this process, but also ended by transferring important portions of parks to former owners, who turned them into restaurants or other facilities. Another consequence was that the price of real estate increased. It is even hard to know the total area of green spaces. A study carried out under the framework of the Swiss–Romanian cooperation program points out that the area of green spaces per capita "increased" from 3 m^2 in 2007 to 23.21 m^2 in 2011 without creating any new green space, but accounting for the private ones [39]. Overall, the problem of Bucharest is that its Master Plan, initiated in 2013 and due in 2016, was not finalized; however, the previous (and obsolete) plan was modified by numerous plans for smaller areas, disregarding the requirements providing for green spaces, among others [40]. As a result, in the lack of clear-cut regulations, new developments are continuously shrinking the existing green spaces; numerous examples are presented by the media, which depict the situation as disastrous [41,42].

A somewhat different situation is found in Cluj. Here, during the first stage, the city began losing an increasing number of landscaped squares or green open spaces to the erection of filling stations, churches or bank offices, with land leases increasing throughout the 1990s. A second stage followed closely in the early 2000s, when relatively large swathes of parks were transformed into housing estates or large-scale shopping centers, thereby almost completely voiding the provisions of planning documents endorsed during the socialist period. This second stage resulted in an overall 30-hectare decrease in the total green area of the city. During the 2010s, a new generation of planning documents struggled to increase the green area ratio per capita, in an attempt to compensate for the losses experienced during previous decades. When artificial increases were sought out, such as declaring green spaces on private properties, litigations soon followed, and the practice was dropped. A third stage became manifest a few years ago, when the Municipality began organizing urban design competitions, a practice that has by now become customary. This latest period is by far the most interesting, as it tackled not only the overhaul of existing green open space network, but it tentatively began pursuing its gradual expansion. However, the status of the city in the media with respect to the GI is controversial; two articles published in 2018 stated that, based on the data from the statistical office, Cluj-Napoca was well positioned with respect to the green space per capita, with 25.8 m^2/inhabitant, close to the European Union limit (26 m^2/inhabitant), surpassing Bucharest, with only 21 m^2/inhabitant, but continues to state the need for urban green spaces, as only two of them, landscaped in 2014, were functional [43]. The same lack of landscaped green spaces is pointed out in a second article, showing that some of the landscaping projects were not implemented, and the city lacks true and functional green spaces [44], and by a study carried out under the framework of the Swiss–Romanian cooperation program, which points out the lack of interest for the landscaped green spaces [45].

The situation of these last two cities must be understood against the background of the general trend during the study period in Romania, especially in large cities. The year 2008 was the peak of a real estate bubble that had started earlier, followed by a drop in property prices. For example, based on real estate data, the price of a three-room apartment from the socialist period in Bucharest increased from 63,805 euros in 2006 to 146,678 in 2008 and then dropped down to 70,783 until the end of 2012 [46]. This increase has to be interpreted in the particular context of Romania, where 96% of people own the homes they live in, this being the highest share in Europe, and only 4% (compared to 30% EU average) live in a rented house or apartment [47], and prices in Bucharest are higher than the Romanian average (1413 euro/m^2 compared to 1341 euro/m^2 in 2019—[48]), with prices in central areas equaling those in Paris or London [49]. At the same time, prices in Cluj-Napoca are even higher on average (1900 euro/m^2 [48]). Because of the real estate boom, former owners turned the restituted property, usually a house with some green space or a parcel of land, into a larger unit, thereby using up almost the entire land.

3.2.2. Poland

In the case of Polish cities, Lublin and Konin were selected for a more detailed analysis. These two cities differ in the nature of the socio-economic processes taking place, including the level of involvement of the local community.

Lublin is one of the least green large Polish cities with a rather limited access of inhabitants to greenery and a relatively high level of GI fragmentation, mainly due to the small share of forests and tall greenery (about 11% of the city area) [50]. This is probably the reason why GI is perceived as one of the most valuable elements of the city's ecosystem. Shaping a coherent system of greenery and improving residents' access to it are goals included in the city's development strategy [51]. The city authorities yield a great importance not only to the protection of forests and revitalization and development of urban GI (in the last decade a total of 72 hectares of green areas have been created and revitalized, including, for example, the construction of Park Czuby), but also to the social participation in the process. Therefore, residents, supported by the activities of non-governmental organizations for the protection of trees and development of urban GI (e.g., the Lublin City Movement—City for People, fighting for the preservation of every tree in the city), play an extremely important role in shaping the urban GI [52]. Activating positive GI processes is not easy, as Lublin is currently the most rapidly developing metropolis in Eastern Poland and one of the most attractive places to live and do business in the whole country. However, at the beginning of the period of economic and social transformation initiated in Poland in 1989, the city was characterized by insufficient industrial development, which resulted in a comparatively low level of performance, as pointed out by economic indicators. However, this backwardness turned out to be an asset in the post-industrial transformation. Although in the period under study, similar to most Polish cities, Lublin lost population (population loss was 1.74% from 2006 to 2012 and 2.38% from 2012 to 2018) and had an average level of economic development; in 2006–2010, it became one of the fastest developing cities in Poland. Despite the global crisis, economy increased significantly in the city, and the growth of the number of businesses was one of the largest among major Polish cities [53]. The effect of economic changes was a gradual increase in the wealth of residents and in the accompanying investment pressure. This pressure, combined with weaknesses of the spatial planning legislation, caused a loss of GI. This loss was primarily related to the felling of trees accompanying the implementation of different types of investments (e.g., road reconstruction, construction of roundabouts, new housing estates, shopping malls etc.) [54]. The loss of fallen trees was unfortunately not fully compensated by planting new ones. For this reason, Lublin lost green areas and stopped the fragmentation processes.

Konin, in turn, is a typically industrial town, which owes its development to the exploitation and processing of brown coal. The city has experienced a particular transformation of the political system, manifested by a progressive, dynamic decline in population (by 3.26% in 2006–2012 and 4.75% in 2006–2018), steady number of businesses, high unemployment, and a significant decrease in the average salary compared to the national average (from 96,3% in 2006 to 93% in 2012). The unfavorable trends during this period were caused by a progressive restructuring of its economy and group mass layoffs among the city's inhabitants. As a result of these demographic and economic processes, in Konin the development pressure was not too high (only slightly more than 500 new apartments were added between 2006 and 2012, and between 2012 and 2018—830). The attention of local authorities was primarily focused on the modernization and expansion of the road network and inclusion of new areas for economic activation in the communication system. These activities have been major contributors to the loss and progressive fragmentation of green spaces, which was not counteracted by the passive and poorly organized residents, connected only to a small extent with the inhabited city (largely immigrant population, recruited to work in industry). However, the positive side consists of the creation of new GI. As a result of reclamation and revitalization activities, operations parks (Park 700-lecia, Park Ojców), recreational amenities (tennis courts, playgrounds, playgrounds, and sport halls), allotments, arable fields and afforested areas were created in the regions of three former open-pit mines, which to some extent blurred the mining history of the city [55]. These activities resulted in an increase in forests (123%), street alignments (by 72 ha), and parks, greeneries, and private house green spaces (by 30 ha) [56].

Therefore, the processes taking place in Lublin and Konin are different, and conditioned by the specificity of economic development, but also, and perhaps most importantly, resulting from the characteristics of local communities. Everything seems to indicate that the positive example of Lublin is largely due to its inhabitants forming a community, and not just a collection of residents. Only a community was able to quickly "find itself" in new political conditions, aware of the role of GI in the city, capable to organize itself and take bottom-up initiatives to protect the GI, and to build up an important social capital in sustainable urban development. However, local authorities also played a role in this success, attaching great importance to both the protection of green spaces from fragmentation and the public participation processes. Conversely, the case of Konin, representing a center of traditional mining industry preferred during the socialist period, reveals a community passively waiting for the public authorities to provide them with jobs and a higher life quality. The inhabitants of Konin are largely characterized by the lack of activity and entrepreneurship, and an unwillingness to undertake bottom-up initiatives. In turn, local authorities, wanting to improve the image of the city and make it a more attractive place to live, focused their attention not only on job creation and the implementation of "hard" investments, but also on the rehabilitation and organization of new GI elements.

4. Discussion

4.1. *Quantitative Analysis*

Our findings indicate that agricultural and natural areas make up the largest share of urban GI (best seen in Figure 2). With several exceptions (mostly in Polish cities), one of them dominates the other, but the other landscaped areas (green urban areas and sports and leisure facilities) rarely sum up to 25% of the urban GI. The dynamics of GI seem to vary mostly by the country and year, at least with respect to the gain and loss of some categories, but also overall. In more detail (see Table 4), during 2006–2012, Polish cities lost more landscaped GI and natural areas, but Romanian cities lost more agricultural land, and during 2012–2018, the trends reverted. The city area seems to influence only the total gain and loss, which is consistent with previous findings [3]. However, the loss of GI seems to be influenced to a great extent, for each category, by the share of this category in the beginning of the period. This means that the well-represented categories tend to lose more than the less-represented ones. In addition, the analyses did not reveal many significant predictors for the landscaped green spaces (green urban areas and sports and leisure facilities). We suspect that their reduced share makes it hard to distinguish the drivers of their loss or gain, compared to the dominant categories (agricultural and natural areas). On a similar note, the loss seems to depend on some of the predictors accounted for (population and density), while the gain seems to be less influenced by them. This might indicate that the urban social metabolism tends to consistently eliminate the GI, while its gain is mostly due to random causes.

Our analysis differentiates between large cities and small ones, the latter having a closer status to rural settlements. Large cities have a well-represented landscaped GI, corresponding to the third type of city nature described by Kowarik [57]. Small cities preserve in their structure agricultural areas [58], and these are the first to be lost when the city increases; some are compensated for by the newly created landscaped areas. However, the loss is proportional to the share of each category, suggesting that the process is irrespective to the type of GI.

Similar to our previous study [3], we also found many inconsistencies. To cite an example for each analysis, natural and (semi-) natural areas are influenced by the density of population in Romania, but not in Poland and overall; natural and (semi-) natural areas seem to be significantly inversely correlated with the total area of the city in Romania, marginally significantly positively correlated to it in Poland, and not correlated overall; the loss of agricultural areas is significantly influenced by density in Romania and marginally significantly overall, but not influenced in Poland; and the transformation of GI is significantly positively influenced by population in Romania and overall, but not in Poland. While it would be easy to blame the small sample size for such differences, all the examples above show significant values in Romania, where the sample size is the smallest (14 cities). Therefore, it is sounder to explain them by the variability of phenomena rather than considering them statistical fluctuations.

Similarly, there seem to be variables which correlate with almost all others (e.g., population, for the analysis of the structure of GI, and total gain of GI, least in Romania, for the analyses looking at its change), and variables that do not correlate with any others, in the case of analyses looking at the changes of GI, e.g., gain of green urban areas). Particularly in the case of the latest analyses, the predictors (area, population, and density) and overall transformation, gain and loss seem to be the most correlated (almost all possible pairwise correlations are significant). Again, these results indicate that it is easier to obtain the overall picture, similar to our previous study [3], but the detailed mechanisms of change are more contextual and harder to generalize.

In brief, despite the many variations, indicating the contextual nature of changes, we were able to show that natural and agricultural areas are the most represented components of the urban GI. The latest tends to diminish its size on a rural-urban gradient. The size, population, and density of cities can play an important role in explaining the loss of different GI categories, but the gain and transformation are less predictable. Most importantly, the loss of GI, regardless of the category, is a generalized phenomenon, and its intensity varies, for each category, with its share in the total urban GI.

4.2. Qualitative Analysis

4.2.1. Romania

Against the individual cases, we can distinguish some salient features of planning practices: in a first phase, socialist planning documents fall into disrepute and their provisions are either eroded gradually or voided completely, through a plethora of lower-rank planning documents explicitly geared toward exacting derogations from higher-ranking plans. In a second stage, planning documents aim at protecting the existing network of open green spaces, while at the same time experimenting with various methods for increasing the overall surface of green areas. When litigations follow, experiments are halted. In a third stage, the provisions within planning documents become subject to national or international competition, thereby seeking a higher degree of validation from the general public. It is definitely too soon to tell, but the practice of organizing urban design and urban planning competitions might well engender a reform within the current architecture of the Romanian planning system.

It is also important to analyze another particular aspect of Romania. One could question the role of people, communities and organizations involved in the process. All these stakeholders do not play an important role in Romania, and our discussion focuses mainly on local administrations and business pressure. Citizens do not play an important role, and their reactions seem limited to protesting against trees being cut off by municipalities in Bucharest [59] or Iași [60], or initiating online petitions in Bucharest [61], with limited success. The reasons are a combination of the fact that, in Romania, citizens are less organized in making up one single voice for their issues, and authorities are generally less prone to listen to them. Nevertheless, if this is the case of population alone, NGOs were more successful; "Save Bucharest" (Salvaţi Bucureştiul in Romanian), an NGO founded in 2008, adopted the legal path and announced in 2020 winning its 166th lawsuit against Bucharest City Hall [62].

Three lessons can be learned from the Romanian case studies. One is that the "winning" cases for GI are those where public administrations got involved or the process occurred naturally, due to the lack of intervention; however, these are the exceptions. The second one shows that the general "rule" of greed for profit applies, especially when analyzing phenomena through the lens of land price, generated a spontaneous urbanization. The real estate crisis slowed down the process after 2008 [25], but overall effects could not be diminished in the least. The third is that planning documents adapt, as Municipalities face inherent weaknesses of the Romanian planning system and search for new ways of gaining legitimacy for their planning actions.

4.2.2. Poland

Unfortunately, in all analyzed Polish cities, the loss and, consequently, a negative balance of GI occurred, and the vast majority of cities also experienced its fragmentation. The functional and spatial conflicts between the needs of public interest (represented by the city) and those of individual interests (often including development companies interested only in short-term profit, and not the well-being of residents) were, among others, due to the expectation of destination for development, mainly residential, of all categories of areas in the city, including the agricultural and green ones. The main source of conflict was the lack of adequate coverage of the city with spatial development plans (their drafting is optional) and launch of further investment areas through non-planning tools, such as decisions on development conditions, often detrimental to the city (interference with land reserved for roads, green areas necessary to maintain a high quality of life, areas designated for conservation, etc.) [63–65]. From 1 January 2017, the difficult situation was significantly worsened by the amendment to the provisions of the Nature Conservation Act, which abolished, among other things, obligation to obtain a permit to cut trees on private real estate (the so-called "Lex Szyszko"—the name of the minister responsible at that time for environmental protection). The law was tightened again on June 17 of the same year, but for half a year, there was massive felling of trees, causing significant losses in tree cover in practically all Polish cities [66].

In Poland, planning conditions are identical for all cities. Therefore, changes in the scope of urban greenery are determined by actions of local authorities (first of all, they are responsible for deciding which areas will be covered by spatial development plans allowing for the protection of GI) and bottom-up actions taken by inhabitants. For this reason, all Polish cities need a substantial improvement of their spatial policy. Without it, they are at risk of increasing spatial chaos [67], of a gradual loss of natural assets, and of fragmentation of ecological corridors under the pressure of short-term goals of the investors.

4.3. Lessons Learned from the Study

Before proceeding to sketching our recommendations, we need to pause for a moment and frame the limitations that have become prominent while conducting this research exercise. Generally speaking, the data and methodology produced significant results. Nevertheless, it is local variations and the failure to ascertain the exact degree of influence these driving forces have on GI that render these limitations explicit. In short, they are the following:

1. The methodology needs more fine tuning to improve statistical significance. This is, however, easier said than done. To illustrate this point, please consider the following data-related issue: the frequency of land cover and land use data are low. Hence, while the Urban Atlas demonstrated its suitability for performing comparative analyses of changes, its temporal span does not allow for highlighting the intrinsic mechanisms of these changes.
2. To complicate things even further, there are some additional limitations that affect the methodological exercise proper. They are the following:
    - Mathematical constraints apply to modeling complex processes describing the dynamics of the GI. Hence, it is difficult to achieve both intuitive clarity and statistical power.
    - The ecological fallacy characterizing ecological studies could cast a doubt on the validity of results. More precisely, even if local influence of the driving forces could be determined if available data were available, there will be always a doubt that aggregated data reflect appropriately each particular case of fragmenting, losing, or changing of the GI.

5. Conclusions and Recommendations

In summary, we found out that the loss of GI is a generalized phenomenon, but its intensity varies with the different categories. Large cities have a GI dominated by natural areas, which are parts of the surrounding natural areas incorporated by them as they grow. Smaller, rural-like towns are dominated by agricultural areas, which tend to shrink as the towns evolve into large and complex cities. Consequently, most urban areas have an infrastructure consisting mainly of agricultural or natural areas, while the landscaped spaces make up only a small share. The loss of GI is uniform and generalized to all investigated cities, and most likely common to other cities not included in our study, but the loss is proportional with the share of each category. While the most important drivers seem to be the city population and its density, the process is influenced, as the case studies indicate, by a local "culture" of administration or citizens in terms of dealing with the GI. The main actors seem to be public administrations in Romania and city dwellers in Poland. However, civil society and city dwellers seem to be more organized in Romania in recent years too, while adopting legal tools in order to make public authorities listen to their voice. Finally, plan coverage seems to play an important role in Poland too, namely, the insufficient level of coverage of cities by plans. Overall, the study confirmed previous results in terms of the overall findings (generalized loss of the GI) but revealed the fact that inner mechanisms are contextual and therefore hard to generalize, as indicated by the statistically significant fluctuations by country and period, although we were able to reveal that most losses account for natural and agricultural areas.

Against the background defined by our findings, we can now venture to make a series of recommendations for planning practitioners. We have conceived these recommendations in a generalized way, as they are applicable to other countries experiencing similar planning issues, such as the central-eastern European ones. However, they apply first to Poland and Romania, and, due to this reason, we have phrased specific recommendations for these countries wherever appropriate. In short, in both Romanian and Polish planning systems,

GI elements are not adequately protected. All planning studies should introduce provisions and solutions for protecting and restoring the environment.

However, the two countries fare differently: In Poland, there is no national document defining directions for spatial planning, at least for the time being. Hence, there is no framework for buttressing spatial policies relevant to GI. In addition, plan coverage at the local level is low, with only about one- third of the country covered by plans. At the time of writing, plans remain optional. Against this background, urban development is managed via administrative decisions, which tend to encourage circumvention. Hence, with plans remaining optional, there are both substantive and procedural barriers to successful GI implementation in the planning practice. Any recommendation will therefore require either a structural change within the Polish planning system, or an incremental improvement in the performance of administrative decisions.

In Romania, the legal framework is there, but the problem is practical: There are no dedicated and precise planning instruments, aimed at ensuring GI critical mass and continuity in real estate development. Simply put, planners and developers do not have readymade prescriptions for creating viable and scalable GI. In addition, there is no comprehensive analysis of Romanian zoning and building regulations aimed at highlighting provisions that work against the creation of GI. Hence, the problem here is merely procedural and could easily be solved by producing scientifically and legally informed planning and design manuals.

Consequently, we propose the following preliminary recommendations, which require a considerable amount of good will from planning practitioners:

1. Consistent formation of the urban GI system, which would ensure the maximum area and compactness (compact shape) of its elements, their spatial continuity, and connection with suburban green areas. The process is possible in Romania, where guidelines for drafting plans exist, but they are outdated; updating them is the easiest solution. However, Poland lacks such tools; the study revealed their need, and they can be developed accounting for the GI from the beginning.
2. Preserving the system's continuity by integrating remnants of natural environment in the urban structure, usually including the water system, and preserving adjacent natural and semi-natural areas. This can also be achieved via the guidelines for developing plans described above.
3. Protecting the remnants of natural and semi-natural urban green areas (meadows, wetlands, etc.). Again, spatial plans are a solution, but this measure needs the cooperation of environmentally aware local communities, and their stronger participation in the process of drafting the plans.
4. Developing properly landscaped green areas of the estates and accompanying greenery which should reduce the pressure on compact alignments of GI. Similarly, this recommendation depends on the cooperation of environmentally aware local communities.
5. Appropriate local and regional spatial policies, considering public participation. Against this background, we would like to highlight the lack of a Polish national document defining directions for spatial planning. Without such a document, spatial policies are bound to remain patchy.
6. Adopting spatial development plans that are not only growth-oriented, but also have a protection component, maintaining the spatial continuity of GI. Again, we face a procedural issue in Poland, as there is no legal obligation to adopt urban plans. Hence, without such obligations, relevant spatial policies are rendered moot. What is left to tend to for Polish GI are development decisions of the public administrations, which usually foster bypassing. An incremental improvement in these decisions seems possible, but it rests on public pressure.
7. Eliminating the instruments (i.e., building conditions) leading to the fragmentation of GI. This requires an in-depth analysis of both zoning and building regulations in both countries. Depending on the results, some legal provisions might need updating.

8. Building developments allowed only in areas already equipped with a technical infrastructure. Unfortunately, enforcement requires a separate discussion, via "derogatory planning", but can be addressed while revising planning provisions in both countries.
9. Educational and awareness activities focused on the importance of GI for a city and its residents. Several other points stressed the need for cooperation of environmentally aware local communities. However, public involvement needs to be responsible, efficient, and positive. For that, both countries need to expand environmental education, even by informative means, enabling it to reach not only city managers and planners, but also local actors.

Supplementary Materials: The following supporting information can be downloaded at: https://www.mdpi.com/article/10.3390/land11050592/s1: Annex S1. Additional tables analyzing the correlations between variables describing and influ-encing the structure and dynamics of GI categories in Romanian and Polish cities covered by Urban Atlas data, containing: Table S1—Correlations between variables describing and influenc-ing the structure of GI categories in Romanian and Polish cities covered by Urban Atlas data, and Table S2— Correlations between variables describing and influencing the dynamics of GI catego-ries in Romanian and Polish cities covered by Urban Atlas data.

Author Contributions: All the authors have equally contributed to the article. Conceptualization, A.-I.P., L.M. and A.M.; methodology, A.-I.P., L.M. and A.M.; formal analysis, A.-I.P., L.M. and A.M.; investigation, A.-I.P., L.M. and A.M.; resources, A.-I.P., L.M. and A.M.; data curation, A.-I.P., L.M. and A.M.; writing—original draft preparation, A.-I.P., L.M. and A.M.; writing—review and editing, A.-I.P., L.M. and A.M.; validation, A.-I.P., L.M. and A.M.; visualization, A.-I.P., L.M. and A.M.; supervision, A.-I.P.; project administration, A.-I.P. All authors have read and agreed to the published version of the manuscript.

Funding: This research received no external funding.

Institutional Review Board Statement: Not applicable.

Informed Consent Statement: Not applicable.

Data Availability Statement: This study used existing data. Processed data cannot be easily stored due to the size but can be obtained by demand from the corresponding author.

Conflicts of Interest: The authors declare no conflict of interest.

References

1. Artmann, M.; Chen, X.; Ioja, I.C.; Hof, A.; Onose, D.-A.; Ponizy, L.; Zavodnik Lamovšek, A.; Breuste, J.H. The role of urban green spaces in care facilities for elderly people across European cities. *Urban For. Urban Green.* **2017**, *27*, 203–213. [CrossRef]
2. Raymond, C.M.; Frantzeskaki, N.; Kabisch, N.; Berry, P.; Breil, M.; Nita, M.R.; Geneletti, D.; Calfapietra, C. A framework for assessing and implementing the co-benefits of nature-based solutions in urban areas. *Environ. Sci. Policy* **2017**, *77*, 15–24. [CrossRef]
3. Petrișor, A.-I.; Mierzejewska, L.; Mitrea, A.; Drachal, K.; Tache, A.V. Dynamics of Open Green Areas in Polish and Romanian Cities During 2006–2018: Insights for Spatial Planners. *Remote Sens.* **2021**, *13*, 4041. [CrossRef]
4. Badiu, D.-L.; Ioja, I.C.; Patroescu, M.; Breuste, J.H.; Artmann, M.; Nita, M.R.; Grădinaru, S.R.; Hossu, C.A.; Onose, D.-A. Is urban green space per capita a valuable target to achieve cities' sustainability goals? Romania as a case study. *Ecol. Ind.* **2016**, *70*, 53–66. [CrossRef]
5. Artmann, M.; Kohler, M.; Meinel, G.; Gan, J.; Ioja, I.C. How smart growth and green infrastructure can mutually support each other—A conceptual framework for compact and green cities. *Ecol. Ind.* **2017**, *96*, 10–22. [CrossRef]
6. Nita, M.R.; Năstase, I.-I.; Badiu, D.-L.; Onose, D.-A.; Gavrilidis, A.-A. Evaluating urban forests connectivity in relation to urban functions in Romanian cities. *Carpathian J. Earth Env. Sci.* **2018**, *13*, 291–299. [CrossRef]
7. Turkelboom, F.; Leone, M.; Jacobs, S.; Kelemen, E.; García-Llorente, M.; Baró, F.; Termansen, M.; Barton, D.N.; Berry, P.; Stange, E.; et al. When we cannot have it all: Ecosystem services trade-offs in the context of spatial planning. *Ecosyst. Serv.* **2018**, *29*, 566–578. [CrossRef]
8. Pătru-Stupariu, I.; Nita, A. Impacts of the European Landscape Convention on interdisciplinary and transdisciplinary research. *Landsc. Ecol.* 2022, *in press*. [CrossRef]
9. Petrisor, A.-I.; Andronache, I.C.; Petrisor, L.E.; Ciobotaru, A.-M.; Peptenatu, D. Assessing the fragmentation of the green infrastructure in Romanian cities using fractal models and numerical taxonomy. *Procedia Env. Sci.* **2016**, *32*, 110–123. [CrossRef]

10. Puşcaşu, V. Some considerations about urban planning and urban landscape relationship in Romania. *Public Admin Reg. Stu* **2011**, *4*, 50–60.
11. Díaz-Palacios-Sisternes, S.; Ayuga, F.; García, A.I. A method for detecting and describing land use transformations: An examination of Madrid's southern urban-rural gradient between 1990 and 2006. *Cities* **2014**, *40*, 99–110. [CrossRef]
12. Grădinaru, S.R.; Iojă, C.I.; Onose, D.A.; Gavrilidis, A.A.; Pătru-Stupariu, I.; Kienast, F.; Hersperger, A.M. Land abandonment as precursor of built-up development at the sprawling periphery of former socialist cities. *Ecol. Indic.* **2015**, *57*, 305–313. [CrossRef]
13. Ianoş, I.; Sîrodoev, I.; Pascariu, G.; Henebry, G. Divergent patterns of built-up urban space growth following post-socialist change. *Urban Stud.* **2016**, *53*, 3172–3188. [CrossRef]
14. Kupková, L.; Bičík, I.; Najman, J. Land Cover Changes along the Iron Curtain 1990–2006. *Geografie* **2013**, *118*, 95–115. [CrossRef]
15. Iaţu, C.; Eva, M. Spatial profile of the evolution of urban sprawl pressure on the surroundings of Romanian cities (2000–2013). *Carpathian J. Earth Environ. Sci.* **2016**, *11*, 79–88.
16. Petrişor, A.-I.; Petrişor, L.E. Transitional dynamics based trend analysis of land cover and use changes in Romania during 1990–2012. *Present Environ. Sust. Dev.* **2018**, *12*, 215–231. [CrossRef]
17. Tascu-Stavre, M.; Banica, C. Old and New in Vama Veche and 2 Mai. *Urban Arhit Constr.* **2014**, *5*, 73–82.
18. Hamma, W.; Petrişor, A.-I. Urbanization and risks: Case of Bejaia city in Algeria. *Human Geogr.* **2018**, *12*, 97–114. [CrossRef]
19. Stoica, I.V.; Vîrghileanu, M.; Zamfir, D.; Mihai, B.A.; Săvulescu, I. Comparative Assessment of the Built-Up Area Expansion Based on Corine Land Cover and Landsat Datasets: A Case Study of a Post-Socialist City. *Remote Sens.* **2020**, *12*, 2137. [CrossRef]
20. Stoica, I.V.; Zamfir, D.; Vîrghileanu, M. Evaluating the Territorial Impact of Built-Up Area Expansion in the Surroundings of Bucharest (Romania) through a Multilevel Approach Based on Landsat Satellite Imagery. *Remote Sens.* **2021**, *13*, 3969. [CrossRef]
21. Mierzejewska, L. *Tereny Zielone w Strukturze Przestrzennej Poznania (Green Areas in the Spatial Structure of Poznań)*; Wyd. PTPN: Poznań, Poland, 2001.
22. Mierzejewska, L. Green areas as a determinant of living conditions in the city (on the example of Poznań). In *Diversification of the Living Conditions of the Population in the City*; Jażdżewska, I., Ed.; Wydawnictwo Uniwersytetu Łódzkiego: Łódź, Poland, 200■; pp. 321–332. (In Polish)
23. Morgenstern, H. Ecologic studies in epidemiology: Concepts, principles, and methods. *Annu. Rev. Public Health* **1995**, *16*, 61–8■ [CrossRef] [PubMed]
24. Petrişor, A.-I.; Ianoş, I.; Tălângă, C. Land cover and use changes focused on the urbanization processes in Romania. *Environ Eng Manag. J.* **2010**, *9*, 765–771. [CrossRef]
25. Petrişor, A.-I.; Sîrodoev, I.; Ianoş, I. Trends in the national and regional transitional dynamics of land cover and use changes in Romania. *Remote Sens.* **2020**, *12*, 230. [CrossRef]
26. Loney, T.; Nagelkerke, N.J. The individualistic fallacy, ecological studies and instrumental variables: A causal interpretation. *Emerg. Themes Epidemiol.* **2014**, *11*, 18. [CrossRef] [PubMed]
27. Petrişor, A.-I. Assessment of the Green Infrastructure of Bucharest using CORINE and Urban Atlas data. *Urban Archit. Constr.* **2015**, *6*, 19–24.
28. Petrişor, A.-I.; Petrişor, L.E. Assessing microscale environmental changes: CORINE vs. the Urban Atlas. *Present Environ. Sustain. Dev.* **2015**, *9*, 95–104. [CrossRef]
29. Hagenauer, J.; Helbich, M. Mining urban land-use patterns from volunteered geographic information by means of genetic algorithms and artificial neural networks. *Int. J. Geogr. Inf. Sci.* **2012**, *26*, 963–982. [CrossRef]
30. Prastacos, P.; Chrysoulakis, N.; Kochilakis, G. Spatial metrics for Greek cities using land cover information from the Urban Atlas. In *Multidisciplinary Research on Geographical Information in Europe and Beyond, Proceedings of the AGILE'2012 International Conference on Geographic Information Science "Bridging the Geographic Information Science", Avignon, France, 24–27 April 2012*; Gensel, J., Josselin, D., Vandenbroucke, D., Eds.; Springer: Berlin, Germany, 2012; pp. 261–266.
31. Local Development Agency of Oradea. *Regulamentul de Funcţionare a Parcului Industrial Eurobusiness II (Regulation on the Functioning of the Industrial Park Eurobusiness II)*. Available online: http://www.adlo.ro/download/REGULAMENT-PARC%202.pdf (accessed on 28 April 2020).
32. Oradea în Direct. *Suprafaţa de Spaţiu Verde al Municipiului Oradea a Crescut în Ultimii Ani. Comunicat (Communicate: The Green Space Area of Oradea Increased in the Recent Years)*. Available online: https://oradeaindirect.ro/suprafata-de-spatiu-verde-al-municipiului-oradea-a-crescut-in-ultimii-ani-comunicat/ (accessed on 28 April 2020).
33. Oradea Press. *Primăvara Vine în Oradea cu Amenajări de Spaţii Verzi (Spring Comes to Oradea with the Landscaping of Green Spaces)*. Available online: https://oradeapress.ro/2011/03/15/primavara-vine-in-oradea-cu-amenajari-de-spatii-verzi/ (accessed on 28 April 2020).
34. F.A. *Oradea: 8 ha noi de Spaţii Verzi (8 New Hectares of Green Spaces)*. Available online: https://www.arenaconstruct.ro/oradea-8-ha-noi-de-spatii-verzi/?cn-reloaded=1 (accessed on 12 November 2021).
35. Finantare.ro. *Zece Hectare de Terenuri Degradate, Transformate de Primaria Oradea in Spatii Verzi, cu Fonduri europene (Ten Hectare of Degraded Land, Transformed by the City Hall of Oradea in Green Spaces, with European Funding)*. Available online: https://www.finantare.ro/zece-hectare-de-terenuri-degradate-transformate-de-primaria-oradea-in-spatii-verzi-cu-fonduri-europene.html (accessed on 12 November 2021).

36. Administratie.ro. *Spatii Verzi din Fondul de Mediu in Municipiul Giurgiu (Green Speces (Financed) from Environmental Funds in Giurgiu)*. Available online: https://www.administratie.ro/spatii-verzi-din-fondul-de-mediu-in-municipiul-giurgiu/ (accessed on 12 November 2021).
37. Vârlan, C. *Giurgiu. De ce nu se Trăieşte Bine în Giurgiu. Cu ce Probleme se Confruntă Locuitorii Acestui Judeţ (Giurgiu. Why One Does Not Live Well in Giurgiu. What Problems Are Faced by the Dwellers of this County)*. Available online: https://www.gandul.ro/financiar/giurgiu-de-ce-nu-se-traieste-bine-in-giurgiu-cu-ce-probleme-se-confrunta-locuitorii-acestui-judet-10675450 (accessed on 12 November 2021).
38. Vătămanu, V. *Arhitectura Spatiilor Verzi din Romania Dupa 1989 (Architecture of Green Spaces in Romania after 1989)*. Available online: https://www.agrimedia.ro/articole/arhitectura-spatiilor-verzi-din-romania-dupa-1989 (accessed on 12 November 2021).
39. Culescu, D.; Răducu-Lefter, A. *Studiu de Caz: Bucureşti, România (Case Study: Bucharest, Romania)*. Available online: http://asop.org.ro/lan-net/studii-de-caz/rezumat-studiu-de-caz-bucuresti.pdf (accessed on 12 November 2021).
40. Ivanov, C. *Odiseea Noului Plan Urbanistic General: De ce în Bucureşti se Construieşte După Planurile de Acum 20 de Ani (The Odyssey of the New Master Plan: Why is Bucharest Building by 20 Years Old Plans)*. Available online: https://www.hotnews.ro/stiri-administratie_locala-23699451-odiseea-noului-plan-urbanistic-general-bucuresti-construieste-dupa-planurile-acum-20-ani.htm (accessed on 12 November 2021).
41. Petrescu, A. *Bucureşti, Capitala cu Cele mai Puţine Spaţii Verzi. Suntem cu Mult sub Limita Minimă Impusă de OMS (Bucharest, the Capital with the Least Geen Spaces. We are Far away below the Minimum Limit Imposed by WHO)*. Available online: https://observatornews.ro/eveniment/bucuresti-capitala-cu-cele-mai-putine-spatii-verzi-suntem-cu-mult-sub-limita-minima-impusa-de-oms-428211.html (accessed on 12 November 2021).
42. Petrescu, R. *Dezastrul din Bucureşti: Capitala Este Oraşul European cu Cea mai Mică Suprafaţă de Spaţii Verzi pe Cap de Locuitor. Bulgarii şi Ungurii au de Două ori mai Mult Verde, iar Suedezii de 6 ori (The Disaster of Bucharest: The Capital is the European City with the Least Geen Space Area Per Capita. Bulgarians and Hungarians have Twice More, and Swedish 6 Times More)*. Available online: https://www.zf.ro/companii/dezastrul-din-bucuresti-capitala-este-orasul-european-cu-cea-mai-20179614 (accessed on 12 November 2021).
43. Preda, B. *Municipiul Cluj-Napoca, Din Punct de Vedere Statistic, stă Destul de Bine în Ceea ce Priveşte Spaţiul Verde Alocat Fiecărui Locuitor, Fiind Foarte Aproape de Limita Uniunii Europene (Statistically Speaking, Cluj Napoca is Well Positioned with Respect to the Green Space Per Capita, Close to the European Union Limit)*. Available online: https://www.monitorulcj.ro/actualitate/65824-cat-de-bine-sta-clujul-la-capitolul-spatii-verzi (accessed on 12 November 2021).
44. Reştea, K. *Clujul şi Criza de Spaţii Verzi (Cluj and the Green Space Crisis)*. Available online: https://actualdecluj.ro/clujul-si-criza-de-spatii-verzi/ (accessed on 12 November 2021).
45. Boancă, P.I.; Bors-Oprişa, S. *Studiu de Caz: Cluj-Napoca, România (Case Study: Cluj-Napoca, Romania)*. Available online: http://asop.org.ro/lan-net/studii-de-caz/rezumat-studiu-de-caz-cluj-napoca.pdf (accessed on 12 November 2021).
46. Anunţul telefonic. *Media si Evolutia Preturilor Imobiliare (Average and Evolution of Real Estate Prices)*. Available online: https://www.anuntul.ro/statistici-imobiliare/bucuresti/bucuresti/vanzari/apartamente-3-camere/evolutie-lunara/#top (accessed on 28 April 2020).
47. Romania-Insider.com. *96% of Romanians Own the Homes They Live in—Highest Share in Europe*. Available online: https://www.romania-insider.com/romania-high-share-homeowners (accessed on 28 April 2020).
48. Delmendo, L.C. *Romanian Housing Market Growing Strongly*. Available online: https://www.globalpropertyguide.com/Europe/Romania (accessed on 28 April 2020).
49. Sfetcu, N. *Ghid Turistic Bucureşti: Ediţia de Buzunar. Ghid Bilingv Engleză/Română (Bucharest Tourist Guide: Pocket Edition. Bilingual English/Romanian Guide)*; Rakuten Kobo: Toronto, ON, Canada, 2015.
50. Łachowski, W.; Łęczek, A. Tereny zielone w dużych miastach Polski. Analiza z wykorzystaniem Sentinel 2. *Urban Dev Issues* **2020**, *68*, 77–90.
51. *Strategia Rozwoju Lublina na Lata 2013–2020 (Lublin's Development Strategy for 2013–2020)*. Available online: http://www.flop.lublin.pl/media/newsy/strategia_rozwoju_lublina_2020_konsultacje.pdf (accessed on 4 May 2020).
52. *Miasto dla Ludzi—Lubelski Ruch Miejski (A City for People—Lublin City Movement)*. Available online: https://lubelskiruchmiejski.wordpress.com/ (accessed on 4 May 2020).
53. *Raport na Temat Wielkich Polskich Miast (Report on Major Polish Cities)*. Available online: https://www.pwc.pl/pl/pdf/miasta/raport_lublin_2011.pdf (accessed on 4 May 2020).
54. Dunajska, A. *Co Ważniejsze w Mieście: Inwestycje czy Drzewa? Raport NIK (What Is More Important in the City: Investments or Trees? NIK Report)*. Available online: https://lublin.naszemiasto.pl/co-wazniejsze-w-miescie-inwestycje-czy-drzewa-raport-nik/ar/c1-3311387 (accessed on 4 May 2020).
55. PAK Kopalnia Węgla Brunatnego Konin. *Tu Była Odkrywka (There Was an Outcrop Here)*. Available online: http://www.kwbkonin.pl/index.php/krajobraz-po-odkrywce/ (accessed on 4 May 2020).
56. Urząd Statystyczny w Poznaniu. *Konin 2013: Podstawowe Dane Statystyczne (Konin 2013: Basic Statistical Data)*. Available online: https://bip.konin.eu/tmpl/sesje_rady/sesja_132_26_02_1390401082.pdf (accessed on 4 May 2020).
57. Kowarik, I. Cities and Wilderness. A New Perspective. *Int. J. Wilderness* **2013**, *19*, 32–36.
58. Panagopoulos, T.; Jankovska, I.; Boştenaru Dan, M. Urban green infrastructure: The role of urban agriculture in city resilience. *Urban Arhit Constr.* **2018**, *9*, 55–70.

59. Vice News. *Protestul din Parcul Tineretului a Fost ca o Plimbare Romantică Prin Ploaie (The Protest in Tineretului Park Was Like a Romantic Walk in the Rain)*. Available online: https://www.vice.com/ro/read/protestul-din-parcul-tineretului-a-fost-ca-o-plimbare-romantica-prin-ploaie (accessed on 29 April 2020).
60. Ştirile PRO TV. *Protest in Lasi ca Teii din Centrul Orasului sa nu fie Taiati (Protest Is Lasi agaisnt Cutting off the Lime Trees)*. Available online: https://stirileprotv.ro/stiri/actualitate/protest-in-iasi-pentru-ca-teiul-lui-eminescu-sa-nu-fie-taiat-autoritatile-vor-aduce-salcami-japonezi.html (accessed on 28 April 2020).
61. Petitieonline.com. *Stop Taierii Copacilor de pe Sos. Pantelimon (Stop Cutting off the Trees on Pantelimon St.)*. Available online: https://www.petitieonline.com/stop_taierii_copacilor_de_pe_sos_pantelimon (accessed on 29 April 2020).
62. DIGI 24 HD. *Nicușor Dan Anunță că a Câștigat al 166-Lea Proces. Tribunalul Suspendă Construirea în „Grădinile Versailles ale Bucureștiului" (Nicușor Dan Announces that He Won the 166th Lawsuit. Bucharest City Hall Suspends the Constructions in the "Versailles Gardens of Bucharest")*. Available online: https://www.digi24.ro/stiri/nicusor-dan-anunta-ca-a-castigat-al-166-lea-proces-tribunalul-suspenda-construirea-in-gradinile-versailles-ale-bucurestiului-1240241 (accessed on 29 April 2020).
63. Mierzejewska, L. Urban planning in Poland in the context of European standards. *Quaest. Geogr.* **2009**, *28*, 29–38.
64. Kolipiński, B. Planowanie przestrzenne w Polsce w minionym 25-leciu (Spatial planning in Poland in the last 25 years). *Maz. Studia Reg.* **2014**, *15*, 109–118.
65. Śleszyński, P.; Nowak, M.; Brelik, A.; Mickiewicz, B.; Oleszczyk, N. Planning and Settlement Conditions for the Development of Renewable Energy Sources in Poland: Conclusions for Local and Regional Policy. *Energies* **2021**, *14*, 1935. [CrossRef]
66. Kronenberg, J.; Łaszkiewicz, E.; Sziło, J. Voting with one's chainsaw: What happens when people are given the opportunity to freely remove urban trees? *Landsc Urban Plan.* **2021**, *209*, 104041. [CrossRef]
67. Nowak, M.L.; Śleszyński, P.; Legutko-Kobus, P. *Spatial Planning in Poland. Law, Property Market and Planning Practice*; SpringerBriefs in Geography; Springer Nature: Cham, Switzerland, 2022.

land

MDPI

Article

The Impact of Fine-Scale Present and Historical Land Cover on Plant Diversity in Central European National Parks with Heterogeneous Landscapes

Radomír Němec [1], Marie Vymazalová [2] and Hana Skokanová [2,*]

[1] South Moravian Museum in Znojmo, Přemyslovců 8, 66902 Znojmo, Czech Republic; nemec@znojmuz.cz
[2] Silva Tarouca Research Institute for Landscape and Horticulture, Lidická 25/27, 60200 Brno, Czech Republic; meri@seznam.cz
[*] Correspondence: hana.skokanova@vukoz.cz

Abstract: As the human population grows, the transformation of landscapes for human uses increases. In recent homogeneous and predominantly agricultural landscapes, land-cover and management changes are considered the main drivers of vascular plant diversity. However, the specific effects of land-cover classes across whole heterogeneous landscapes are still insufficiently explored. Here, we investigated two floristic surveys realised in 1997 and 2021, accompanied by fine-scale land-cover classes detected in 1950, 1999 and 2018, to reveal the impact of historical and present land cover on the pattern of species composition and species richness in the bilateral Podyjí and Thayatal National Parks. Multi-dimensional analyses revealed that the species composition was driven by the fine-scale historical land cover, the overall species richness was mostly affected by the river phenomenon and the present richness was mostly affected by increased soil nutrients. In well-preserved protected areas, it is especially desirable to restore disappearing land-cover classes with traditional or compensatory management to retain plant species richness, which is a key factor of biodiversity. However, management plans should also take into account the increasing amount of nitrogen in soils from long-term continual deposition, which can strongly impact the species richness, even in national parks with low current deposition.

Keywords: species richness; species composition; land cover; Podyjí/Thayatal National Parks; vascular plants

1. Introduction

In central Europe, humans have affected landscapes and plant diversity since the Neolithic times [1]. However, with growing populations and the consequent pressure on land use, this impact has steadily increased over time [2,3]. This pressure has been most pronounced during the past two hundred years, after the modern agricultural revolution and industrial revolution at the end of the 18th and beginning of the 19th century [4] that introduced, among others, new crops, new agricultural systems, mechanisation and artificial fertilisation [5]. Besides food, the growing population also demanded more space for living, resulting in the spread of settlements and the accompanying transportation infrastructure [6]. While more fertile regions started to be overexploited over this period, less fertile and accessible regions, typically mountainous regions, started to be abandoned, mainly leading to the spread of forest [2,3]. The land overexploitation of fertile regions and abandonment of mountainous regions accelerated after the Second World War in western parts of Europe, connected with the so-called productivist agriculture and market economies [5], while in eastern parts of Europe, the introduction of the so-called socialist agriculture, connected with forced collectivisation, were the main drivers [7]. Both types were characterised by the heavy use of fertilisers, pesticides and mechanisation, leading to the loss and fragmentation of traditionally managed habitats [8]. Therefore, the long historical land-use/cover continuity in large areas was greatly disrupted, leading to homogenisation and sharp declines in biodiversity [6]. Indeed, many studies showed that regions with continuous land-cover and/or traditional land-use management enable the accumulation of many species at specific individual sites and can contribute to reducing the chance of random extinctions of rare species [9–11]. Such regions are usually protected and harbour one of the most species-rich plant communities, e.g., Bílé Karpaty (White Carpathian) Mts. in the Czech Republic [12–15].

However, historical continuity is only one of the factors influencing the diversity of plant communities. Diversity naturally varies across landscapes, with environmental factors, such as precipitation, soil pH, nutrient and light availability; biomass productivity [16–19]; and geological diversity causing some regions to have unusually high diversity and others to be much poorer. In addition, high topographic and meso-climatic diversity [20] can also affect patterns of species diversity, as well as alien plant invasions [21]. Both topographic and meso-climatic diversity may be even more accentuated in deep river valleys, where they are known as a 'river phenomenon' [22]. Nevertheless, the main drivers of species composition and species richness are considered to be land-use and management changes (in connection with historical continuity), which affect not only predominantly lowland agricultural landscapes [11,23] and settlements [24] but also mountain regions [25,26]. In general, land-cover studies have usually focused on only one type of habitat, such as grasslands [11,24] and forests [6,27], in landscapes with relatively homogeneous environmental conditions. They have generally confirmed that the combined historical and current land use is reflected in current patterns of diversity [28]. A few studies have investigated more than one land-cover category, e.g., open rural habitats [24]. A study by Michalcová [10] found that the land-cover diversity of the whole landscape contributes to between-site similarities in species composition, with the co-occurrence of many species increasing the species richness. However, complex studies of fine-scale land-cover classes at a regional scale in heterogeneous landscapes are still lacking.

In this study, we focused on two aspects of the relationships between fine-scale land-cover history and plant diversity at a regional scale in two central European national parks. First, we examined spatial heterogeneity and compared two types of landscapes with distinctive geomorphological features: a deep river valley and surrounding rather flat plains. Second, we focused on all types of habitats in the studied landscape. Our analyses, therefore, aimed to answer the following questions: (1) How does fine-scale land cover influence plant species composition in naturally well-preserved areas with long-term human impacts, but with very distinctive geomorphologies? (2) Is the species composition dependent on present or rather historical land cover? (3) Which types of land cover and relevant landscape heterogeneity indicators are important for current species richness compared with more than twenty years ago?

2. Materials and Methods

2.1. Study Area

This study was performed in the bilateral Podyjí and Thayatal National Parks (NPs) and buffer zone of the Podyjí NP, situated between the towns of Vranov nad Dyjí (48°54′ N, 15°49′ E) and Znojmo (48°52′ N, 16°03′ E) along the state border between the Czech Republic and Austria (Figure 1). The total area is 105 km^2, with NP Podyjí covering 63 km^2, Thayatal NP covering 13 km^2 and the Podyjí NP buffer zone covering 29 km^2. A protected landscape area on the Czech side was declared in 1978 and was then transformed into a national park in 1991. On the Austrian side of the Dyje (Thaya) river, a national park was declared in 2000. The main part of the protected areas is formed by the river Dyje, which flows through a deep valley with varied morphology [29] and a diverse mosaic of habitats on slopes of different aspects [22]. The Dyje valley is surrounded by more or less flat relief of adjacent plateaus. The altitudinal range is from 210 m a.s.l. (river Dyje in Znojmo) to 536 m a.s.l. (Býčí hora hill near Vranov n. D.), and the maximum valley depth is almost 200 m. From a geological perspective, the crystalline bedrock of the Bohemian massif is mainly formed by gneiss, schist and granite, with restricted occurrences of crystalline limestone [30]. The eastern part of the study area is filled with Neogene sediments and the plateaus are covered by loess. Other Quaternary sediments can be found in the Dyje river and many stream valleys.

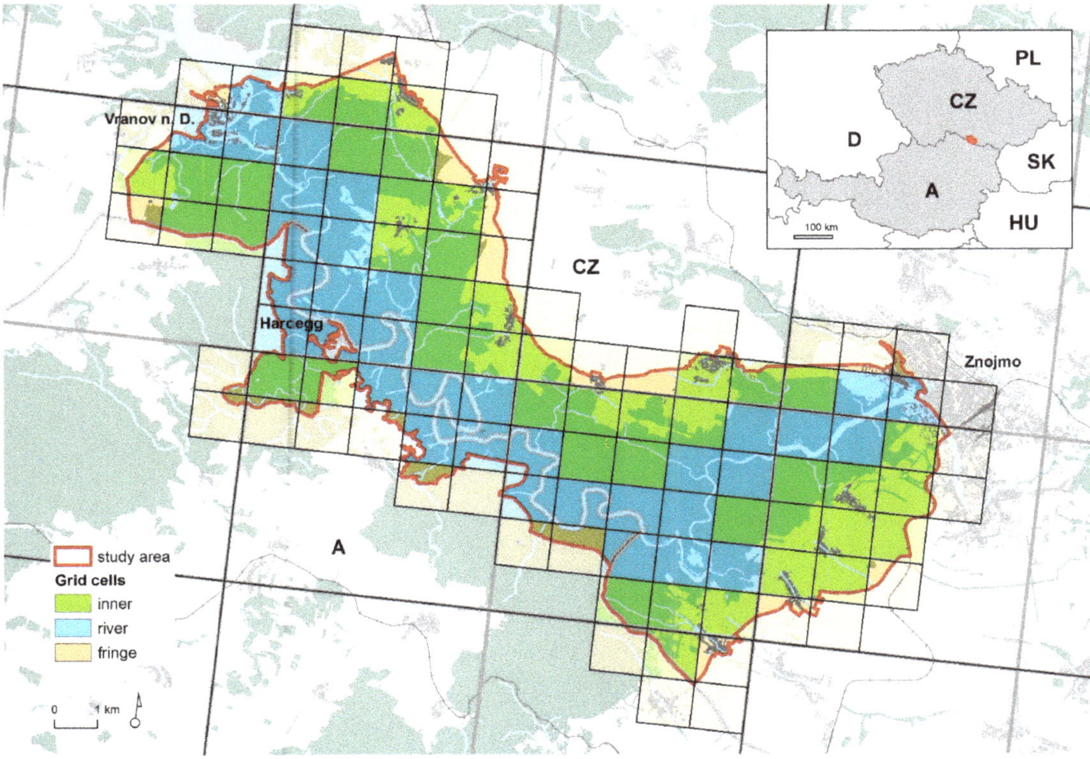

Figure 1. Study area in Austria and the Czech Republic divided into 115 grid cells within three groups according to the geomorphology and surveyed area of each cell (see Section 2.2): inner, river and fringe. The background map shows forests (dark green), watercourses (blue) and settlements (grey).

The study area is classified as a warm and dry region, with an average yearly temperature between 8 and 9 °C, mean annual precipitation of 550–600 mm and a distinct NW–SE gradient from a moderately warm to warm climate [31]. The area is located in the transition zone between the Hercynian and Pannonian floristic regions [32] and, therefore, has a significant proportion of thermophilous and continental species [22]. These varied environmental conditions have also been accompanied by specific human impacts, including the location along the so-called Iron Curtain in the second half of the 20th century, characterised by restricted movement along the Czech–German and Czech–Austrian borders. Therefore, the study area can be considered a biodiversity hotspot [33–35], with 1555 historically recorded species of vascular plants [36], i.e., much more species than the individual national parks in adjacent Austria (e.g., the area richest in species is the Hohe Tauern harbour with 1077 species, [37]).

The Podyjí and Thayatal NPs and their surroundings have been continuously inhabited since 5000–6000 BC [38–42]. For more than 800 years, mainly before the industrial revolution in the 19th century, human impacts were reflected in deforestation, pasture, burning, litter raking and coppicing [43], as well as in the cultivation of slope terraces for wine production. After the industrial revolution and the population explosion in the 19th century, anthropogenic pressure spread to the non-forested, agricultural landscape. This was accomplished by using more efficient tillage systems and the introduction of fertilisers in the 19th century and by agricultural mechanisation, the use of pesticides and fungicides, and land consolidation in the 20th century [5]. These practices have completely changed the landscape configuration and contributed to its degradation, especially in the intensively used agricultural surroundings of both national parks. While forests significantly dominated in Thayatal NP for at least the last 200 years, Podyjí NP was characterised by quite a large proportion of open landscape (24–35%) with permanent grassland and later arable land between the 1840s and 1930s, though forests have dominated since the second half of the 20th century. In contrast, the Podyjí NP buffer zone has been characterised by a greater variety in the form of vineyards, orchards, grasslands and settlements, and forests were (and still are) quite scarce [44]. Long-term continuity of forests and grasslands based on topographic maps has shown that nearly 50% of forests in the NP were preserved over the last 200 years, while less than 1% of grasslands were preserved [44].

2.2. Vascular Plant, Land Cover and Supplementary Data

Our data set consisted of a floristic grid cell survey of vascular plants conducted in two periods: 1992–1993 (first floristic survey; [45]) and 2019–2020 (second floristic survey; [36]). The grid used was derived from the grid of Central European floristic mapping by dividing the area into basic grid cells that are 1′ × 0.6′ (1.2 × 1.1 km) in size. Accordingly, the study area comprised 115 grid cells, with each cell containing a list of vascular plant taxa. The standardised methodology was applied with one exception. While during the first floristic survey, plant data were collected for the whole grid cell, regardless of the borders of the study area, during the second floristic survey, plant data were collected only for parts of the grid cell lying inside the study area. The nomenclature of plant taxa follows Kaplan et al. [46], where some hybrids or records identified only at the genus level were removed. The resulting data set included 1412 species, which is still higher than in individual Austrian National Parks (see above, [37]).

We assessed the land cover based on aerial photos, orthophotos and other digital sources in three time steps: 1954–1955 (pixel resolution 0.5 m), 1999–2001 (pixel resolution 0.2 m) and 2018–1019 (pixel resolution 0.2 m). Other digital sources were in a vector format and included agricultural data from the Land Parcel Information System (from 2020); digital cadastre data (from 2020); biotope mapping (from 2013); and settlements, roads, railroads and watercourses from the Fundamental Base of Geographic Data on the Czech Republic (from 2020). Roads, railroads and watercourses were transformed to polygon layers by buffering them with preset criteria (first-class roads 22 m wide, second-class roads 12 m, third-class roads 10 m, paved roads and streets 6 m, unpaved roads and railroads 4 m, watercourses 2 m). For the Czech part of the study area, we combined the vector data and used the result as a basis for the present state, which was adjusted and classified with the help of orthophotos from 2018–2019 [44]. We derived the present land cover in the vector format for the Austrian part by vectorizing each orthophoto. The minimum mapping unit was set to 50 m^2, but in the case of settlement areas, it was lowered to 15 m^2 in order to capture individual buildings. We verified the resulting land-cover layer vector in the field and adjusted it accordingly. Then, we used the present land-cover layer vector as a source for the 1999–2001 layer, where the boundaries of individual polygons were adjusted via backward editing [5] and reclassified according to the relevant orthophoto. We derived the land cover for the 1954–1955 layer via the same procedure, using the 1999–2001 layer as a basis. However, since the orthophoto from 1954–1955 had a coarser resolution, some features, namely, settlements and, to some degree, small agricultural plots, were characterized. The land-cover layers in the following text are referred to as the 1950s (from period 1954–1955), 1990s (from period 1999–2001) and present (from period 2018–2020).

In total, we distinguished 30 land-cover classes. They included forest (closed canopy woody vegetation); sparse forest (open canopy woody vegetation, characterized according to [47] as a forest with trees or shrubs 15–40 m apart); clearings (including shrubs and forest nurseries); grass-forb (grassland with scattered trees); meadows (including pastures); fallow land (land not cultivated for at least two years); natural bare surfaces (mainly rocks); natural water bodies (in the form of pools); artificial water bodies; streams; swamps (wetlands); smallholdings; small orchards (<1 ha); small vineyards (<1 ha); vineyards with trees, arable fields with trees; small arable fields (<1 ha); large orchards (>1 ha); large vineyards (>1 ha); large arable fields (1–30 ha); very large arable fields (giant) (>30 ha); gardens; public greenery; recreational areas; mining areas; waste dumps; residential built-up areas (with public buildings); industrial, agricultural and commercial areas; and roads and railroads. Two land-cover classes (natural water bodies and railroads) were omitted from the analyses due to their negligible areas.

For the land-cover analysis, for each grid cell, we calculated the proportion of land-cover classes from the three periods. We also calculated the number of patches for each grid cell, which served as the landscape heterogeneity indicator.

As supplementary variables, we used five characteristics derived from maps that could probably influence the species richness or species composition: the area belonging to Podyjí National Park (NP), the area of the Podyjí NP buffer zone, the area belonging to Thayatal NP, the area of unchanged land cover and the geological diversity. The last was computed as the Shannon diversity index based on the proportions of bedrock types from Geologische Karte der Republik Österreich 1:50,000, sheet Retz [48]. The geological diversity was counted for each grid cell, with original bedrock types merged into seven broader categories (gneiss, granite, schist, calcareous rocks, neogene sediments, loess, other Quaternary sediments). Further, we used the number and characteristics of alien plant species according to Pyšek et al. [49] and the total number of all four (C1–C4) categories of endangered species and the number of critically endangered (C1) and endangered (C2) species evaluated using The Red List of vascular plants of the Czech Republic [50]. Based on the species composition we computed mean Ellenberg indicator values calibrated for the Czech flora (EIVs; light, temperature, moisture, soil reaction, nutrients; [51]).

2.3. Statistical Analyses

To address question 1, first, we had to calculate the Bray–Curtis dissimilarity for the land cover and the Sörensen dissimilarity for floristic presence/absence data. Then, the composition and changes in land cover and floristic data were explored using the following analyses: (1) principal coordinates analysis (PCoA) embedded mentioned dissimilarities and were combined with the Hellinger distance (supplementary variables were tested using the multiple regression with 999 permutations ($p < 0.001$), and Ellenberg indicator values were tested using the modified permutation test according to Zelený et Schaffers [52]), (2) permutational multivariate analysis of variance using distance matrices (PERMANOVA) [53,54] and (3) multivariate homogeneity of the groups' dispersions (PERMDISP) [54,55]. Further, the floristic beta diversity between particular surveys was computed as the Sörensen dissimilarity.

To address questions 1 and 3, changes in species richness, alien species, neophytes, EIVs and beta diversity were displayed in boxplots and analysed using parametric pairwise or unpaired t-tests accordingly, while changes in individual classes of land cover (used as relative area, i.e., percentages of the whole grid cell) were assessed using the non-parametric Friedman test (if $p < 0.001$, then analysis with post hoc Wilcoxon pair rank-sum tests with Bonferroni correction followed). Further changes in land-cover heterogeneity were evaluated using ANOVA repeated measures analysis (if $p < 0.001$, then analysis with Tukey's post hoc tests followed) and changes in the proportions of unchanged land were analysed using the non-parametric Wilcoxon rank-sum test.

To address question 2, but also necessary for question 1 to reveal the limit value of included/surveyed grid cells area, which shows the similar pattern of species richness as the whole area of grid cells, and to identify the major land-cover classes influencing the species richness, we performed an analysis of regression trees based on the CART concept (classification and regression trees [56]; libraries rpart and rpart.plot [57]). First, we used the species richness per grid cell recorded during the second floristic survey as a dependent variable, while the proportions of each land-cover class detected in the second floristic survey, included/surveyed area of grid cells, landscape heterogeneity and proportions of individual NPs and the buffer zone were used as explanatory variables.

The regression trees based on the second floristic survey specified a limit value of 72 ha (53%) of the surveyed grid cell area (see Section 3), with grid cells surveyed in a larger area that was comparable in species richness to grid cells surveyed in the whole area. For subsequent analyses, we also divided the whole number of grid cells (n = 115) into three groups: (1) fringe cells (n = 37 for the second floristic survey, while n = 32 for the first floristic survey since five fringe cells were not surveyed) comprising grid cells on the border of the study area with smaller surveyed areas; and other cells (n = 78), which were separated into (2) inner cells (n = 44), within which the flow of the Dyje river is not present, and (3) river cells around and including the Dyje river (n = 34). Then, using regression trees, we analysed the corresponding data from the first floristic survey, examining only the inner and river cells (n = 78) according to another survey methodology. To better understand the environmental patterns, we then performed another regression tree with the addition of mean Ellenberg indicator values among the explanatory variables for the first and second surveys separately. All four trees were pruned according to expert knowledge into a maximum of eight leaf nodes.

To address questions 1 and 2, i.e., to examine the impact of the fine-scale land cover of three time periods on vegetation composition, we first classified grid cells based on the land-cover data. The proportions of 28 land-cover types were square-rooted and the Bray–Curtis dissimilarity was computed; then, the beta-flexible method ($\beta = -0.25$) was used to create three distinctive groups. Such a classification was performed on land-cover data from all periods, while the classification from the 1950s land cover was displayed on floristic data in the principal coordinates analysis (PCoA; Sörensen dissimilarity, Hellinger distance). The impact of these land-cover classifications, together with the landscape heterogeneity and geomorphology coupled with the meso-climate (river phenomenon), on species composition was tested using distance-based redundancy analysis (db-RDA; Sörensen dissimilarity, Hellinger distance). Because the high geological diversity of the Podyjí and Thayatal NPs could also affect the species richness and the species composition [10,22,30], we included this diversity in the db-RDA as a covariate to remove its impact. First, we calculated the significance and adjusted the R^2 of the whole model, and then used a forward selection method appropriate for particular variable testing in db-RDA.

We used ArcGIS software (ESRI 2016–2021), version 10.3–10.5, for the production of land-cover layers and all maps, and R software (R Core Team 2018) for statistical analyses with the libraries tidyverse [58], ggpubr [59], rstatix [60], vegan [61], cluster [62], rpart [57] and rpart.plot [63].

3. Results

The PERMANOVA analysis showed that the land cover significantly differed ($p < 0.001$) between the 1950s and 1990s within both the inner and river cells, while there was no significant difference between the 1990s and the present (Table 1). No difference in land cover was found within fringe cells. At the same time, the PERMDISP analysis revealed no changes in the overall landscape diversity between the examined periods.

Table 1. The permutational multivariate analysis of variance using distance matrices (PERMANOVA) and multivariate homogeneity of groups dispersions (PERMDISP) for the land-cover and floristic data were based on the Bray–Curtis dissimilarity for land cover and Sörensen dissimilarity for floristic presence/absence data. n.s.—not significant.

Characteristics	Regions	PERMANOVA		PERMDISP	
		F	p	F	p
First vs. second floristic mapping	Whole model	7.33	<0.001	4.66	0.032
	River	4.12	0.002		n.s.
	Inner	4.56	<0.001	3.87	0.03
Land cover 1950 vs. 1997	Whole model	8.04	<0.001		n.s.
	River	10.29	<0.001		
	Inner	6.01	<0.001		
	Fringe		n.s.		
Land cover 1997 vs. 2018	Whole model		n.s.		n.s.
Land cover 1950 vs. 2018	Whole model	8.71	<0.001		n.s.
	River	12.77	<0.001		
	Inner	5.69	<0.001		
	Fringe		n.s.		

The large shift in land cover between the 1950s and 1990s (Figure A1) was clear, especially in the huge decrease in sparse forest (Figure 2b), meadow (Figure 2d) and grass-forb vegetation (Figure 2e) within both the inner and river cells; in contrast, there was an increase in fallow land (Figure 2f) within inner cells and closed canopy forest within river cells (Figure 2a). Another significant decrease was documented in inner cells within orchards, vineyards with trees, fields with trees and small fields (<1 ha; Figure 2c); in contrast, there was an increase in gardens and giant fields (>30 ha).

The same analyses for floristic data revealed a significant difference in the species composition of inner ($p < 0.001$) and river ($p = 0.002$) cells between the first and second floristic surveys, as well as in the gamma diversity of inner cells ($p < 0.03$), which was not significant in river cells (Table 1).

The PCoA analysis of species composition with regressed supplementary variables (Figure 3) showed a slow decrease in the species richness of both inner and river cells (but according to the pairwise t-test, the differences within these groups were not significant, Figure 4a), and there was a decrease in the light conditions (Figure A2b). Conversely, we found an increase in soil nutrients within both inner and river cells (EIV for nutrients; Figure A2a), while river cells showed an obvious shift to higher moisture conditions. Fringe cells revealed an increase in soil nutrients. The beta diversity comparing both surveys was significantly higher within inner cells (Figure 4b).

The impact on the species composition in the study area showed a similar pattern in both examined floristic surveys (Table 2). The highest impact was caused by the historical land cover (identified in the previous period; Table 1 and Figure 5), followed by the river phenomenon (geomorphology combined with meso-climate) and present land cover. In contrast, the landscape heterogeneity (Figure A2c) was not significant, and the land cover of the 1950s applied to the second floristic survey had only a negligible impact (under 1% of the explained variability).

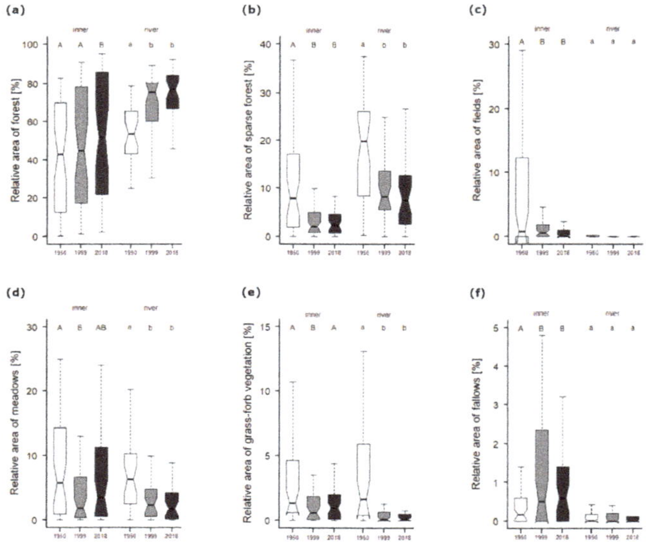

Figure 2. Changes in the relative areas (percentages of the whole grid cell area) of (**a**) forest, (**b**) sparse forest, (**c**) fields (<1 ha), (**d**) meadows, (**e**) grass-forb vegetation and (**f**) fallows were assessed using the Friedman test, and when significant ($p < 0.001$), with the post hoc Wilcoxon pairwise test with a Bonferroni correction ($p < 0.01$), for three periods: 1950, 1999 and 2018. Whiskers indicate the non-outlier range, with the median as the middle; boxes show 25–75% percentiles.

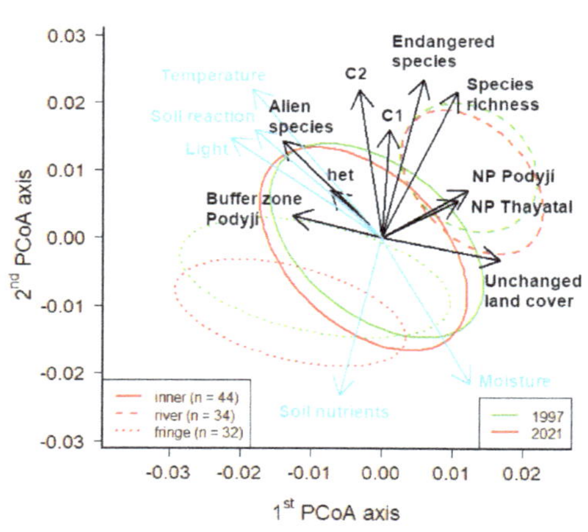

Figure 3. The principal coordinates analysis (PCoA) of the floristic data (Sörensen dissimilarity) of 110 grid cells in the Podyjí and Thayatal NPs and the Podyjí buffer zone from both floristic surveys (1997 and 2021). Ellipses represent groups of inner, river and fringe grid cells within both floristic surveys and are drawn around the medoids of the respective cells (see Methods); if they do not overlap, the groups differed. Only those supplementary variables that were significant in the multiple regression are shown ($p < 0.001$; het—landscape heterogeneity; C1—number of critically endangered species; C2—number of endangered species category C2; endangered species—number of all endangered species of categories C1–C4 according to [50]). The blue arrows represent Ellenberg indicator values, which were tested using a modified permutation test according to Zelený and Schaffers ([52]; $p < 0.001$ for light, temperature and moisture; $p < 0.002$ for soil nutrients and $p < 0.003$ for soil reaction).

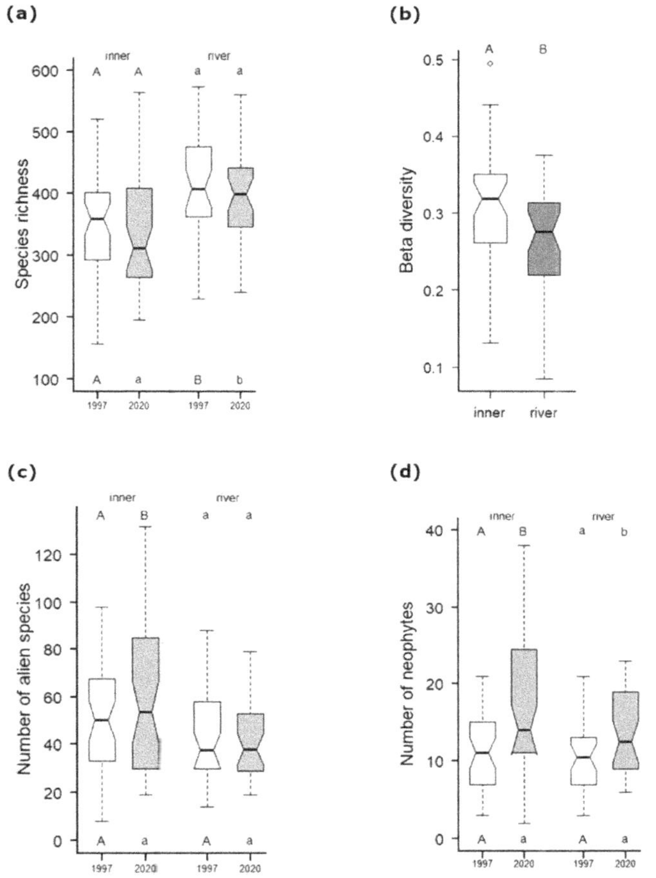

Figure 4. Changes and differences in (**a**) species richness, (**b**) beta diversity (Sörensen dissimilarity; non-pair *t*-test), (**c**) number of alien species and (**d**) number of neophytes per grid cell within both the first (1997) and the second survey (2021) separated into inner (n = 44) and river (n = 34) group cells (see Methods). The letters above boxes indicate the results of pairwise *t*-tests between surveys within each group, while the letters below boxes show results of the non-pair *t*-test between the groups (**a**,**c**,**d**). Whiskers indicate the non-outlier range with the median as the middle; boxes show the 25–75% percentiles.

Among the present land-cover classes, the most important for species richness was streams, i.e., those grid cells including parts of the Dyje river (Figure 6b). This was followed by recreation, then sparse forest and meadows. Similarly, in the first floristic survey, the major impact on species richness was caused by streams (Figure 6a), which was also an important variable in the secondary branch level, together with meadows (Figure 6a; only 78 grid cells analysed; see Section 2.3).

Table 2. Variables tested in the distance-based redundancy analysis (db-RDA; Sörensen dissimilarity, Hellinger distance; geological diversity filtered as a covariate), assessed the variation of the whole model and then tested using forward selection with an adjusted R^2 and Holm's correction. n.s.—not significant.

Survey	Variables	Adj.R^2	p	Holm's Correctec p
Second floristic survey	Whole model	0.126	0.001	0.126
	Land-cover classification 1999	0.059	0.001	0.059
	River phenomenon	0.034	0.001	0.034
	Land-cover classification 2018	0.023	0.001	0.023
	Land-cover classification 1950	0.008	0.001	0.008
	Landscape heterogeneity 2018		n.s.	
	Landscape heterogeneity 1999		n.s.	
First floristic survey	Whole model	0.135	0.001	0.135
	Land-cover classification 1950	0.058	0.001	0.058
	River phenomenon	0.047	0.001	0.047
	Land-cover classification 1999	0.021	0.001	0.021
	Landscape heterogeneity 1999		n.s.	

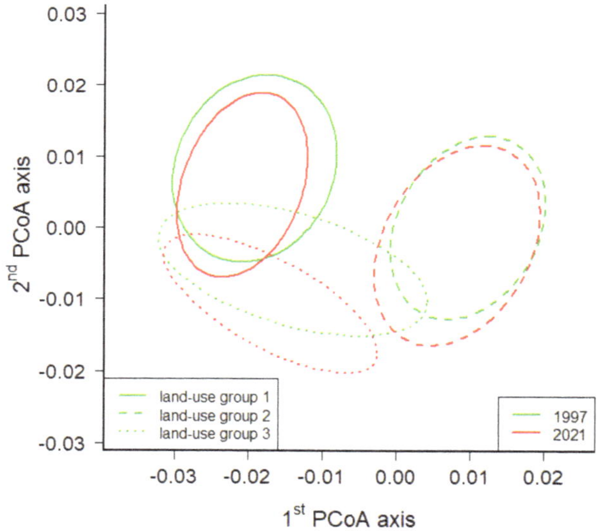

Figure 5. The principal coordinates analysis (PCoA) of the floristic data (Sörensen dissimilarity) of 110 grid cells in Podyjí and Thayatal NPs and the Podyjí buffer zone from both floristic surveys (1997 and 2021) shows the shift in plant species composition within three groups of grid cells clustered on land cover observed in 1950. Ellipses are drawn around the medoids of three groups of cells based on the classification of land-cover data detected in the 1950s (see Methods); if they do not overlap, the groups differed.

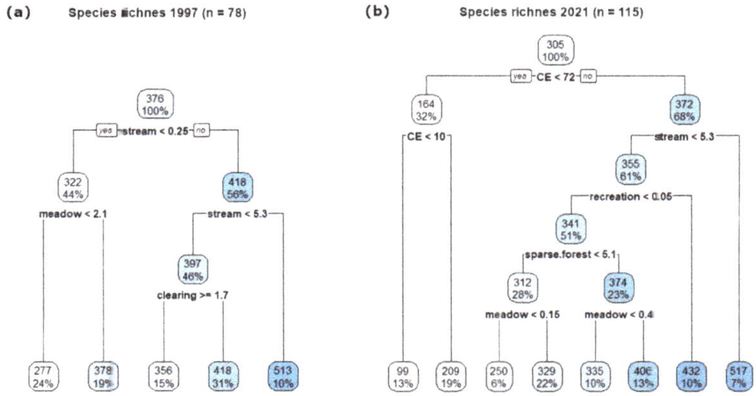

Figure 6. The regression trees for species richness in surveys in (**a**) 1997 (n = 78) and (**b**) 2021 (n = 115) based on the proportions (%) of individual land-cover categories per grid cells, landscape heterogeneity and proportions (%) of individual NPs and surrounding area, resp. surveyed area per grid cells (CE; within 2021). Numbers in each node represent the average of species richness in this node.

When adding the Ellenberg indicator values (EIVs) to the explanatory variables (Figure 7b), the lower content of nutrients (EIV for nutrients) was the most important for species richness in the second floristic survey, while the next node level comprised both the stream and EIVs for soil reaction (the most species-rich grid cells had a higher value of EIV for soil reaction). The EIV for light was also important. Similarly, when adding the EIVs into the first floristic survey analysis (Figure 7a), the major impact remained from streams, while in the secondary branch level, both meadows and lower EIVs for nutrients were important. Again, the EIV for light was also important.

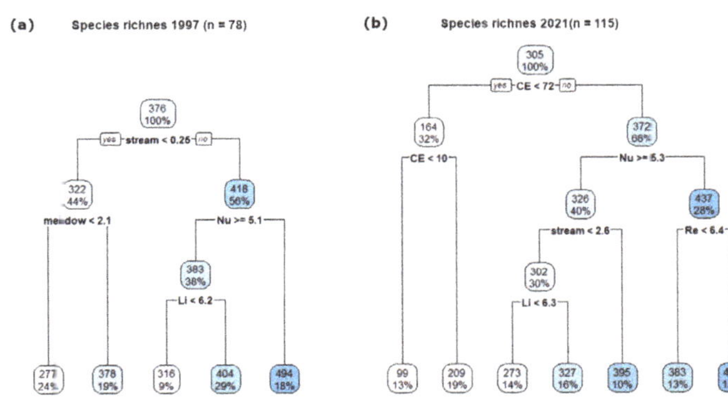

Figure 7. The regression trees for species richness in surveys in (**a**) 1997 (n = 78) and (**b**) 2021 (n = 115) based on the proportions (%) of individual land-cover categories per grid cells, landscape heterogeneity and proportions (%) of individual NPs and Podyjí buffer zone, as well as Ellenberg indicator values (see Section 2.2; Nu—nutrients, Re—soil reaction, Li—light), resp. surveyed area per grid cells (CE; within 2021). Numbers in each node represent an average of species richness in this node.

4. Discussion

Our analyses might seem to have been led by partitioning based on the different beta diversity patterns (Figure 4b) and geomorphology into (1) river cells characterised by high geomorphological and meso-climatic diversity [21] (river phenomenon) accompanied by a diverse mosaic of habitats on slopes with different aspects, and (2) inner cells representing mainly gently undulating landscape with settlement accompanied by a rich mosaic of land cover, including rare heathlands, resulting in high plant diversity ([64]; Figure 3). However, the strong river phenomenon described in the studied area [22] and documented in our analyses (Figures 4a, 6 and 7) did not result in major impacts on species composition (Table 1). The major impact on species composition was found to be due to the historical land cover (Table 1 and Figure 5). This effect of historical land-cover continuity was demonstrated mainly for traditionally managed grasslands [9,11,65]. However, continually used grasslands (between 1840 and 2018) in the study area covered only 0.6%. Therefore, any separate effect of grassland continuity in our study area would be rather weak. This is different from, e.g., the White Carpathians, where continual meadows cover 3.8% of their area. On small plot sizes, these meadows are the most species-rich in the world [12,13]. On the other hand, forest continuity was found in 47.5% of our study area, and therefore, might be considered to have a strong impact, as it influences the species richness and composition mainly through the dispersal limitation of many forest specialists [66]. It should be stressed that our study reflected the land-cover composition of the whole landscape because our analyses included fine-scale information on land cover with 28 categories and their classification resulted in robust partitions (Figure 5). In both analysed floristic surveys, the historical land cover explained more variability than the cover in the respective period of the floristic survey, even though roughly 20 years (in the first survey) or even 40 years (in the second survey) had passed. This corresponded in particular with the overgrowing and abandonment of some habitat types because vegetation generally responds with some delay [67,68] and some species specialists may still survive there. Moreover, the transformation into new habitats could also permit the persistence of some species from the original habitats. These new habitats include (1) near-natural land covers inside settlements, e.g., public greenery, some managed road verges or some recreation areas [24], or (2) near-natural land covers outside settlements [6].

The main driver of species richness in the study area in both surveys was found to be the river phenomenon (Figure 6a,b), which was likely caused by sharp environmental gradients within relatively small areas [22]. The species richness was also strongly influenced by meadow areas in both surveys, although the area of historically continuous meadows (1840–2018) was very low compared with other well-preserved natural areas (e.g., White Carpathians, see above). However, these fragments still represent valuable biodiversity hotspots [29] that are important for the maintenance of species richness [24]. This is true even though the meadows within river cells declined in 1999 and continued to decline until the present, while meadows within inner cells have been partly restored (Figure 2d). In the second floristic survey, the recreation land-cover area was also important for species richness. This finding was the major difference between the two surveys and was partly due to an increase in alien species near settlements (inner cells; Figure 4c), especially escapes of cultivated plants (e.g., *Muscari armeniacum, Tulipa gesneriana, Primula vulgaris, Rudbeckia hirta, Lunaria annua*) in combination with the perseverance of some rare weeds (e.g., *Asperugo procumbens, Chenopodium vulvaria, Urtica urens*), and the occurrence of disturbed surfaces harbouring some rare endangered species (e.g., *Trifolium retusum, T. striatum, Filago germanica*). Another important land cover that was distinguished in the second floristic survey was the sparse forest area (Figure 2b), which still harbours some rare species of historically continually managed forests ([69]; the majority of unchanged land cover, Figure A2d). These sparse forests represent just remnants of species-rich traditionally managed forests [43], which were gradually abandoned, like in many other parts of Europe [8] until the establishment of the so-called Iron Curtain (the restricted military area on the state border, 1951–1990). This led to the total isolation of the river valley and adjacent areas and resulted in a huge decrease in sparse forests, as well as meadows and grass-forb vegetation within both river and inner grid cells (Figure 2b,d,e). Whereas sparse forests in river cells (Figure 2b) closed their canopy (;with the exception of habitats on extreme slopes with shallow soil) and meadows with grass-forb vegetation were mostly overgrown by forests (Figure 2a), inner cells were only partly abandoned (increasing of fallows, Figure 2d–f) and accessible/open areas of highly productive and easily accessible plains [26] were mostly transformed into giant fields (>30 ha; [44]). Between 1999 and 2018, the forest abandonment and canopy closing within river cells substantially slowed down, but significantly increased in settlement surroundings (inner cells; Figure 2a,b and Figure A2b). Our results, therefore, showed that the loss of the historical state of dynamic equilibrium between traditional management and natural dynamics was followed by abandonment, similar to the adjacent area of Lower Austria [70], and later by some regeneration processes, which reduced ecological complexity at the landscape scale [6] and negatively affected biodiversity (Figure 3).

When we included environmental characteristics in the form of Ellenberg indicator values (EIVs), the major driver of species richness in the first survey was still the river phenomenon (Figure 7a). Additional important variables were represented by meadows and nutrients. Meadows were important outside the river valley because grassland specialist species are rare in agricultural [24] and forested landscapes. On the other hand, nutrients played a big role in the river valley, where the most species-rich grid cells contained lower nutrients. This was typical, especially for grid cells with the majority of grassland and wetland species pools [16]. Within the nutrient-rich river cells containing fewer grassland and wetland species, light availability was likely decisive in limiting the richness of forest species [18,71]. The second survey analysis showed a substantial shift, with the nutrients content (EIV) becoming the major driver of species richness, while the river phenomenon dropped into the second branch, together with soil reaction (Figure 7b). It is alarming that the availability of nutrients (the C:N ratio was a proxy for calibrated EIVs; [51]) became the major species richness driver, even in the bilateral Podyjí and Thayatal NPs, where N deposition is one of the lowest in the whole Czech Republic [72]. These findings are nonetheless in agreement with Gallego-Zamorano et al. [73], who found a considerable impact of N deposition on species richness in some regions of Europe. Their findings also show that Europe is currently the most impacted continent, with an average decline in species richness of 34% caused by a combination of land-cover changes and N deposition, despite the trend of decreasing N deposition in Europe (for about the last 30 years; [74]). However, influences of N deposition are not easy to disentangle because of strong linkages to abandonment within the landscape scale since N deposition accelerates overgrowing. The negative N deposition impact on species richness in the Podyjí/Thayatal NPs was documented by the decline of heathlands and acidic dry grasslands [75]. Their occurrence is conditioned by nutrient limitations [76] and their species richness decline was observed across Europe [77]. Although nutrient availability (EIV) increased in both the inner and river grid cells (Figure A2a), the grid cell species richness did not significantly decrease (Figure 4a), despite indications of a declining trend (Figure 3). This could have been caused by a combination of the increase in neophytes in both inner and river cells (Figure 4d), appropriate conservation management and several biodiversity hotspots with well-buffered soils of high pH (Figure 7b). In such hotspots, the N-deposition-driven decline of sensitive species can be slow and additional species might even colonise them [78]. The well-buffered high-pH soils are also responsible for the fact that the most species-rich grid cells in the second survey are those with lower nutrient content and higher soil reaction (Figure 7b). In nutrient-rich grid cells mostly outside the river valley, light availability is likely important in determining species richness (see above; Figure 7b).

5. Conclusions

We found that in a heterogeneous landscape, historical land cover had the largest impact on species composition. At the same time, the effect of historical land cover was higher than the present land cover; therefore, we can expect that the present land cover will affect the species composition in the near future.

The species richness was driven by the river phenomenon more than 20 years ago and also at present, even though the landscape around settlements in the protected areas was under stronger human impact than in the river valley and resulted in a rich mosaic of land covers supporting species richness. Some near-natural land-cover classes with appropriate management could harbour and ease the spread of many valuable native species. Enlarging their area of occurrence is important for preventing genetic corrosion and partly substitute for overgrowing and the loss of their original habitats. On the other hand, many endangered species are very limited in their spread and are closely connected to unique localities with a continuous history, and thus, require appropriate protection and management.

Especially in well-preserved protected areas, it is desirable to restore disappearing land-cover classes with traditional or compensatory management to retain their plant species richness, which is mostly a key factor to preserve biodiversity. However, management plans should take into account the increasing amount of nitrogen in soil derived from long-term continual deposition because, combined with land-cover changes, it has a negative impact on species richness. It is necessary to highlight that soil nutrients strongly affect the species richness, even in NPs with low current N deposition, because the effects of long-term human impacts through land use/cover and N deposition are not only restricted to local scales but also operate mostly on regional and even global scales.

Author Contributions: Conceptualization—R.N., H.S. and M.V.; methodology—M.V. and R.N.; data collection of vascular plants (1997, 2021)—R.N.; analyses—M.V. and R.N.; map visualization—R.N.; paper writing—M.V., R.N. and H.S.; review and editing—R.N. and H.S. All authors have read and agreed to the published version of the manuscript.

Funding: This paper was prepared with support from the Síťové mapování cévnatých rostlin NP Podyjí/Thayatal (ConNat ATCZ045 v Programu INTERREG V-A Rakousko—Česká republika) at the Silva Tarouca Research Institute for Landscape and Horticulture with the help of Technological Agency of the Czech Republic (grant no. SS02030018—Centre for Landscape and Biodiversity).

Institutional Review Board Statement: Not applicable.

Informed Consent Statement: Not applicable.

Data Availability Statement: Not applicable.

Acknowledgments: The authors would like to thank Petr Filippov, Martin Musil, Zuzana Němcová, Radomír Řepka and the anonymous reviewers who have made significant contributions to the quality of the paper.

Conflicts of Interest: The authors declare no conflict of interest.

Appendix A

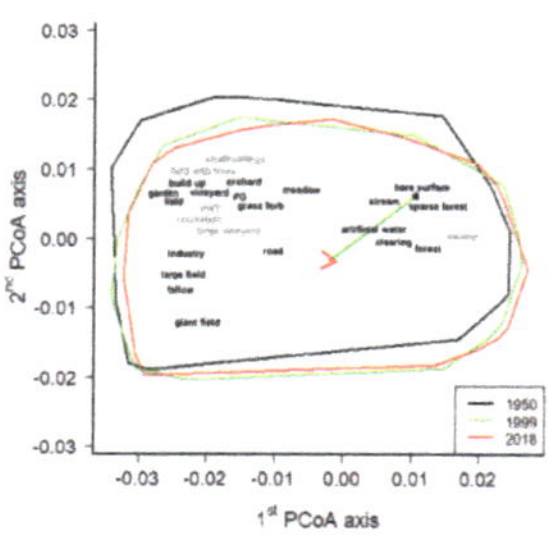

Figure A1. The principal coordinates analysis (PCoA) of land-cover data (proportions were square-rooted and the Bray–Curtis dissimilarity and Hellinger distance used) of all 115 grid cells in the Podyjí/Thayatal NPs detected in the years 1950, 1999 and 2018. Envelopes were drawn around the marginal grid cells in the PCoA diagram. The coloured arrow shows the shift of land-cover medoids between the searched years (1950, 1999 and 2018). Individual land covers are projected as weighted averages; grey coloured land covers occurred in less than 30 grid cells; VWT—vineyards with trees, PG—public greenery.

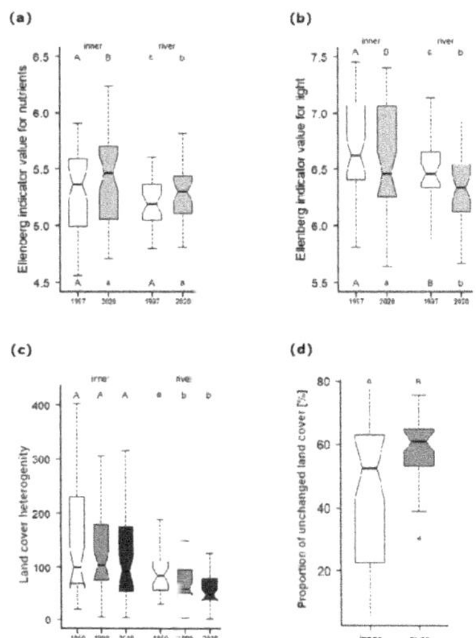

Figure A2. Changes and differences in the (**a**) Ellenberg indicator value (EIV) for nutrients, (**b**) EIV for light, (c) land-cover heterogeneity and (**d**) proportions of unchanged land cover (1950–2018) per grid cell within both the first (1997) and the second floristic survey (2021); or three land-cover searched periods (1950, 1999 and 2018) separated into inner (n = 44) and river (n = 34) group cells (see Methods). The letters above boxes (in (a) and (b)) indicate the results of pairwise t-tests between surveys within each group, while the letters below boxes show the results of non-pair *t*-tests between the groups. However, the letters above the land-cover heterogeneity boxes (**c**) show results based on the Friedman test, and when significant ($p < 0.001$), the post hoc Wilcoxon pairwise test with a Bonferroni correction ($p < 0.01$) was applied for the three periods: 1950, 1999 and 2018. Differences between proportions of unchanged land cover (**d**) were assessed using ANOVA repeated measures analysis (if $p < 0.001$, then Tukey's post hoc tests followed). Whiskers indicate the non-outlier range with the median as the middle; boxes represent 25–75% percentiles.

References

1. Magyari, E.K.; Chapman, J.; Fairbairn, A.S.; Francis, M.; Guzman, M. Neolithic human impact on the landscapes of North-East Hungary inferred from pollen and settlement records. *Veg. Hist. Archaeobotany* **2012**, *21*, 279–302. [CrossRef]
2. Dahlström, A.; Cousins, S.A.O.; Eriksson, O. The history (1620–2006) of land use, people and livestock, and the relationship to present plant species diversity in a rural landscape in Sweden. *Environ. Hist.* **2006**, *12*, 191–212. [CrossRef]
3. Vives, G.S.; Miras, Y.; Riera, S.; Julia, R.; Allée, P.; Orengo, H.; Paradis-Grenouillet, S.; Palet, J.P. Tracing the land use history and vegetation dynamics in the Mont Lozere (Massif Central, France) during the last 2000 years: The interdisciplinary study case of Countrasts peat bog. *Quart. Int.* **2014**, *353*, 123–129. [CrossRef]
4. Berglund, B.E.; Gaillard, M.J.; Björkman, L.; Persson, T. Long-term changes in floristic diversity in southern Sweden: Palynological richness, vegetation dynamics and land-use. *Veg. Hist. Archaeobotany* **2008**, *17*, 573–583. [CrossRef]
5. Skokanová, H.; Havlíček, M.; Unar, P.; Janík, D.; Šimeček, K. Changes of Ortolan Bunting (*Emberiza hortulana* L.) habitats and implications for the species presence in SE Moravia, Czech Republic. *Pol. J. Ecol.* **2016**, *64*, 98–112.
6. Amici, V.; Landi, S.; Frascaroli, F.; Rocchini, D.; Santi, E.; Chiarucci, A. Anthropogenic drivers of plant diversity: Perspective on land use change in a dynamic cultural landscape. *Biodivers. Conserv.* **2015**, *24*, 3185–3199. [CrossRef]
7. Bezák, P.; Petrovič, F. Agriculture, landscape, biodiversity: Scenarios and stakeholder perceptions in the Poloniny National Park (NE Slovakia). *Ekológia Bratisl.* **2006**, *25*, 82–93.
8. Hejcman, M.; Hejcmanová, P.; Pavlů, V.; Beneš, J. Origin and history of grasslands in Central Europe—A review. *Grass Forage Sci.* **2013**, *68*, 345–363. [CrossRef]
9. Hájková, P.; Roleček, J.; Hájek, M.; Horsák, M.; Fajmon, K.; Polák, M.; Jamrichová, E. Prehistoric origin of extremely species-rich semi-dry grasslands in the Bílé Karpaty Mts. *Preslia* **2011**, *83*, 185–204.
10. Michalcová, D.; Chytrý, M.; Pechanec, V.; Hájek, O.; Jongepier, J.W.; Danihelka, J.; Grulich, V.; Šumberová, K.; Preislerová, Z.; Ghisla, A.; et al. High plant diversity of grasslands in a landscape context: A comparison of contrasting regions in Central Europe. *Folia Geobot.* **2014**, *49*, 117–135. [CrossRef]

11. Johansson, L.J.; Hall, K.; Prentice, H.C.; Ihse, M.; Reitalu, T.; Sykes, M.T.; Kindström, M. Semi-natural grassland continuity, long-term land-use change and plant species richness in an agricultural landscape on Öland, Sweden. *Landsc. Urban Plan.* **200**, *84*, 200–211. [CrossRef]
12. Wilson, J.B.; Peet, R.K.; Dengler, J.; Pärtel, M. Plant species richness: The world records. *J. Veg. Sci.* **2012**, *23*, 796–802. [CrossRef]
13. Chytrý, M.; Dražil, T.; Hájek, M.; Kalníková, V.; Preislerová, Z.; Šibík, J.; Ujházy, K.; Axmanová, I.; Bernátová, D.; Blanár, D.; et al. The most species-rich plant communities in the Czech Republic and Slovakia (with new world records). *Preslia* **2015**, *87*, 217–278.
14. Škodová, I.; Hájek, M.; Chytrý, M.; Jongepierová, I.; Knollová, I. Vegetace (Vegetation). In *Louky Bílých Karpat (Grasslands of the White Carpathian Mountains)*; Jongepierová, I., Ed.; ZO ČSOP Bílé Karpaty: Veseli nad Moravou, Czech Republic, 2008; pp. 128–177.
15. Škodová, I.; Devánová, K.; Senko, D. Subxerophilous and mesophilous grasslands of the Biele Karpaty Mts (White Carpathian Mts) in Slovakia. *Tuexenia* **2011**, *31*, 235–269.
16. Cornwell, W.K.; Grubb, P.J. Regional and local patterns in plant species richness with respect to resource availability. *Oikos* **2003**, *100*, 417–428. [CrossRef]
17. Merunková, K.; Chytrý, M. Environmental control of species richness and composition in upland grasslands of the southern Czech Republic. *Plant Ecol.* **2012**, *213*, 591–602. [CrossRef]
18. Axmanová, I.; Chytrý, M.; Zelený, D.; Li, C.F.; Vymazalová, M.; Danihelka, J.; Horsák, M.; Kočí, M.; Kubešová, S.; Lososová, Z.; et al. The species richness–productivity relationship in the herb layer of European deciduous forests. *Glob. Ecol. Biogeogr.* **2012**, *21*, 657–667. [CrossRef]
19. Vonlanthen, C.; Kammer, P.; Eugster, W.; Bühler, A.; Veit, H. Alpine vascular plant species richness: The importance of daily maximum temperature and pH. *Plant Ecol.* **2006**, *184*, 13–25. [CrossRef]
20. Chytrý, M.; Tichý, L. Phenological mapping in a topographically complex landscape by combining field survey with an irradiation model. *Appl. Veg. Sci.* **1998**, *1*, 225–232. [CrossRef]
21. Pyšek, P.; Prach, K. Plant invasions and the role of riparian habitats: A comparison of four species alien to central Europe. *J. Biogeogr.* **1993**, *20*, 413–420. [CrossRef]
22. Zelený, D.; Chytrý, M. Environmental control of vegetation pattern in deep river valleys of the Bohemian Massif. *Preslia* **2007**, *79*, 205–222.
23. Williams, J.J.; Newbold, T. Local climatic changes affect biodiversity responses to land use: A review. *Divers. Distrib.* **2020**, *26*, 76–92. [CrossRef]
24. Cousins, S.A.; Eriksson, O. The influence of management history and habitat on plant species richness in a rural hemiboreal landscape, Sweden. *Landsc. Ecol.* **2002**, *17*, 517–529. [CrossRef]
25. Spiegelberger, T.; Matthies, D.; Müller-Schärer, H.; Schaffner, U. Scale-dependent effects of land use on plant species richness of mountain grassland in the European Alps. *Ecography* **2006**, *29*, 541–548. [CrossRef]
26. Tasser, E.; Tappiner, U. Impact of land use changes on mountain vegetation. *Appl. Veg. Sci.* **2002**, *5*, 173–184. [CrossRef]
27. Amici, V.; Santi, E.; Filibeck, G.; Diekman, M.; Geri, F.; Landi, S.; Scoppola, A.; Chiarucci, A. Influence of secondary forest succession on plant diversity patterns in a Mediterranean landscape. *J. Biogeogr.* **2013**, *40*, 2335–2347. [CrossRef]
28. Gustavsson, E.; Lennartsson, T.; Emanuelsson, M. Land use more than 200 years ago explains current grassland plant diversity in a Swedish agricultural landscape. *Biol. Conserv.* **2007**, *138*, 47–59. [CrossRef]
29. Chytrý, M.; Vicherek, J. *Lesní Vegetace NP Podyjí*; Academia: Praha, Czech Republic, 1995; p. 166.
30. Roetzel, R.; Fuchs, G.; Batík, P.; Čtyroký, P.; Havlíček, P. *Geologische Karte der Nationalparks Thayatal und Podyjí / Geologická Mapa Národních Parků Thayatal a Podyjí, 1:50,000*; Geologische Bundesanstalt: Wien, Austria, 2005.
31. Tolasz, R. *Atlas Podnebí Česka*; Czech Hydrometeorological Institute: Praha, Czech Republic, 2007; p. 255.
32. Chytrý, M.; Grulich, V.; Tichý, L.; Kouřil, M. Phytogeographical boundary between the Pannonicum and Hercinicum: A multivariate analysis of landscape in the Podyji/Thayatal National Park, Czech Republic/Austria. *Preslia* **1999**, *71*, 23–41.
33. Stejskal, R. Nosatcovití Brouci (Coleoptera, Curculionoidea) ve Vybraných Lesních Geobiocenózách Národního Parku Podyjí. Ph.D. Thesis, MENDELU, Brno, Czech Republic, 2006; p. 123.
34. Šumpich, J. *Motýli Národních parků Podyjí a Thayatal. Die Schmetterlinge der Nationalparke Podyjí und Thayatal*; Správa Národního parku Podyjí: Znojmo, Czech Republic, 2011; p. 428.
35. Myšák, J.; Horsák, M. Biodiversity surrogate effectiveness in two habitat types of contrasting gradient complexity. *Biodivers. Conserv.* **2014**, *23*, 1133–1156. [CrossRef]
36. Němec, R. (Ed.) *Rozšíření Cévnatých Rostlin Národních Parků Podyjí a Thayatal*; Správa Národního parku Podyjí: Znojmo, Czech Republic, 2021; p. 399.
37. Zulka, K.P.; Gilli, C.; Paternoster, D.; Banko, G.; Schratt-Ehrendorfer, L.; Niklfeld, H. *Die Bedeutung der Österreichischen Nationalparks für den Schutz, Die Bewahrung und das Management von Gefährdeten, Endemischen und Subendemischen Arten und Lenensräumen*; Umweltbundesaumt: Wien, Austria, 2021; p. 259.
38. Podborský, V.; Vildomec, V. *Pravěk Znojemska*; Jihomoravské muzeum ve Znojmě: Znojmo, Czech Republic, 1972; p. 282.
39. Kovárník, J. Výsledky terénního archeologického průzkumu na Znojemsku (okr. Znojmo). *Přehled Výzkumů* **1983**, *28*, 100–102.
40. Kovárník, J. Nové archeologické nálezy ze Znojemska a Třebíčska (okr. Třebíč, Znojmo). *Přehled Výzkumů* **1993**, 128–131.
41. Kovárník, J. Další archeologické nálezy ze Znojemska a Třebíčska (okr. Třebíč, Znojmo). *Přehled Výzkumů* **1993**, *35*, 115–126.

2. Čižmář, Z. Mašovice—"Pšeničné" (okr. Znojmo). In *Život a Smrt v Mladší Době Kamenné*; Čižmář, Z., Ed.; Ústav archeologické a památkové péče: Brno, Czech Republic, 2008; pp. 126–143.
3. Vrška, T.; Janík, D.; Pálková, M.; Adam, D.; Trochta, J. Below-and above-ground biomass, structure and patterns in ancient lowland coppices. *iFor. BiogeoSci. For.* **2016**, *10*, 23–31. [CrossRef]
4. Skokanová, H.; Musil, M.; Havlíček, M.; Pokorná, P.; Svobodová, E.; Zýka, V. Analýza vývoje krajiny národních parků Podyjí a Thayatal a jejich okolí. *Thayensia* **2021**, *18*, 3–46.
5. Grulich, V. *Atlas Rozšíření Cévnatých Rostlin Národního parku Podyjí/Thayatal. Verbreitungsatlas der Gefäßpflanzen des Nationalparks Podyjí/Thayatal*; Masarykova Univerzita: Brno, Czech Republic, 1997; p. 297.
6. Kaplan, Z.; Danihelka, J.; Chrtek, J.; Kirschner, J.; Kubát, K.; Štech, M.; Štěpánek, J. (Eds.) *Klíč ke Květeně České Republiky*, 2nd ed.; Academia: Praha, Czech Republic, 2019; p. 1168.
7. Miklín, J.; Miklínová, K.; Čížek, L. Změny krajinného krytu na území národního praku Podyjí mezi lety 1938 a 2014. *Thayensia* **2016**, *13*, 59–80.
8. Geologische Bundesanstalt. Geologische Karte der Republik Österreich 1:50.000. Available online: https://gisgba.geologie.ac.at/gbaviewer/?url=https://gisgba.geologie.ac.at/arcgis/rest/services/KM50/AT_GBA_KM50_GE_LS99/MapServer (accessed on 16 February 2022).
9. Pyšek, P.; Danihelka, J.; Sádlo, J.; Chrtek, J., Jr.; Chytrý, M.; Jarošík, V.; Kaplan, Z.; Krahulec, F.; Moravcová, L.; Pergl, J.; et al. Catalogue of alien plants of the Czech Republic (2nd edition): Checklist update, taxonomic diversity and invasion patterns. *Preslia* **2012**, *84*, 155–255.
0. Grulich, V. The Red List of vascular plants of the Czech Republic. *Příroda* **2017**, *35*, 75–132.
1. Chytrý, M.; Tichý, L.; Dřevojan, P.; Sádlo, J.; Zelený, D. Ellenberg-type indicator values for the Czech flora. *Preslia* **2018**, *90*, 83–103. [CrossRef]
2. Zelený, D.; Schaffers, A.P. Too good to be true: Pitfalls of using mean Ellenberg indicator values in vegetation analyses. *J. Veg. Sci.* **2012**, *23*, 419–431. [CrossRef]
3. Anderson, M.J. A new method for non-parametric multivariate analysis of variance. *Austral Ecol.* **2001**, *26*, 32–46.
4. Anderson, M.J.; Crist, T.O.; Chase, J.M.; Vellend, M.; Inouye, B.D.; Freestone, A.L.; Sanders, N.J.; Cornell, H.V.; Comita, L.S.; Davies, K.F.; et al. Navigating the multiple meanings of β diversity: A roadmap for the practicing ecologist. *Ecol. Lett.* **2011**, *14*, 19–28. [CrossRef] [PubMed]
5. Anderson, M.J. Distance-based tests for homogeneity of multivariate dispersions. *Biometrics* **2006**, *62*, 245–253. [CrossRef] [PubMed]
6. Breiman, L.; Friedman, J.; Olshen, R.; Stone, C. *Classification and Regression Trees*; CRC Press: Boca Raton, FL, USA, 1984.
7. Therneau, T.; Atkinson, B. rpart: Recursive Partitioning and Regression Trees. *R Package Version* **2019**, *4*, 1–15.
8. Wickham, H.; Averick, M.; Bryan, J.; Chang, W.; McGowan, L.D.; François, R.; Grolemund, G.; Hayes, A.; Henry, L.; Hester, J.; et al. Welcome to the tidyverse. *J. Open Source Softw.* **2019**, *4*, 1686. [CrossRef]
9. Kassambara, A. ggpubr: "ggplot2" Based Publication Ready Plots. R Package Version 0.4.0. Available online: https://CRAN.R-project.org/package=ggpubr (accessed on 1 March 2021).
0. Kassambara, A. rstatix: Pipe-Friendly Framework for Basic Statistical Tests. R package version 0.7.0. Available online: https://CRAN.R-project.org/package=rstatix (accessed on 1 March 2021).
1. Oksanen, J.; Guillaume, F.B.; Friendly, M.; Kindt, R.; Legendre, P.; McGlinn, D.; Minchin, P.; O'hara, R.; Simpson, G.; Solymos, P.; et al. Vegan: Community Ecology Package. R Package Version 2.2-0. Available online: https://www.scirp.org/(S(351jmbntvnsjt1aadkposzje))/reference/ReferencesPapers.aspx?ReferenceID=1778707 (accessed on 16 February 2022).
2. Maechler, M.; Rousseeuw, P.; Struyf, A.; Hubert, M. *Cluster: Cluster Analysis Basics and Extensions, R Package Version 2.0.7-1*; R Package Vignette: Madison, WI, USA, 2021.
3. Milborrow, S. rpart.plot: Plot 'rpart' Models: An Enhanced Version of 'plot.rpart', R package version 3.0.9, 2020. Available online: https://cran.r-project.org/web/packages/rpart.plot/index.html (accessed on 16 February 2022).
4. Gilbert, O.L. *The Ecology of Urban Habitats*; Chapman & Hall: London, UK, 1989; p. 369.
5. Kuhn, T.; Domokos, P.; Kiss, R.; Ruprecht, E. Grassland management and land use history shape species composition and diversity in Transilvanian semi-natural grassland. *Appl. Veg. Sci.* **2021**, *24*, 1–12. [CrossRef]
6. Nordén, B.; Dahlberg, A.; Brandrud, T.E.; Fritz, Ö.; Ejrnaes, R.; Ovaskainen, O. Effects of ecological continuity on species richness and composition in forests and woodlands: A review. *Ecoscience* **2014**, *21*, 34–45. [CrossRef]
7. Galvánek, D.; Lepš, J. Changes of species richness pattern in mountain grasslands: Abandonment vs. restoration. *Biodivers. Conserv.* **2008**, *17*, 3241–3253. [CrossRef]
8. Plieninger, T.; Hui, C.; Gaertner, M.; Huntsinger, L. The Impact of Land Abandonment on Species Richness and Abundance in the Mediterranean Basin: A Meta-Analysis. *PLoS ONE* **2014**, *9*, e98355. [CrossRef]
9. Dreslerová, D.; Hájek, P.; Sádlo, J.; Pokorný, P.; Cílek, V. *Krajina a Revoluce. Významné Přelomy ve Vývoji Kulturní Krajiny Českých Zemí*; Malá skála: Praha, Czech Republic, 2008; p. 248.
0. Grass, A.; Tremetsberger, K.; Hössinger, R.; Bernhardt, K.-G. Change of species and habitat diversity in the Pannonian region of eastern Lower Austria over 170 years: Using herbarium records as a witness. *Nat. Resour.* **2014**, *5*, 583–596. [CrossRef]
1. Hofmeister, J.; Hošek, J.; Modrý, M.; Roleček, J. The influence of light and nutrient availability on herb layer species richness in oak-dominated forests in central Bohemia. *Plant Ecol.* **2009**, *205*, 57–75. [CrossRef]

72. Záhora, J.; Chytrý, M.; Holub, P.; Fiala, K.; Tůma, I.; Vavříková, J.; Fabšičová, M.; Keizer, I.; Filipová, L. The Effect of Nitrogen Accumulation on Heathlands and Dry Grasslands in the Czech Podyjí National Park. *Životné Prostr.* **2016**, *2*, 97–107.
73. Gallego-Zamorano, J.; Huijbregts, M.A.J.; Schipper, A.M. Changes in plant species richness due to land use and nitrogen deposition across the globe. *Divers. Distrib.* **2022**, *28*, 745–755. [CrossRef]
74. Ackerman, D.; Millet, D.B.; Chen, X. Global estimates of inorganic nitrogen deposition across four decades. *Glob. Biogeochem Cycles* **2019**, *33*, 100–107. [CrossRef]
75. Chytrý, M.; Vicherek, J. Travinná, keříčková a křovinná vegetace Národního parku Podyjí/Thayatal. *Thayensia* **2003**, *5*, 11–84.
76. Krahulec, F.; Chytrý, M.; Härtel, H. Smilkové trávníky a vřesoviště (*Calluno-Ulicetea*). In *Vegetace České Republiky. 1. Travinná a Keříčková Vegetace*; Chytrý, M., Ed.; Academia: Praha, Czech Republic, 2007; pp. 281–319.
77. Stevens, C.J.; Duprè, C.; Dorland, E.; Gaudnik, C.; Gowing, D.J.G.; Bleeker, A.; Diekmann, M.; Alard, D.; Bobbink, R.; Fowler, D.; et al. Nitrogen deposition threatens species richness of grasslands across Europe. *Environ. Pollut.* **2010**, *158*, 2940–2945. [CrossRef]
78. Van den Berg, L.J.L.; Vergeer, P.; Rich, T.C.G.; Smart, S.M.; Guest, D.; Ashmore, M.R. Direct and indirect effects of nitrogen deposition on species composition change in calcareous grasslands. *Glob. Change Biol.* **2011**, *17*, 1871–1883. [CrossRef]

land

MDPI

Article

Rapid Reclamation and Degradation of *Suaeda salsa* Saltmarsh along Coastal China's Northern Yellow Sea

Jing Zhang [1,2,3], Yan Zhang [1], Huw Lloyd [4], Zhengwang Zhang [5] and Donglai Li [1,*]

1 Provincial Key Laboratory of Animal Resource and Epidemic Disease Prevention, College of Life Sciences, Liaoning University, Shenyang 110036, China; jingzhangcug@yeah.net (J.Z.); 13842652675@163.com (Y.Z.)
2 State Key Laboratory of Urban and Regional Ecology, Research Center for Eco-Environmental Sciences, Chinese Academy of Sciences, Beijing 100085, China
3 University of Chinese Academy of Sciences, Beijing 100049, China
4 Ecology and Environment Research Centre, Department of Natural Sciences, Manchester Metropolitan University, Chester Street, Manchester M1 5GD, UK; H.Lloyd@mmu.ac.uk
5 Ministry of Education Key Laboratory for Biodiversity Science and Ecological Engineering, College of Life Sciences, Beijing Normal University, Beijing 100875, China; zzw@bnu.edu.cn
* Correspondence: lidonglai@lnu.edu.cn

Abstract: *Suaeda salsa* saltmarshes are an important coastal wetland habitat of China's northern Yellow Sea, which plays a critical role in sequestering carbon (blue carbon), protecting shorelines, maintaining biodiversity, and has substantial economic value (e.g., ecotourism). However, the area of *S. salsa* has been rapidly declining due to several different threats from reclamation and invasive species that impact its natural succession. Here, we map the changes in the distribution of the *S. salsa* saltmarshes along the northern Yellow Sea of China (NYSC) at 5-year intervals by applying the supervised maximum likelihood method to analyze Landsat images from 1988 to 2018 and investigate the potential impact of three important factors on habitat change by analyzing the temporal changes in *S. salsa* saltmarshes with other land covers. *S. salsa* saltmarsh areas have decreased by 63% (264 km^2 ha to 99 km^2), and the average loss of *S. salsa* saltmarshes was 5.5 km^2/year along the NYSC over the past three decades. There have been many dramatic declines in the two main distribution areas of *S. salsa* saltmarshes with a 77% loss of habitat area in Liaodong Bay (from 112 km^2 to 26 km^2) and a 52% loss in the Yellow River Delta wetland-Guangli-Zhima estuarine wetland (from 137 km^2 to 65 km^2). Land reclamation is the most important impact factor in the loss of *S. salsa* saltmarshes, while there have been limited effects of natural succession and smooth cordgrass (*Spartina alterniflora*) invasion. In light of the important ecological services and economic value of the *S. salsa* habitat, emergency conservation actions (e.g., habitat restoration, strictly supervision) are needed to limit the rapid habitat loss, which should include the immediate cessation of extensive land reclamation along the NYSC.

Keywords: *Suaeda salsa* saltmarsh; land cover change; land reclamation; mudflat; *Spartina alterniflora*; Yellow Sea

1. Introduction

Coastal wetlands, which are composed by multiple natural landscapes connecting mainland and oceanic ecosystems, provide hugely significant ecological services (e.g., buffering ocean storms [1], shoreline protection against rising sea-levels, and maintaining biodiversity [2]) and have critical economic value from the contribution of their vast marine biological resources [3]. However, in recent decades, coastal wetlands have been under intensive pressure worldwide from land reclamation [4,5], reduced stream discharge and sediment fluxes from rivers [4,6], erosion, and sea-level rise [7]. Coastal wetland reclamation by seawall construction and other industrial and aquaculture land uses are the primary drivers for coastal wetland loss, resulting in a 16% loss of coastal tidal flats globally from 1984 to 2016 [5]. Reclamation has also indirectly negatively impacted local wetland ecosystems through soil compaction [8], heavy mental contamination, oil pollution, and biodiversity loss [9].

As one of the fifty largest marine ecosystems worldwide [10], the Yellow Sea, located in northeastern Asia, supports an estimated USD 30 billion in ecosystem services and approximately 60 million people across urban, industrial, agricultural and fishery sectors per year [11]. Given the variety of coastal wetland landscapes and the richness of marine benthic invertebrate resources, the Yellow Sea also supports more than 200 species of migratory waterbirds during stopovers and the breeding season along the East Asian–Australia Flyway [12,13]. Consequently, numerous protected areas and Ramsar sites have been established along the coast of the Yellow Sea [14], including the Yancheng National Nature Reserve, which is located near the southern Yellow Sea and is listed as the UNESCO World Heritage site [15]. The unprecedented urban, industrial, and agricultural land reclamation has however, resulted in a loss of 65% of tidal flats during the past five decades [16], increasing the vulnerability of local communities to cope with extreme weather events, especially with predicted climate warming and sea level rise. The reclamation of China's coastal wetlands is estimated to have caused a loss of ecosystem services and internationally shared biodiversity [2,17]. To improve the protection level for coastal wetlands and to provide strict controls for reclamation management, in 2018, the State Council of the People's Republic of China announced a respite to claiming coastal wetland for creating extra land, particularly for the more ecologically sensitive coastal wetlands distributed in the northern Yellow Sea of China (http://www.mnr.gov.cn/gk/tzgg/201807/t20180726_2187020.html; accessed on 1 August 2021).

Tidal mudflats are the dominant coastal wetland habitat in the Yellow Sea region and are among the most extensive coastal wetland habitats globally [16,18]. These mudflats nourish countless macro-benthic invertebrates and are consequently used by millions of migratory waterbirds, including globally threatenrd species (e.g., Spoon-billed Sandpiper (*Calidris pygmaea*) [19], Bar-tailed Godwit (*Limosa lappnica*), and Great Knot (*C. tenuirostris*) [20]) for feeding but are also used by millions of people for clam production [12,13,21]. *Suaeda salsa* saltmarshes, which are distributed in the upper tidal region, represents another important wetland landscape component in the northern Yellow Sea [8,22], serving as a key ecotourism destintion ("red beach tourism") and an economic driver in areas such as Liaodong Bay. As an indication of this, from 2016 to 2020, the local government has paid RMB 0.3 billion to restore *S. salsa* saltmarshes and other natural marine ecosystems in an effort to maintain the "red beach landscape" [8].

S. salsa saltmarshes also provide important ecosystem services, such as carbon sequestration("blue carbon"), shoreline protection, and biodiversity maintenance [23,24], and serves as an important foraging and resting habitat for endangered waterbirds that migrate along the EAAF, including the Red-crowned Crane (*Grus japonensis*) [25,26], the Far Eastern Curlew (*Numenius madagascariensis*), and the Great Knot [22,27]. However, *S. salsa* saltmarshes also face serious threats from reclamation, lower runoff of river discharge, uncontrolled tourist development, oil exploitation, invasive species, and sea-level rise [8,24]. Due to these threats, during 1997 to 2016, the *S. salsa* saltmarsh area on the Liao River Estuary has been dramatically reduced and heavily fragmented [6,24], with similar declines reported from the Yellow River Delta from 1970 to 2015 [28]. Furthermore, because of their proximity to the wetland landscape mosaic in the upper tidal flats, *S. salsa* saltmarshes are naturally exposed to more serious reclamation effects than other mudflats. This loss and fragmentation is also having detrimental impacts on the population of shorebirds that depend on the *S. salsa* saltmarshes. Despite these losses, there is still a surprising yet significant gap in our knowledge regarding the patterns of spatial–temporal variation of these coastal wetland habitats across the northern Yellow Sea over the past several decades, thereby limiting our conservation evidence base, which is necessary for creating viable and realistic *S. salsa* saltmarsh restoration strategies at the regional or/and national government level.

In order to fill these gaps, we mapped the historic spatial–temporal distribution and more recent changes to *S. salsa* saltmarsh distribution along the northern Yellow Sea of China from 1988 to 2018 at intervals of five years and calculated the change in the *S. salsa* saltmarsh distribution change rates from the interpretation of Landsat long time-series images. Moreover, in order to understand the reclamation and degradation of *S. salsa* saltmarshes in this region, we analyzed the temporal transfer into and out of *S. salsa* saltmarshes with other land cover types.

2. Materials and Methods

2.1. Study Site

The northern Yellow Sea of China (NYSC) is a semi-closed shallow sea over the continental shelf of eastern Asia, with an average water depth of 26 m [29]. The NYSC extends from the Yalu River estuary (40°59′ N, 124°19′ E) in the north to the Shandong Peninsula (37°25′ N, 122°41′ E) in the south (Figure 1). The tides around this region are typically semidiurnal and diurnal in some places, ranging from 1.5 to 8 m but are approximately 2–4.6 m. The main coastal types include both cliff and alluvial coasts along the NYSC. The cliff coast is distributed along the Liaodong and Shandong Peninsula with no saltmarsh habitat, with the remaining coast being alluvial coast mudflat nourished by material exchanged between river sediment and the ocean. In this study, only the alluvial coast was investigated. As a species of salt-related vegetation, *S. salsa* can grow on large areas of the upper tidal flat where it is influenced by the influx of freshwater from more than 30 rivers in this area [29,30]. This habitat can frequently be submerged and exposed during floods and ebb tides. In recent decades, smooth cordgrass (*Spartina alterniflora*) has invaded and become established in the *S. salsa* habitat along the south of the NYSC [31,32].

2.2. Landsat Images and Pre-Processing to Analyze Land Cover Changes in NYSC

We took the DESCR conceptual framework [33] as a basis for this work to identify the drivers, exchanges, state of the environment, consequences, and responses of land loss. We downloaded the Landsat-5 Thematic Map (1988–2008, 35 images) and the Landsat-8 Operational Land Imager (2013–2018, 18 images) remote sensing images from the United States Geological Survey website (USGS; https://earthexplorer.usgs.gov; accessed on 1 May 2020) and the Geospatial Data Cloud website (https://www.gscloud.cn; accessed on 1 May 2020), respectively. The highest quality remote sensing images with <2% cloud cover from mid-August to early November were selected because the spectral signatures of *S. salsa* and other vegetation (e.g., reed (*Phragmites australis*), smooth cordgrass, rice) are more similar to each other in other seasons (e.g., spring and summer) [6].

2.3. Extracting *S. salsa* Saltmarsh Distribution Data

A total of 63 images with seven periods (1988, 1993, 1998, 2003, 2008, 2013, 2018) were selected and screened to quantify land cover change. Artificial shorelines are formed by many types of ground features, such as sea dikes, roads, aquaculture ponds, saltpans, and cliffs. Given their complex structures, we delineated the artificial shorelines through visual interpretation with 1988 images and defined reclamation as the change in the area between the artificial shorelines during successive periods. According to the shapefile of the artificial shorelines along the NYSC, we clipped all of the images after Radiometric Calibration and FLAASH atmospheric correction using ENVI 5.3. Given the typical red coloration and distribution along the upper tidal region along the NYSC, *S. salsa* was easily distinguished from the other eight land cover types that were able to be identified: mudflat, reed, bare land, water cover, farmland, smooth cordgrass, reclaimed land, and forest (Figure 2). Finally, the supervised maximum likelihood method and visual interpretation were used to describe the land cover changes according to the Regions of Interests (ROIs) that we marked on the RS images from 1988 to 2018. Overall, 498 *S. salsa* ROIs and 9476 ROIs of other land cover types were randomly selected as training samples. Furthermore, on account of the influence of small patches on statistical power and mapping, Majority Analysis was used to coalesce the results with other land covers. All of the interpretations of the RS images were performed with ENVI 5.3.

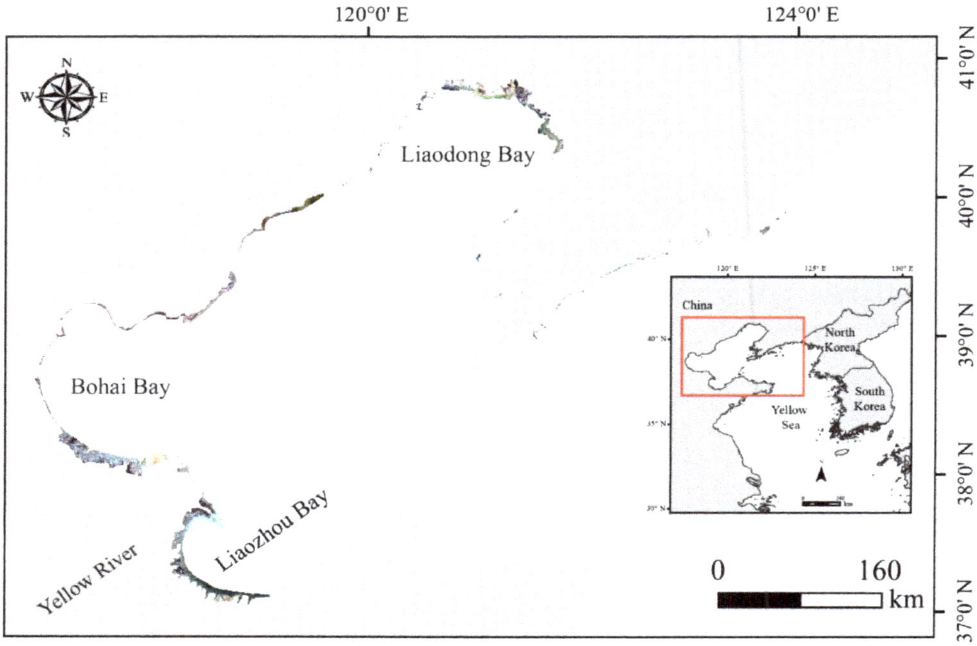

Figure 1. Location of the Northern Yellow Sea of China with subset RS images of 1988.

2.4. Accuracy Assessment

The accuracy of the final classification of *S. salsa* saltmarshes was assessed using geographical coordinates collected from field surveys in 2020 (125 and 2369 points in the *S. salsa* and other land cover types, respectively) and high spectral resolution images from 2013 to 2018 from Google Earth. A confusion matrix consisting of the overall accuracy and the kappa coefficient was created to measure the consistency between our classification and the validation samples. The classification for the estimated overall accuracy was >89.95%, and the kappa coefficient was 0.83.

2.5. Land Cover Change of *S. salsa* Saltmarsh

We used Class Statistics to calculate the area of *S. salsa* vegetation and other land classifications along the NYSC coast in ENVI 5.3 software. To explore the temporal and spatial change of *S. salsa* saltmarshes from 1988 to 2018, Change Detection Statistics were used to calculate the area of land cover transferred into and out of *S. salsa* saltmarshes with reclamation area, mudflat, reed, smooth cordgrass, and other land cover types at five-year periods (1988–1993; 1993–1998; 1998–2003; 2003–2008; 2008–2013; 2013–2018). The average change rates of different land cover types were calculated by dividing the total change in area by the periods in which the changes occurred. Maps of *S. salsa* saltmarshes, reclamation, area and mudflats for all stages were overlaid into one shapefile;the temporal and spatial distribution maps of three land cover types from 1988 to 2018 were then drawn using ArcGIS 10.2 (ESRI, Inc., Redlands, CA, USA).

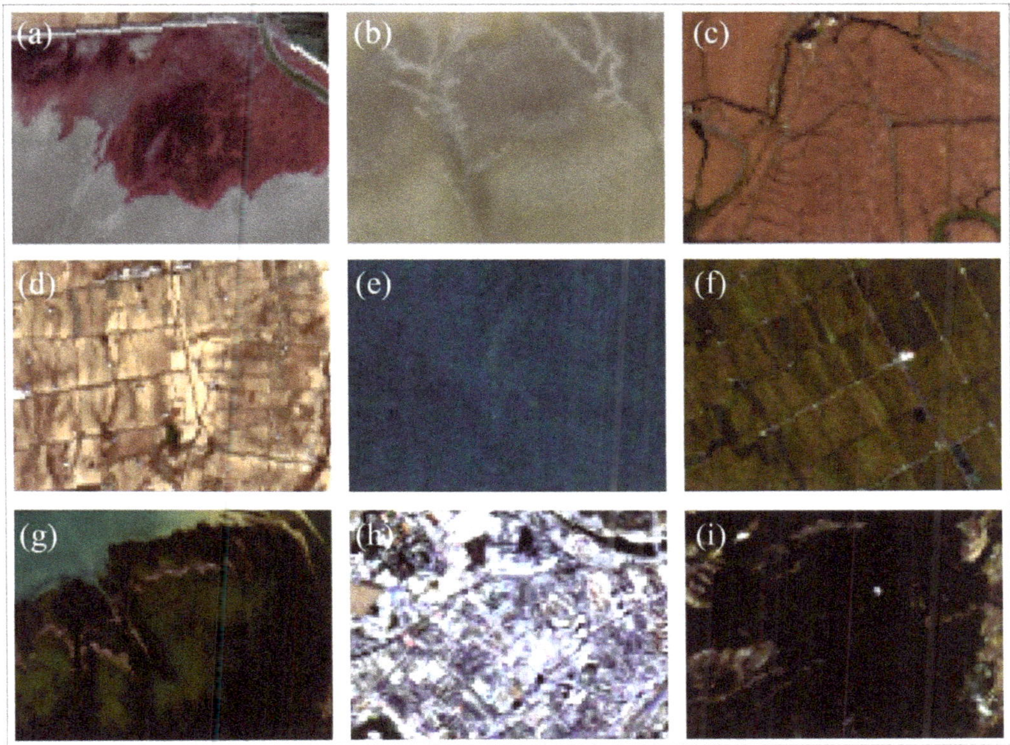

Figure 2. The nine different types of land cover in the Northern Yellow Sea of China: (**a**) *S. salsa*; (**b**) mudflat; (**c**) reed (*Phragmites. australis*); (**d**) bare land; (**e**) water cover; (**f**) farmland; (**g**) smooth cordgrass (*Spartina alterniflora*); (**h**) reclaimed land; (**i**) forest.

3. Results

3.1. Temporal and Spatial Distribution of S. salsa Saltmarsh

In 1988, the extent of the *S. salsa* habitat was estimated to be 263 km^2 (Figure 3), it was and mainly distributed in the Liaodong Bay (Liaohe-Xiaolinghe estuarine wetland, 112 km^2), the Yellow River Delta wetland, and the Guangli-Zhima estuarine wetland (137 km^2) in northern Laizhou Bay. In addition, 6% of the *S. salsa* habitat was located in other estuarine wetlands with lower annual runoff (Figure 4). During the last three decades, the area composed of *S. salsa* saltmarshes dramatically decreased by 63%, from 263 km^2 to 99 km^2, with an average loss rate of 5.5 km^2/year along the NYSC. There were two significant periods of change in the extent of *S. salsa* saltmarsh loss: the *S. salsa* saltmarsh area decreased dramatically by 55% from 263 km^2 to 118 km^2 from 1988 to 1993 and then further declined by 16% from 1993 to 2018 although there were obvious increases detected in 1998 and 2008.

In the two core areas of *S. salsa* saltmarsh distribution, Liaodong Bay suffered the most dramatic decline, with a net loss of 77% from 112 km^2 to 26 km^2 (Figure 5), and in the Yellow River Delta and Guangli-Zhima estuarine wetlands, the *S. salsa* habitat decreased by 52%, from 137 km^2 to 65 km^2 (Figure 6). From 1988 to 2018, the *S. salsa* saltmarsh area in Huludao city, Jinzhou City, Yingkou City, the eastern part of Panjing City, the Guangli-Zhima estuarine wetland, and u thepper tidal flat of Liaohekou National Nature Reserve almost disappeared entirely. However, the *S. salsa* saltmarsh area in the Yellow

River Delta wetland where the Yellow National Nature Reserve is located experienced no significant change (Figures 5 and 6).

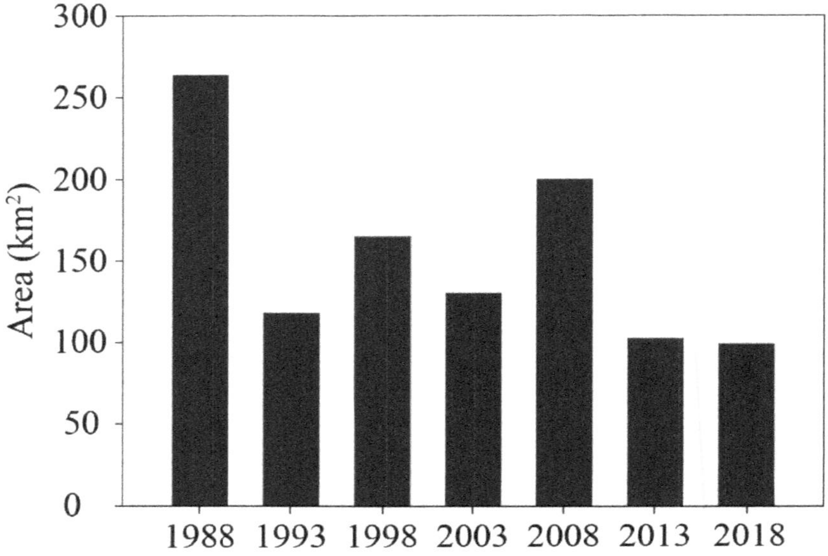

Figure 3. Total *S. salsa* saltmarsh area at different stages between 1988 and 2018 along the NYSC.

3.2. Effects of Reclamation and Natural Succession on S. salsa Saltmarsh Dynamics

Land reclamation was the main reason for the decline of the *S. salsa* saltmarsh areas and is responsible for a loss of 111 km^2 of the whole NYSC region (Figures 4 and 7), corresponding to 78 km^2 in Liaodong Bay (Figure 8) and 19 km^2 in the Yellow River Delta and Guangli-Zhima estuarine wetlands (Figure 9). In turn, there was some degree of natural succession between the *S. salsa* saltmarsh and mudflat areas with the total net area of 3 km^2 the *S. salsa* saltmarsh areas succeeding into mudflats from 1988 to 2018. There were different patterns between regions, with 14 km^2 of mudflat succeeding into *S. salsa* saltmarsh in Liaodong Bay and 35 km^2 of *S. salsa* saltmarsh succeeding into mudflats in the Yellow River Delta and Guangli-Zhima estuarine wetlands. The highly invasive smooth cordgrass was first found in the Yellow River Delta wetland and the Guangli-Zhima estuarine wetlands (19 km^2) and in the Tianjin Coastal wetland (4 km^2) in 1993 and 1998, respectively. During the last several decades, only 1 km^2 *S. salsa* saltmarsh area was entirely lost to smooth cordgrass invasion, which mainly occupied the *S. salsa* saltmarsh area in the Yellow River Delta wetland and the Guangli-Zhima estuarine wetland (0.49 km^2; Figure 9).

4. Discussion

In this study, we first mapped the historical spatial–temporal distribution of *S. salsa* saltmarsh areas near the NYSC from 1988 to 2018 by using a long time-series of Landsat images, which revealed that Liaodong Bay and the Yellow River Delta wetland—Guangli-Zhima estuarine wetlands are the core distribution regions of *S. salsa* saltmarsh areas (95%), which remains true to present day. Consistent with other findings on the loss of coastal wetlands, there has been an extensive loss of *S. salsa* saltmarsh areas on the coast of the NYSC during the past three decades, primarily driven by land reclamation [4,16,24]. During the past several decades, significant areas of coastal wetlands have been lost due to human reclamation and natural changes; for example, tidal flats have declined by

16.02% since 1984, and mangroves have decreased by 1% per year on average, whereas mudflat accretion has converted to erosion during the past decade because of the decline of coastal sediment supplied by the nearby rivers [4,5,34]. Notably, along the coast of the Yellow Sea, 49% of mudflat areas and 67% of saltmarsh areas have been lost from 1984 to 2015. Meanwhile, the land reclamation areas have increased by 7696 km^2, which has resulted in the decrease of numerous waterbird stopover sites in this region [4]. Further declines are likely in the near future if the recent levels of reclamation continue.

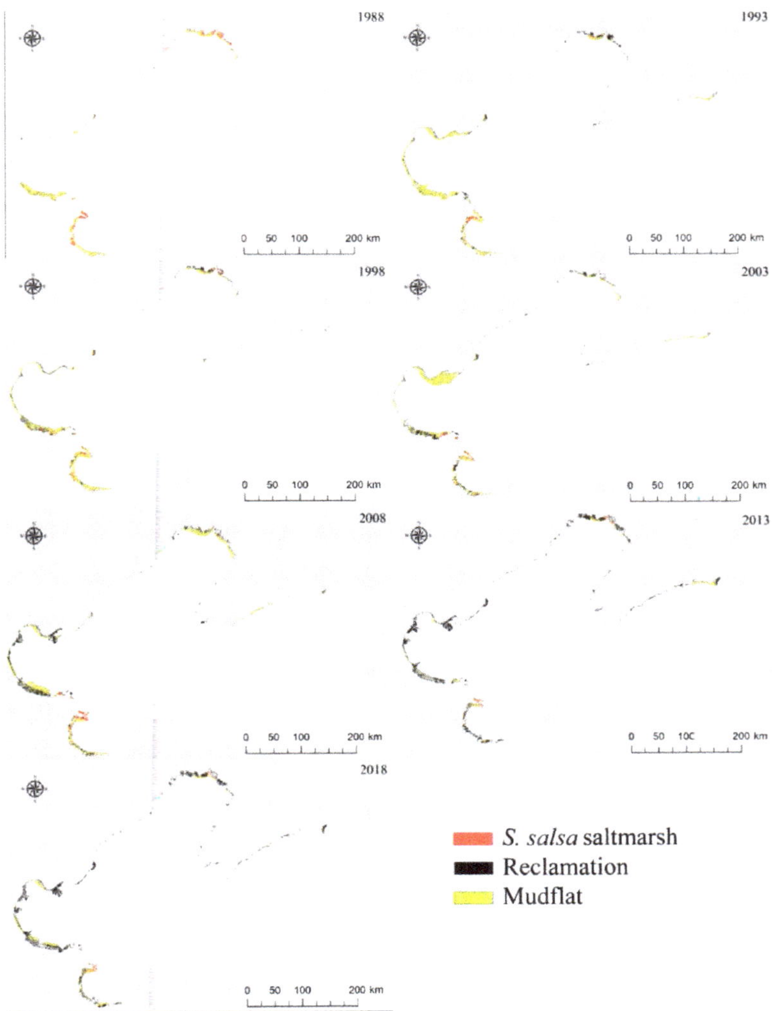

Figure 4. Maps showing the land cover change of the *S. salsa* saltmarsh area and mudflats and the reclaimed land along the NYSC between 1988 and 2018.

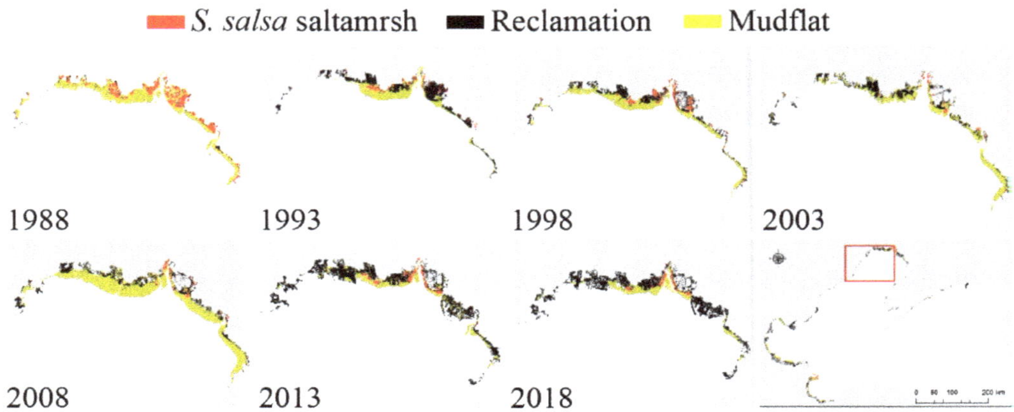

Figure 5. Maps showing the land cover change of *S. salsa* saltmarsh area and mudflats and the reclaimed land in Liaodong Bay between 1988 and 2018.

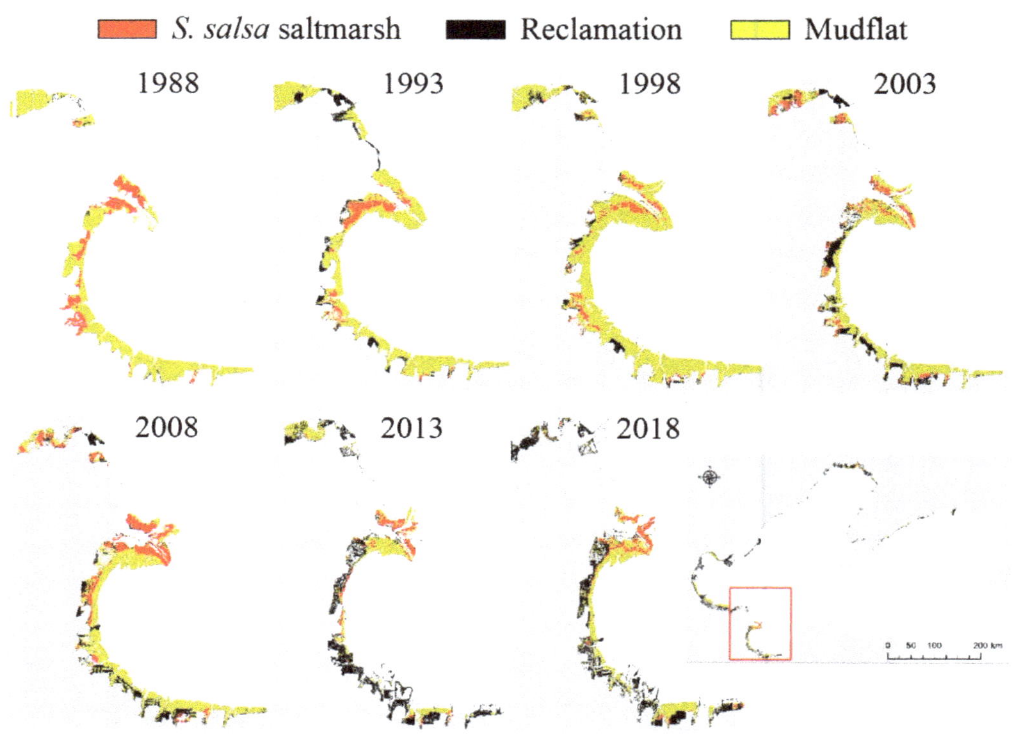

Figure 6. Maps showing the land cover change of *S. salsa* saltmarsh areas and mudflats and the reclaimed land in Yellow River Delta wetland and Guangli-Zhima estuarine wetland between 1988 and 2018.

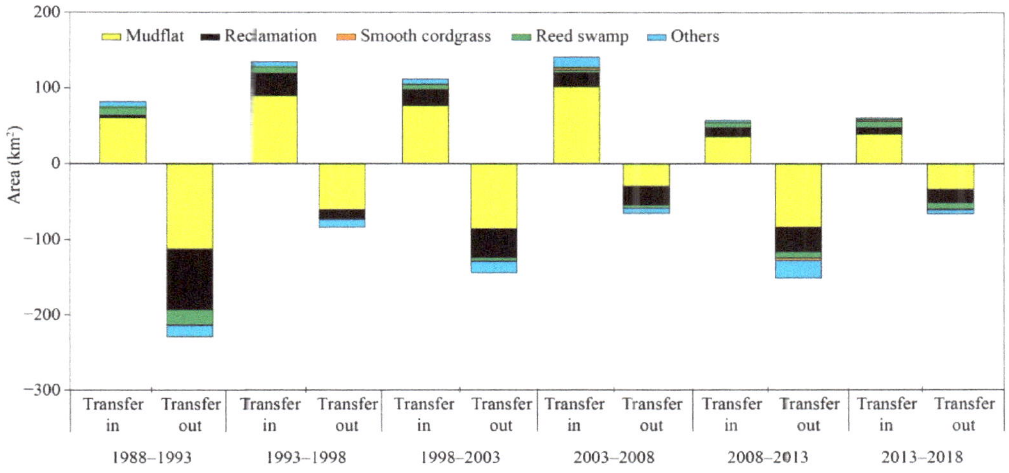

Figure 7. Land change between *S. salsa* saltmarsh area and other habitats near the NYSC from 1988 to 2018.

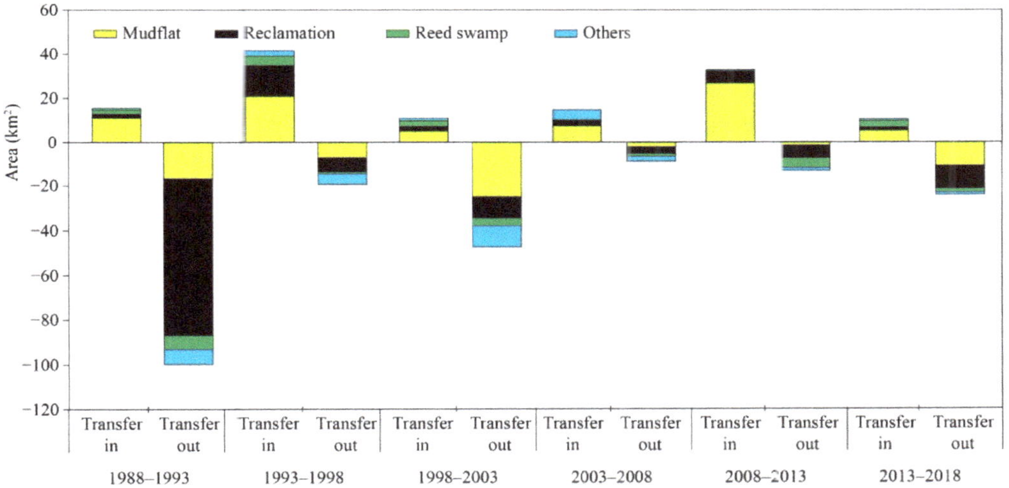

Figure 8. Land change between *S. salsa* saltmarsh area and other habitats near Liaodong Bay from 1988 to 2018.

Our study has revealed variation in the loss of *S. salsa* saltmarsh areas across different regions of the NYSC, which may be related to the differences in the conservation management capacities of different protected areas; the *S. salsa* saltmarsh area in Liaodong Bay was subject to an enormous 77% decline, whereas in the Yellow River Delta wetland and the Guangli-Zhima estuarine wetland has declined by 52%. Variation in land-use practices also contributes to these patterns of habitat loss; Liaodong Bay is the largest bay within the Yellow Sea and is nourished by several rivers, but in order to develop the local economy, a large area of *S. salsa* saltmarsh has been converted into aquaculture ponds, infrastructure such as roads, and reservoirs. Another area of great concern is the near complete loss of the *S. salsa* saltmarsh in Huludao city, Jingzhou City, Yingkou City, and eastern Panjing City. Even in the Liaohekou National Nature Reserve, the *S. salsa* saltmarsh has declined rapidly because of tourism development and the construction of oil production infrastructure, particularly during the period from 1988 to 1993. In contrast, due to the conservation management of the Yellow River Delta National Nature Reserve, little reclamation occurred in this region except for in the Guangli-Zhima estuarine wetland.

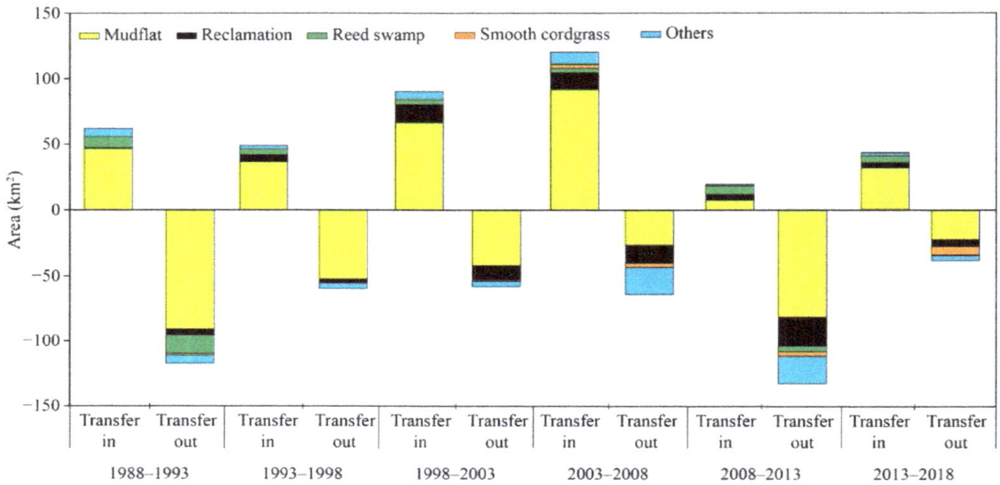

Figure 9. Land change between *S. salsa* saltmarsh area and other habitats near the Yellow River Delta wetland and the Guangli-Zhima estuarine wetland from 1988 to 2018.

Surprisingly, we found that the invasion by smooth cordgrass and natural succession has had limited impact on the decline of *S. salsa* saltmarsh areas near the NYSC. This contrasts with findings from the nearby Yancheng National Nature Reserve, where a large area of *Suaeda* was invaded by smooth cordgrass [31], which was initially introduced to control shoreline erosion and to promote the reclamation of China's coastal tidal flats in 1979 [4,32]. Significant areas of coastal wetlands have been invaded by the rapid expanding smooth cordgrass along the natural shoreline, which may dramatically change the ecological functions of the native coastal wetland landscape. For example, smooth cordgrass can rapidly proliferate to occupy large areas of mudflats, which could impact migratory shorebirds that strongly depend on these habitats as feeding grounds [35,36]. In recent years, there is some evidence to suggest that smooth cordgrass had expanded to the upper tidal flat and into the *S. salsa* saltmarsh in the north of China's coastal wetlands [32], and there are concerns that this may already be having an effect on the numerous endangered waterbird species that use these habitats.

After the Reform and Opening-up policy, China has become the second largest economy, which is part due to to rapid economic growth along its coastal region [2]. However, such rapid economic development would come at the inevitable expense of ecological security. During the past three decades, over 53% of coastal wetland areas have been lost due to land reclamation, developing aquaculture, and agriculture in order to promote economic development along the Yellow Sea [4]. Generally, *S. salsa* saltmarshes, which are located in the upper tidal flats, suffered the most extensive reclamation, which has been confirmed by this and two other previous studies [6,24]. Moreover, other factors (e.g., oil pollution, sea-level rise, lower runoff of river discharge) have prevented the growth and expansion of the *S. salsa* saltmarshes [28]. Fortunately, since the idea of "clear waters and green mountains are as good as mountains of gold and silver" advocated by Xi Jinping, the President of the People's Republic of China, and the strict ban on coastal wetland land reclamation since 2018 by the State Council of the People's Republic of China, numerous developments such as reclaiming land from the sea and aquaculture ponds have been prevented, leading to opportunities for habitat restoration efforts; for example, the abandoned aquaculture ponds, roads, and tourist facilities in Liaohekou National Nature Reserve have recently been restored to mudflats and *S. salsa* habitat, increasing the distribution region of *S. salsa* saltmarshes in this reserve (albeit slowly). However, illegal reclamation continues in the region, and further efforts are needed from the government to halt these and to provide greater funding for wetland restoration.

In summary, we quantified the spatial–temporal distribution and changes of *S. salsa* saltmarshes along the NYSC using Landsat long-time series images. We found that human land reclamation was the main reason for the loss and fragmentation of *S. salsa* saltmarshes from 1988 to 2018. Although different protection policies have been announced since the dramatic degradation of these vital coastal wetland ecosystems, which have made some significant progress regionally, urgent action is still needed to reverse these losses and to assess the ecological consequences, especially for the rapid loss of the *S. salsa* saltmarshes along the NYSC. Only two national nature reserves and a handful of provincial nature reserves are currently established along the NYSC [14,37]. Such a paucity of suitably sized and correctly situated protected areas along the NYSC may further accelerate the loss of this critical coastal wetland ecosystem and its biodiversity.

Author Contributions: D.L. and Z.Z.; conceived and designed the study. J.Z., Y.Z. and D.L.; performed experiment. J.Z. and Y.Z.; analyzed the data. J.Z. and D.L.; wrote the manuscript. J.Z., H.L., Z.Z. and D.L.; modified, edited, and finalized the manuscript. All authors reviewed and approved the manuscript. All authors have read and agreed to the published version of the manuscript.

Funding: This work was financially supported by the National Natural Science Foundation of China (No. 31911540468 and 31672316 to D.L.), the Natural Science Foundation of Liaoning Province of China (2019-MS-154 to D.L.), and the Non-profit Foundation of Marine Environment and Ecological Conservation of CNOOC (CF-MEEC/TR/2020-20 to Z.Z.).

Institutional Review Board Statement: Not applicable.

Informed Consent Statement: Not applicable.

Data Availability Statement: The data subsets used and/or analysed during this study are available from the corresponding author on request.

Acknowledgments: We thank the Liaohekou National Nature Reserve, the Yellow River Delta National Natural Reserve, Yu Bai, and Hang Zhang for their assistance with data collection in the field and Lei Cao and Jianqiang Li for the useful discussions about the manuscript. We thank the Landsat data supplied by the USGS, the Geospatial Data Cloud website, Google Earth, and Google Maps to provide analysis and data validation. We also thank two anonymous reviewers for their constructive comments.

Conflicts of Interest: The authors declare no conflict of interest.

References

1. Arkema, K.K.; Guannel, G.; Verutes, G.; Wood, S.A.; Guerry, A.D.; Ruckelshaus, M.; Kareiva, P.; Lacayo-Emery, M.; Silver, J.M. Coastal habitats shield people and property from sea-level rise and storms. *Nat. Clim. Chang.* **2013**, *3*, 913–918. [CrossRef]
2. Ma, Z.; Melville, D.S.; Liu, J.; Chen, Y.; Yang, H.; Ren, W.; Zhang, Z.; Piersma, T.; Li, B. Rethinking China's new great wall. *Science* **2014**, *346*, 912–914. [CrossRef]
3. Barbier, E.B.; Hacker, S.D.; Kennedy, C.; Koch, E.W.; Stier, A.; Silliman, B. The value of estuarine and coastal ecosystem services. *Ecol. Monogr.* **2011**, *81*, 169–193. [CrossRef]
4. Chen, Y.; Dong, J.; Xiao, X.; Ma, Z.; Tan, K.; Melville, D.; Li, B.; Lu, H.; Liu, J.; Liu, F. Effects of reclamation and natural changes on coastal wetlands bordering China's Yellow Sea from 1984 to 2015. *Land Degrad. Dev.* **2019**, *30*, 1533–1544. [CrossRef]
5. Murray, N.; Phinn, S.R.; DeWitt, M.; Ferrari, R.; Johnston, R.; Lyons, M.B.; Clinton, N.; Thau, D.; Fuller, R.A. The global distribution and trajectory of tidal flats. *Nature* **2018**, *565*, 222–225. [CrossRef] [PubMed]
6. Wang, Y.; Ji, Y.; Sun, Z.; Li, J.; Zhang, M.; Wu, G. Analysis of Suaeda heteroptera cover change and its hydrology driving factors in the Liao River Estuary wetlands, China. *IOP Conf. Ser. Earth Environ. Sci.* **2020**, *467*, 012150. [CrossRef]
7. Passeri, D.L.; Hagen, S.C.; Medeiros, S.C.; Bilskie, M.V.; Alizad, K.; Wang, D. The dynamic effects of sea level rise on low-gradient coastal landscapes: A review. *Earth's Futur.* **2015**, *3*, 159–181. [CrossRef]

8. Lu, W.; Xiao, J.; Lei, W.; Du, J.; Li, Z.; Cong, P.; Hou, W.; Zhang, J.; Chen, L.; Zhang, Y.; et al. Human activities accelerated th degradation of saline seepweed red beaches by amplifying top-down and bottom-up forces. *Ecosphere* **2018**, *9*, e02352. [CrossRef]
9. Studds, C.E.; Kendall, B.E.; Murray, N.; Wilson, H.B.; Rogers, D.I.; Clemens, R.S.; Gosbell, K.; Hassell, C.J.; Jessop, R.; Melvill D.S.; et al. Rapid population decline in migratory shorebirds relying on Yellow Sea tidal mudflats as stopover sites. *Nat. Commu* **2017**, *8*, 14895. [CrossRef]
10. Barter, M. *Shorebirds of the Yellow Sea: Importance, Threats and Conservation Status*; Wetlands International Global Series International Wader Studies 12; Wetlands International: Canberra, Australia, 2002.
11. MacKinnon, J.; Verkuil, Y.I.; Murray, N. *IUCN Situation Analysis on East and Southeast Asian Intertidal Habitats, with Particula Reference to the Yellow Sea (Including the Bohai Sea)*; Occasional Paper of the IUCN Species Survival Commission No. 47, IUCN Gland, Switzerland; Cambridge, UK, 2012.
12. Bamford, M.; Watkins, D.; Bancroft, W.; Tishchler, G.; Wahl, J. *Migratory Shorebirds of the East Asian—Australasian Flyway*; Wetlan International: Canberra, Australia, 2008.
13. China Coastal Waterbird Census Group; Bai, Q.; Chen, J.; Chen, Z.; Dong, G.; Dong, J.; Dong, W.; Fu, V.W.K.; Han, Y.; Lu, G et al. Identification of coastal wetlands of international importance for waterbirds: A review of China Coastal Waterbird Survey 2005–2013. *Avian Res.* **2015**, *6*, 12. [CrossRef]
14. Ma, Z.; Chen, Y.; Melville, D.S.; Fan, J.; Liu, J.; Dong, J.; Tan, K.; Cheng, X.; Fuller, R.A.; Xiao, X.; et al. Changes in area and numbe of nature reserves in China. *Conserv. Biol.* **2019**, *33*, 1066–1075. [CrossRef]
15. UNESCO. Migratory Bird Sanctuaries along the Coast of Yellow Sea-Bohai Gulf of China (Phase I). Available online: http //whc.unesco.org/en/list/1606/ (accessed on 13 November 2019).
16. Murray, N.; Clemens, R.S.; Phinn, S.R.; Possingham, H.; Fuller, R. Tracking the rapid loss of tidal wetlands in the Yellow Se *Front. Ecol. Environ.* **2014**, *12*, 267–272. [CrossRef]
17. Moores, N.; Rogers, D.I.; Rogers, K.; Hansbro, P. Reclamation of tidal flats and shorebird declines in Saemangeum and elsewher in the Republic of Korea. *Emu* **2016**, *116*, 136–146. [CrossRef]
18. Salman, A. Muddy coasts of the world: Processes, deposits and functions. *J. Coast. Conserv.* **2002**, *8*, 103. [CrossRef]
19. Peng, H.-B.; Anderson, G.Q.A.; Chang, Q.; Choi, C.-Y.; Chowdhury, S.U.; Clark, N.A.; Gan, X.; Hearn, R.D.; Li, J.; Lappo, E.G et al. The intertidal wetlands of southern Jiangsu Province, China—globally important for Spoon-billed Sandpipers and othe threatened waterbirds, but facing multiple serious threats. *Bird Conserv. Int.* **2017**, *27*, 305–322. [CrossRef]
20. Choi, C.-Y.; Battley, P.; Potter, M.; Rogers, K.G.; Ma, Z. The importance of Yalu Jiang coastal wetland in the north Yellow Sea t Bar-tailed Godwits Limosa lapponica and Great Knots Calidris tenuirostris during northward migration. *Bird Conserv. Int* **201** *25*, 53–70. [CrossRef]
21. Duan, H.; Xia, S.; Jackson, M.; Zhao, N.; Liu, Y.; Teng, J.; Meng, Z.; Yu, X.; Shi, J. Identifying new sites of significance to waterbird conservation and their habitat modification in the Yellow and Bohai Seas in China. *Glob. Ecol. Conserv.* **2020**, *22*, e01031. [CrossRef]
22. Zhang, J.; Bai, Y.; Huang, Z.; Zhang, Z.; Li, D. Community composition and behavioral differences of migrating shorebird between two habitats within a *Suaeda salsa* saltmarsh-mudflat wetland mosaics. *Biodivers. Sci.* **2021**, *29*, 351–360. [CrossRef]
23. Guan, B.; Yu, J.; Lu, Z.; Xie, W.; Chen, X.; Wang, X. The ecological effects of *Suaeda salsa* on repairing heavily degraded coasta saline-alkaline wetlands in the Yellow River Delta. *Acta Ecol. Sin.* **2011**, *31*, 4835–4840.
24. Tian, Y.; Luo, L.; Mao, D.; Wang, Z.; Li, L.; Liang, J. Using Landsat images to quantify different human threats to the Shuangta Estuary Ramsar site, China. *Ocean Coast. Manag.* **2017**, *135*, 56–64. [CrossRef]
25. Li, D.; Liu, Y.; Sun, X.; Lloyd, H.; Zhu, S.; Zhang, S.; Wan, D.; Zhang, Z. Habitat-dependent changes in vigilance behaviour o Red-crowned Crane influenced by wildlife tourism. *Sci. Rep.* **2017**, *7*, 1–11. [CrossRef]
26. Li, D.; Zhang, J.; Chen, L.; Lloyd, H.; Zhang, Z. Burrow ambient temperature influences Helice crab activity and availability fo migratory Red-crowned cranes Grus japonensis. *Ecol. Evol.* **2020**, *10*, 11523–11534. [CrossRef]
27. Li, D.; Chen, S.; Guan, L.; Lloyd, H.; Liu, Y.; Lv, J.; Zhang, Z. Patterns of waterbird community composition across a natural an restored wetland landscape mosaic, Yellow River Delta, China. *Estuar. Coast. Shelf Sci.* **2011**, *91*, 325–332. [CrossRef]
28. Lu, X.; Huang, Y.; Hong, J.; Zeng, D.; Yang, L. Spatial and temporal variations in wetland landscape patterns in the Yellow Rive Delta based on Landsat images. *China Environ. Sci.* **2018**, *38*, 4314–4324. [CrossRef]
29. Chen, K. *Study on Migratory Waterbirds and Wetlands in the Yellow Sea*; China Forestry Publish Press: Beijing, China, 2006.
30. Xing, Y.; Xing, J. Some taxonomy errors in studies on China Suaeda. *Mar. Sci.* **2019**, *43*, 97–102. [CrossRef]
31. Fang, R.; Shen, Y. Effects of Coast Beach Reclamation on the Change of Landscape Pattern and Its Spatial Centroids: A Case Stud in Coastal Wetland of Part of Yancheng National Natural Reserve. *J. Nat. Resour.* **2015**, *30*, 772–783. [CrossRef]
32. Liu, M.; Mao, D.; Wang, Z.; Li, L.; Man, W.; Jia, M.; Ren, C.; Zhang, Y. Rapid Invasion of Spartina alterniflora in the Coastal Zon of Mainland China: New Observations from Landsat OLI Images. *Remote Sens.* **2018**, *10*, 1933. [CrossRef]
33. Silva, R.; Martínez, M.; Van Tussenbroek, B.; Guzmán-Rodríguez, L.; Mendoza, E.; López-Portillo, J. A Framework to Manag Coastal Squeeze. *Sustainability* **2020**, *12*, 10610. [CrossRef]
34. Wang, H.; Yang, Z.; Saito, Y.; Liu, P.; Sun, X.; Wang, Y. Stepwise decreases of the Huanghe (Yellow River) sediment loa (1950–2005): Impacts of climate change and human activities. *Glob. Planet. Chang.* **2007**, *57*, 331–354. [CrossRef]
35. Cui, B.-S.; He, Q.; An, Y. Spartina alterniflora invasions and effects on crab communities in a western Pacific estuary. *Ecol. Eng* **2011**, *37*, 1920–1924. [CrossRef]

36. Liu, C.; Jiang, H.; Hou, Y.; Zhang, S.; Su, L.; Li, X.; Pan, X.; Wen, Z. Habitat Changes for Breeding Waterbirds in Yancheng National Nature Reserve, China: A Remote Sensing Study. *Wetlands* **2010**, *30*, 879–888. [CrossRef]
37. Mao, D.; Wang, Z.; Wang, Y.; Choi, C.-Y.; Jia, M.; Jackson, M.V.; Fuller, R.A. Remote Observations in China's Ramsar Sites: Wetland Dynamics, Anthropogenic Threats, and Implications for Sustainable Development Goals. *J. Remote Sens.* **2021**, *2021*, 1–13. [CrossRef]

land

MDPI

Article

Severe Drought Monitoring by Remote Sensing Methods and Its Impact on Wetlands Birds Assemblages in Nuntași and Tuzla Lakes (Danube Delta Biosphere Reserve)

Vasile Jitariu [1], Alexandru Dorosencu [2], Pavel Ichim [1] and Constantin Ion [3,*]

1 Faculty of Geography and Geology, Alexandru Ioan Cuza University, 700506 Iasi, Romania; vasile.jitariu@uaic.ro (V.J.); pavel.ichim@uaic.ro (P.I.)
2 Danube Delta National Institute for Research and Development, 165 Babadag Street, 820112 Tulcea, Romania; alexandru.dorosencu@ddni.ro
3 Faculty of Biology, Alexandru Ioan Cuza University, 700506 Iasi, Romania
* Correspondence: constantin.ion@uaic.ro

Abstract: The present paper aims to highlight the impact of the partial or total drying of the Nuntași and Tuzla lakes (from the Danube Delta Biosphere Reserve) as a result of intense drought phenomena on groups of waterfowl that are encountered in this region. Our analysis combined satellite remote sensing techniques with bird observations that were made monthly during the analyzed period, corroborated with the meteorological context of the time interval that was taken into account. The results of the satellite image processing show a partial drying in 2013 and a total drying in 2020 of the Nuntași and Tuzla lakes, which were caused by both natural factors (drought) and anthropogenic factors (inadequate management of the area—e.g.,: communication channels with surrounding lakes are clogged). These situations have led to repercussions for groups of birds, which behave differently depending on their ecology. Pelicans and swans are the most affected birds, they leave the area in the absence of water, whereas gulls and terns are not affected by the decrease in the water surface, they even increase their numbers in such conditions. Our study also shows that from 2010 to 2020 the largest numbers of birds (total numbers of birds), with the exception of pelicans, were recorded in 2013 and 2020, more precisely in the years when the water surface decreased considerably. Another important feature of this paper involves highlighting how fragile an ecosystem can be in the context of climate change, but also how important it is to involve human society in maintaining the adequate conditions for an ecosystem that is part of one of the most important biodiversity hotspots on the planet, the Danube Delta Biosphere Reserve.

Keywords: remote sensing; drought impact; aquatic birds; Danube Delta Biosphere Reserve

1. Introduction

Droughts are one of the most frequent meteorological disasters that occur on a large scale, with negative implications in several fields of human society [1]. Whether we are talking about meteorological drought, agricultural drought or hydrological drought [2], this phenomenon makes its repercussions manifest on multiple levels, but, nevertheless, it is a phenomenon that is difficult to anticipate [3] and complex when it comes to analysis, which is why a number of indicators have been designed in order to highlight its intensity [4,5].

Drought can be considered the most complex climatic phenomenon [6] because it triggers several factors, such as: atmospheric precipitations, air temperature, humidity, soil water reserves evapotranspiration, wind speed etc.; these being the main climatic parameters that define the state of dry weather. Drought spell occurrence is directly influenced by the characteristics of the active surface, soil characteristics, plants' physiological peculiarities and the consequences of the anthropogenic influence on the environment [6]. Drought is normally classified into one of three types: meteorological drought (characterized by a lack of precipitation), hydrologic drought (a lack of water supply, declining river flow and groundwater supply) and agricultural drought (crop water deficit) [2]. In Romania, the highest frequencies of drought phenomena are specific to the eastern, northeastern and southeastern regions, including the Danube Delta Biosphere Reserve [7,8].

This paper will analyze the influence of hydrological drought on the variation of the water surface (WS) of the Nuntași and Tuzla Lakes and the impact of its drying in two representative periods that were captured on satellite images. Hydrologic drought is defined as a period during which stream flows are inadequate to supply the established uses under a given water management system [9]

The drought phenomenon, depending on the region in which it takes place, can produce changes/anomalies in biodiversity. For example, in arid and semi-arid regions, biodiversity responds very quickly to precipitation [10] and in the case of the Sonoran Desert, Mc Creedy and van Riper, 2014 [11] observed a delay in nest initiation for 15 species of birds in 2006–2007. Magoulick and Kobza (2003) [12] mentioned that, for freshwater areas, severe drought phenomena can reduce the habitat area and can lead to changes in fish behavior (predator and competition relations) because drought leads to shifts in refuge spacing. Batanero et al. (2017) [13] have argued that intense droughts lead to a more pronounced increase in wetland eutrophication, especially where inclusive large colonies of birds can be found. In order to understand drought at different temporal scales, bird abundance and distribution may better reveal which drought factors impact on birds and therefore improve our understanding of how climate change impacts species and the landscapes that they inhabit [14]. Little is known about the responses of birds with different functional and behavioral characteristics to drought, or how these responses vary geographically across large areas [15]. Moreover, microclimate variables are the key factors that influence the distribution, diversity and density of the wetland bird species [16].

Waterfowls' abundance and species richness depends especially on the size, shape and depth of a body water. Moreover, the presence of water could assure the development of the birds' food and create places for birds to roost [17].

Satellite remote sensing is an alternative through which spatio-temporal observations can be made regarding changes in the earth's surface and, by this method, depending on the state of the vegetation [18] or the fluctuation of the WS [19], some conclusions can be drawn regarding the severity of the drying occurrence and the variation of the WS. Monitoring microclimate variables and birds is possible in order to understand the fluctuation of bird assemblages in the context of climatic changes.

Although the literature dealing with aspects of remote sensing, climatology and ecology is still scarce, with this paper we want to outline the importance of multidisciplinarity in science through which, in addition to the classical climate–biodiversity analysis, we can track changes in habitats through satellite remote sensing and underline the impact on aquatic birds.

The broader impact of the study is that the presence of birds could be correlated with remote sensing analysis and the weather parameters' fluctuation. The impact of climate change over the distribution of the precipitation amount in Romania was analyzed by Croitoru et al., 2018 [20]. The climate scenarios of the Representative Concentration Pathways (RCPs) 4.5 were used. Using those climate scenarios, it was observed that, in the future, the Dobruja region will record a decrease in its precipitation amount of approximately 90% as compared to the historical climate data that were recorded during dry spell events, which will intensify drought periods [20]. The fate of birds will be disastrous if the drought is not stopped through adequate management measures regarding the hidrorological system, bykeeping the channels unclogged.

This study aims to highlight the impact of the Nuntași and Tuzla Lakes' WS area variations on their aquatic bird groups in the context of the current climate trends through remote sensing methods combined with data from avifauna field monitoring. Our study focuses on the years 2013 and 2020 because the most intense modifications of the WS were observed from the field information at that time and this was also observed by Serban et al., (2022) [21].

2. Study Area

The Nuntași–Tuzla complex is a natural reserve that is included in the Danube Delta Biosphere Reserve and Natura 2000 network of protected areas (SPA and SCI) and it represents one of the main hotspots for birds in Eastern Europe [22] (Figure 1).

Figure 1. Location of Nuntași–Tuzla complex inside the Danube Delta Biosphere Reserve.

The medium-sized shallow lakes Nuntași–Tuzla are situated between 8–11 km west of the Black Sea shoreline [22–24]. Among the main characteristics concerning these lakes' morphometry, they have a length of about 2 km, a maximum width of 3 km and a depth that varies between 2.15 and 6.15 m [25]. There are marginal lakes of the Razim–Sinoie lagoon complex that resulted when the natural levees had appeared along the former sea's bay. The lakes have been undergoing a slow process of desalination in the last half a century. Now the water is brackish and waves stir up the sediments that cause the lakes to be turbid when it is windy. It is known that most brackish lakes are turbid and that this state may be due to the absence of phytoplankton grazers (*Daphnia* sp.) [26]. Being turbid, the only places that are suitable for submerged vegetation growth are some marginal shallow and covered sites. In some years, floating algal beds cover large parts of the lakes as a response to the high availability of nutrients in the water column. Common reed (*Phragmites australis*) dominates the shore vegetation and forms several islands in the southern half of the lake system. Two rivers (the Nuntași and Săcele) supply the north-western and south-western extremities, respectively, of the Nuntași–Tuzla lakes complex with fresh water that is collected from precipitation and two man-made channels that interconnect it with Lake Istria in the north and Lake Sinoie in the south. The northern channel provides the most significant water supply for the complex.

The lakes are surrounded by extensive cereal and oilseed rape fields to the north, west and south (Figure 1), these are excellent autumn and winter feeding places for the goose and swan populations. The WS represents a safe night-roost and an important drinking source for the waterbird populations of the area [27].

In the spring and autumn, the complex is heavily used by migrating waterbirds. The lakes are an important feeding and staging area for migrating waterbirds in the so-called Via Pontica flyway and one of the key roosting sites for geese, swans, pelicans and shelducks. The shallow water and muddy fields of the lakes attract waders and gulls in great numbers.

The WS varies due to both natural and anthropic factors [24]. In 2013 and 2020, the lake dried (partially and totally, respectively) because of climatic factors and management measures.

The choice of the study area was motivated by the drying phenomenon in 2020, which, due to its visual impact, has also been the subject of a considerable number of press articles and TV reports. Given the fact that this complex of lakes is located in the Danube Delta Biosphere Reserve, a biodiversity hotspot, we considered it necessary to carry out a study in order to examine the implications of this phenomenon on the avifauna of the region.

3. Methodology

3.1. Bird Census

Waterbird species surveys have been carried out once per month using four points that offer the optimal visibility over the lakes. Individual count units have been used in order to reduce bias. The counts have been performed regularly in the morning and evening, including the roosting sites.

The data were collected in such a way as to avoid the double counting of the same individuals from two neighboring points and the resulting numbers were summed in order to obtain the totals per location. The birds were identified with binoculars and field scopes [27,28].

We focused, in this analysis, on aquatic bird assemblages, such as: geese, swans, dabbling ducks (they live in shallow waters and feed by tipping up rather than diving), shelducks, gulls, terns, waders and pelicans. We took into consideration only these birds, because they represent a constant presence on the Nuntași and Tuzla Lakes, which are used mainly for feeding (by gulls, terns, waders, the ruddy shelduck), roosting (by geese and swans) or both (by ducks and the common shelduck).

3.2. Climate Data

To understand the climate characteristics of the region we used the ROCADA gridded daily climatic dataset over Romania (1961–2013) that was developed by Dumitrescu & Bârsan in 2015 [29]. For the latest period (after 2014) we used daily observed meteorological data that were obtained from the National Meteorological Administration of Romania (NMA RO) from SYNOP messages that were issued by official weather stations. The daily data that were analyzed were obtained from 3 weather stations (Gura Portiței, Jurilovca and Constanța). The daily data that we used cover the main weather elements were: the mean daily air temperature, minimum temperature, maximum temperature and precipitation (daily rain amount and number of days with precipitation) in 2013 and 2020.

3.3. Satellite Data

Satellite imagery is an excellent way to track various phenomena, both natural and artificial, but it must be admitted that one of the main disadvantages of this research method is the availability of the scenes, which is why, in some situations, the involvement of other methods is necessary (field data, meteorological context, etc.). The analysis of a phenomenon by satellite remote sensing methods can be hampered either by an overloaded atmosphere, clouds, or too much time between scenes.

From the perspective of satellite images, depending on the analyzed period and the availability of materials, we chose two sets of images. Landsat-5 satellite images were obtained through Google Earth Engine (GEE) for 2013 and, due to the fact that the study area is not very large, we downloaded the scenes without clouds, without using a mosaic of images from several days. Although in 2013 we could use other satellites with better resolution, we continued with Landsat-5 because we identified cloudless sequences for a longer period of time. The Landsat-5 TM sensor contains seven bands: blue, green, red, NIR, TIR, and two TIR bands (Table 1). For the year 2020 we used Sentinel-2A/2B satellite images, images with a very good resolution (10 m, 20 m), 13 spectral bands and with a high frequency of revisitation (Table 1).

Table 1. Specifications of Landsat-5 and Sentinel-2.

Satellite	Sensor	Year	Day/Month	Resolution (m)	Wavelength (μm)
Landsat-5	TM	2013	2/May 6/August 22/August 7/September	30	Band 1: 0.45–0.52 Band 2: 0.52–0.60 Band 3: 0.63–0.69 Band 4: 0.76–0.90 Band 5: 1.55–1.75 Band 6: 10.40–12.50 Band 7: 2.08–2.35
Sentinel-2	2A/2B	2020	26/June 11/July 31/July 25/August 9/September 19/September 24/October 23/November	60 10 10 10 20 20 20 10 20 60 60 20 20	Band 1: 0.443 Band 2: 0.490 Band 3: 0.560 Band 4: 0.665 Band 5: 0.705 Band 6: 0.740 Band 7: 0.783 Band 8: 0.842 Band 8A: 0.865 Band 9: 0.945 Band 10: 1.375 Band 11: 1.610 Band 12: 2.190

3.4. Data Analysis

In order to observe the fluctuations of the water level of Lake Nuntași, we considered the use of two indices. The normalized difference water index (NDWI) is one of the most frequently used indicators to identify open water features being developed in this direction by McFeeters (1996) [30], but later being used in the classification of wetlands [31], the mapping of tidal flats (Murray, 2012) [32] and in the mapping of the spatiotemporal changes of some lakes [33].

The normalized difference vegetation index is a well-known index that has a high degree of applicability, from the identification of land use/cover [34,35] and the monitoring of vegetation health [36,37], to the monitoring of drought phenomena [38,39] and afforestation and deforestation phenomena [40,41]. This index is also used in monitoring wetlands or the water levels of some lakes [42–44].

We used these two indicators in parallel to obtain information on the variation of the water body's surface and to compare the results, so that the final data are as close as possible to the real situation, especially since both indicators reveal extremely useful information about the presence/absence of water.

The *NDWI* can be obtained from a division that is based on the *NIR* band and green band as in Equation (1) as shown below [30] from which we will distinguish the WS as the areas where we identify values that exceed the threshold of "0" [33,45].

$$NDWI = \frac{(Green - NIR)}{Green + NIR} \tag{1}$$

The *NDVI*, shown in the following Equation (2), is explained by the fact that the reflectance of the vegetation in the *NIR* band is higher than that which is found in the *Red* band and, in this way, the vegetation is easily observed. Water bodies are associated with *NDVI* values that are less than 0 [46].

$$NDVI = \frac{NIR - Red}{NIR + Red} \tag{2}$$

After obtaining the aforementioned indices, we reclassified the rasters and extracted only the "water" values (mentioned above), resulting in a "mask" of the water surface for each scene that we used.

Before modeling the birds, meteorological data and the WS, we plotted the total number of birds in those years with severe droughts, 2013 and 2020. Consequently, we analyzed the total number of birds for the 2010–2020 period, in order to see the differences between those years when the lakes had water and those with a lack of water. The differences were checked using a Kruskal–Wallis test as is mentioned by Marusteri and Bacarea, 2009 [47]. The statistical analyses were computed using Poisson regression, which is a generalized linear model that has been used in order to elucidate the relationships between biological assemblages of species and their environment, as suggested by Van Strien et al., 2004 [48]. Given that the histograms regarding the number of birds that were analyzed per month indicated a deviation from the normal distribution and that the abundance values for each group of birds are inclined to the right in the Poisson distribution, we used Poisson regression [49,50]. We computed, in XLSTAT software, the biological assemblages (the average number of each group) which were represented by a series of aquatic birds as dependent variables. The explanatory variables consisted of the meteorological data per each month and year, such as the frequency of precipitation, average precipitation, temperature mean, minimum temperature, maximum temperature and the WS in those months for which we have access to satellite images.

4. Results

4.1. Climatic Background

Although the study area is relatively small, during the analyzed time interval the mean temperature and the amounts of monthly precipitation highlighted strong fluctuations. According to Köppen climate classification, Dobrogea is located in type BSk (semiarid steppe region) as a transition between a Mediterranean climate on its poleward margin and a cooler climate with a mild winter, low precipitation, great variability in the precipitation from year to year, low relative humidity and high evaporation rates [51].

The study area is characterized, from a climatic point of view, by a mean air temperature that can vary between 11.5 and 12 °C (according to the data from the stations that were mentioned above) with the highest temperatures encountered in July and August and the minimum temperatures manifesting in January and February. The average multiannual precipitations are distinguished by quite significant differences in the territory, which fluctuate from 366.7 mm at Gura Portitei station to 424.6 mm at Constanta station. The minimum quantity of rainfalls are typically recorded in February, March and August, while the highest amount of precipitation falls in May and June (Figure 2; Table 2). The low amount of precipitation in August, in the context of high temperatures, leads to an intense process of evaporation and in atypical, particularly dry years this can extend over a longer period. The climatic elements of this study broadly follow the topoclimatic and microclimatic characteristics that formed the basis of the episodes in which Lake Nuntași dried up. A climate description of the region was made by associating the climate elements with the climate region that is specified by the Koppen climate classification.

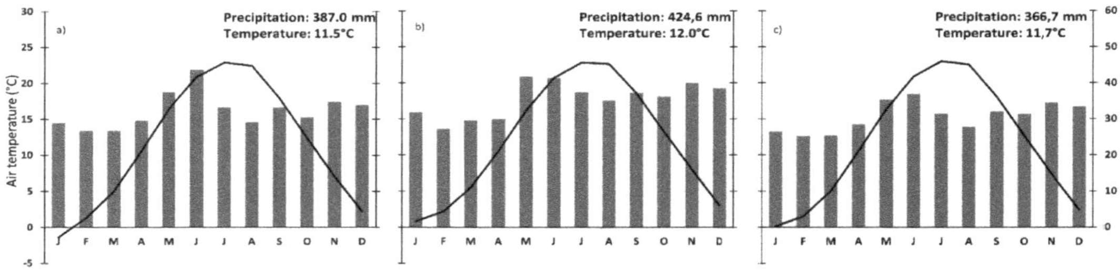

Figure 2. The multiannual mean variation of air temperature and precipitations at (**a**) Jurilovca, (**b**) Constanța and (**c**) Gura Portiței weather station from 1961 to 2020.

Table 2. Multiannual monthly mean number of days with precipitations over 1 mm and monthly number of days with precipitations over 1 mm in 2013 and 2020 for Jurilovca, Constanța and Gura Portiței weather stations.

		J	F	M	A	M	J	J	A	S	O	N	D
Mean number	**Jurilovca**	4.9	5.0	4.7	5.3	7.2	6.3	5.1	4.4	4.8	4.3	5.3	5.8
	Constanța	5.3	5.0	5.1	5.5	7.0	6.4	5.3	4.0	4.4	4.7	5.7	6.4
	Gura Portiței	4.9	5.1	4.8	5.4	7.3	6.7	5.2	4.4	5.0	4.6	5.6	6.2
2013	**Jurilovca**	5	5	2	7	7	9	6	4	7	6	4	0
	Constanța	9	4	4	6	6	8	5	4	6	4	4	2
	Gura Portiței	6	5	3	7	7	8	6	3	9	6	5	2
2020	**Jurilovca**	1	4	1	2	4	6	2	1	1	5	5	10
	Constanța	1	4	3	1	5	7	3	1	2	4	3	11
	Gura Portiței	2	5	1	2	5	6	1	0	2	6	5	9

4.2. Bird Observation

As a brief overview of the bird inventory, three species of swans were recorded for the area: the mute swan (*Cygnus olor*), whooper swan (*Cygnus cygnus*) and Bewick's swan (*Cygnus columbianus*). The whooper and Bewick's swans winter in this area and feed on the adjacent arable fields, which include winter cereals (wheat and barley) and oilseed rape. They use the lakes only for roosting and drinking water. The mute swan is present all year around. The number of mute swans constantly varies in the study area depending on the water level and vegetation state (submerged and floating plants). Years with a drastic aquatic surface reductions cause a lack of swans on these lakes.

The red-breasted goose (*Branta ruficollis*), greater white-fronted goose (*Anser albifrons*) and greylag goose (*Anser anser*) are observed in the area in large numbers. Other geese species like the lesser white-fronted goose (*Anser erythropus*) and bean goose (*Anser fabalis*) are recorded in very small numbers alongside large flocks of the greater white-fronted goose. Except for the greylag goose which is also a breeding species in the area, all of the other geese species use the lakes only as roosting and drinking sites during the winter (Figure 3).

Waders use the area mainly for feeding, so their number depends on both the WS and the depth and/or the extent of the mudflats.

In order to observe the link between the different biotic and abiotic factors and the presence of birds, satellite imagery is an excellent way to track various phenomena, both natural and artificial, but it must be admitted that one of the main disadvantages of this research method is the availability of scenes. This is why, in some situations, the involvement of other methods is necessary: including field data, meteorological context, etc.

In 2013 there was a narrowing of the water line, but not a complete drying. We identified a decrease in the water's surface area from about 800 ha (Lake Nuntași and Lake Tuzla) to 300 ha in the May–September period. There was also a greater impact of the drought on Lake Tuzla (in the south), which was the first to dry out and it remained so for the longest period of that year (Figure 4).

Figure 3. Agglomeration of birds (red-breasted goose) which roost in Nuntași–Tuzla lakes.

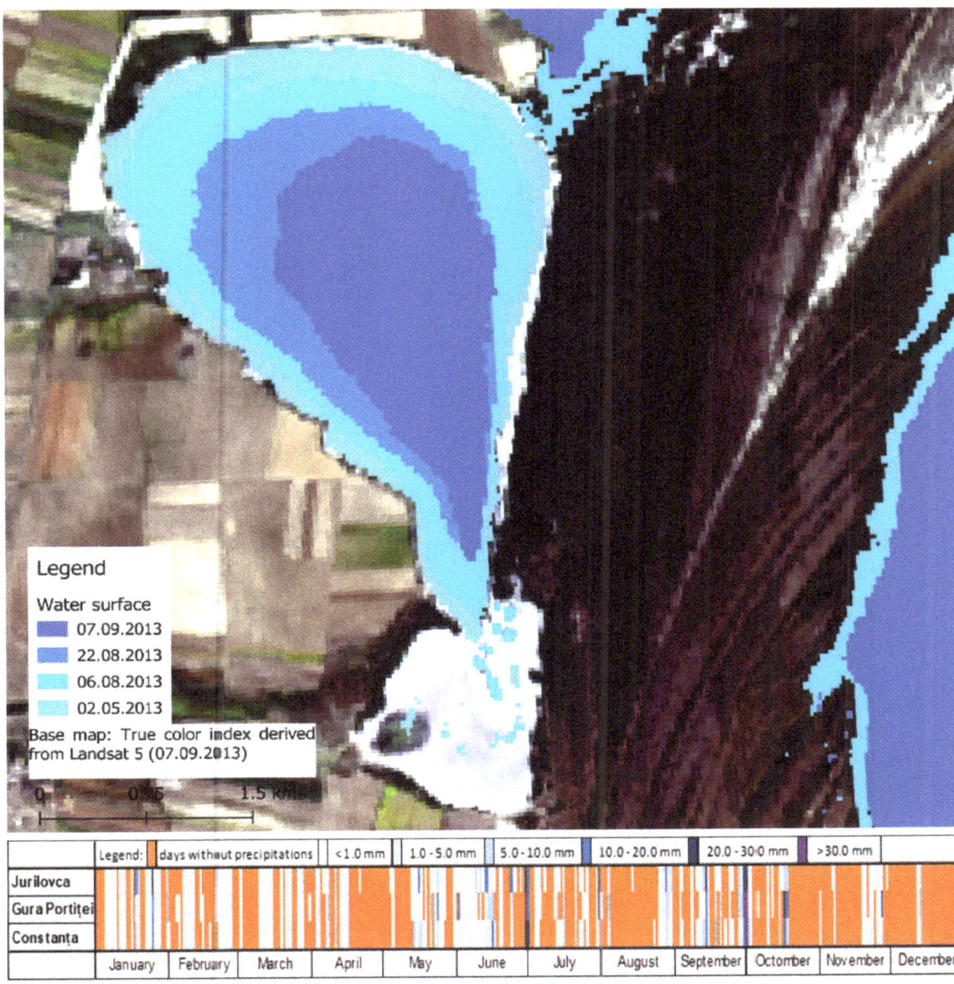

Figure 4. WS in 2013 (Landsat-5) (**above**); Days with representative precipitation amounts at, Jurilovca, Gura Portiței and Constanța weather stations in 2013 (**below**).

Following the reduction of the water's surface in 2013, we identified that the most-affected birds were swans and ruddy shelducks, which were missing in May–August and April–June, respectively (Figure 5). Dabbling ducks were also declining during the droughts. Gulls and terns did not seem to be affected by the restriction of the surface of the water, remaining in large numbers throughout the season. Geese were found in larger numbers during the winter, when specimens come from the north of the continent and are crowded into small areas. Pelicans were found in small numbers in winter because many migrate south, but at the same time their numbers were declining as the WS shrunk, due to their feeding method. Wader birds had low populations due to the droughts. Shelducks seemed to be very versatile, their number decreased both in conditions of drought and when the water level was higher and it was more difficult for them to reach the food at the bottom of the lakes (food that consists of small mollusks, crustaceans and insects) (Figure 4).

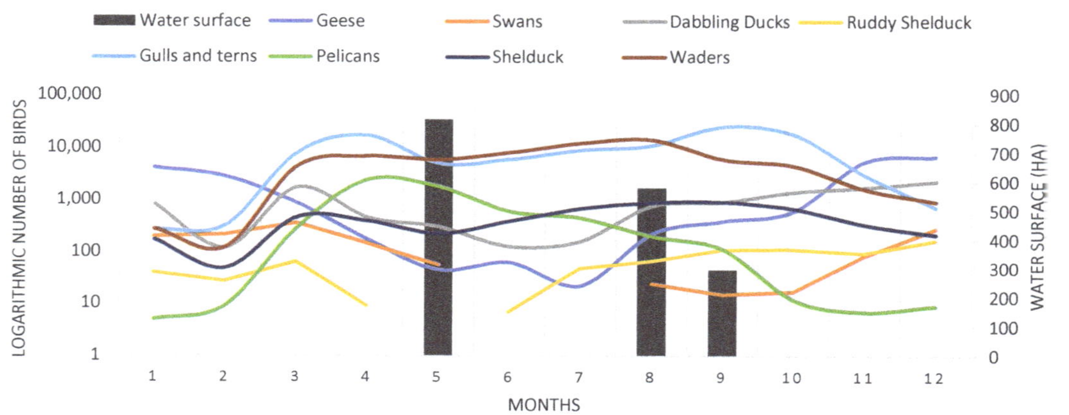

Figure 5. Average number of birds per the months in 2013.

For the 2020 drought phenomenon, we used Sentinel-2 satellite images which helped us to track the phenomenon on a more detailed time scale due to the higher frequency of the scenes. The Nuntași–Tuzla lakes have suffered increases and decreases in their WSs. About 80 consecutive days with less than 5% water were identified during August, September and October. In fact, also based on these satellite images, we could observe the dependence of the studied lakes on the climatic factor through the very fast response of the drying up due to the lack of precipitation (Figure 6). The favorable context of this phenomenon was given also by the fact that the winter of 2019–2020 was the warmest winter in the history of meteorological measurements at the Jurilovca weather station and the amount of precipitation was slightly below the average of the entire range of meteorological observations. During that winter, the North Atlantic Oscillation index was in a positive phase, which generated a mild and dry winter in southern and southeastern Europe [52] by the extension and presence of the high atmospheric pressure over the Azores Islands which extended over the Mediterranean basin [53].

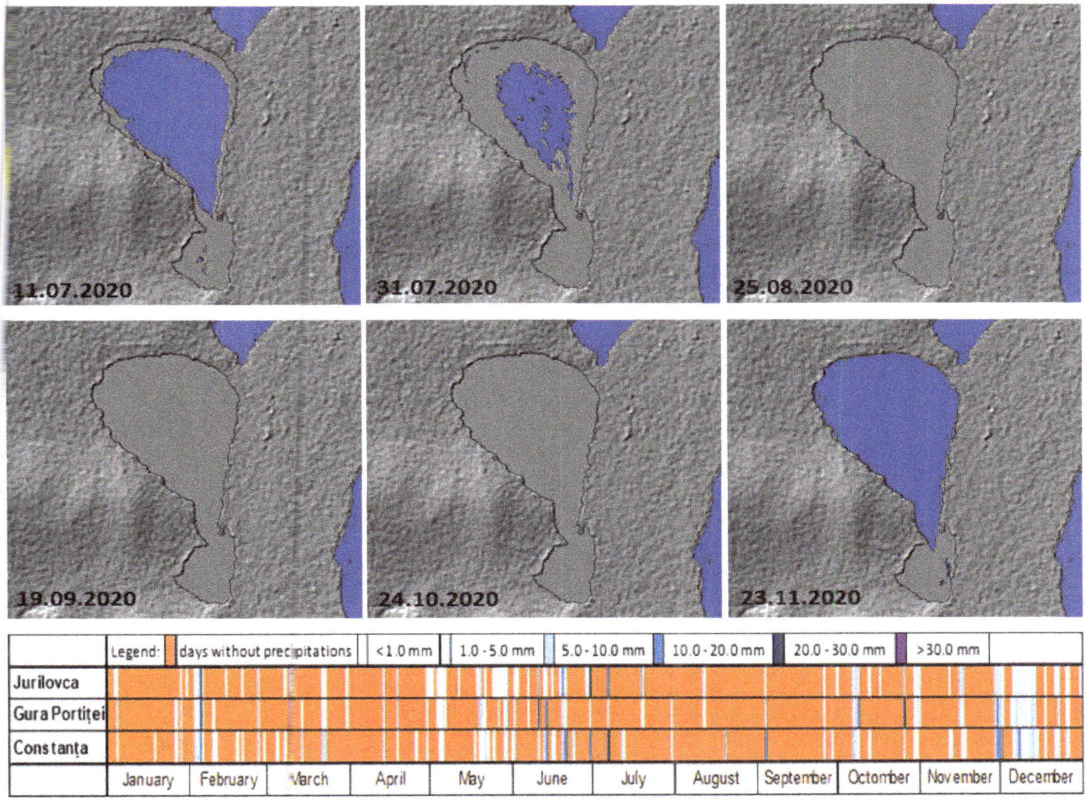

Figure 6. WS in 2020 (Sentinel-2) (**above**); Days with representative precipitation amounts at Jurilovca, Gura Portiței and Constanța weather stations in 2020 (**below**).

Given the entire meteorological context, the drought of 2020 manifested itself at a higher intensity on the two lakes and implicitly on the lakes' birds. In contrast to swans and geese, the low lake surface is accessible as a feeding habitat for waders only at a low water level. Dabbling ducks and the common shelduck use the area for roosting and feeding and their numbers are less influenced by the WS and depth, but for duck species there is a gradual shift in the species' presence and abundance according to the season and water depth. The same situation also applies to the ruddy shelduck. Gulls and terns were present throughout the analyzed period and it seems that the presence of these birds is differently affected by water body variations. Their numbers increased considerably when the WS decreased. Pelicans, on the other hand, prefer areas with a certain level of water that facilitates fish presence, so their number is generally low in the years when the lakes may dry out and for the year 2020 they appeared in an extremely small number and for a very short period. The absence of swans was observed for a longer period of time in 2020 than in 2013, as they were not present throughout the warm season (Figure 7). The number of birds was generally higher in 2020 than in 2013, as a result of the total drying of the lakes, but this difference is not statistically significant as is shown below (Figure 7).

In the third decade of November (in the year 2020) there was an almost complete filling of the lakes, a process that was realized after the clearing of the canal that connects Lake Nuntași to Lake Sinoie through Lake Istria. The importance of human intervention in the well-being of that habitat is undeniable, which is observed after the rapid filling of the lake even in the context of a small amount of precipitation (Figure 8).

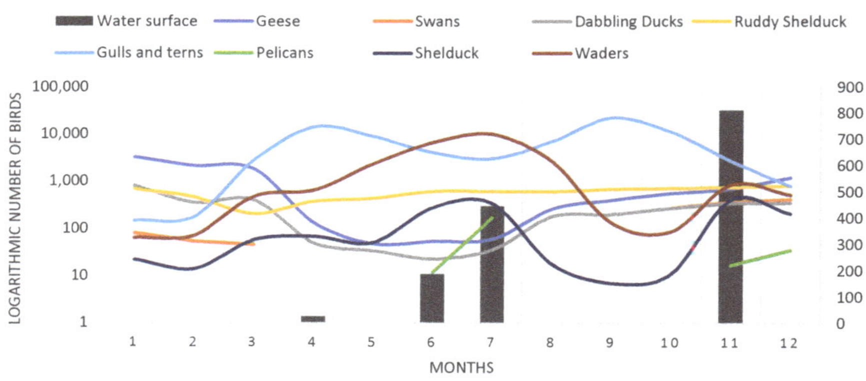

Figure 7. Average number of birds per the months in 2020.

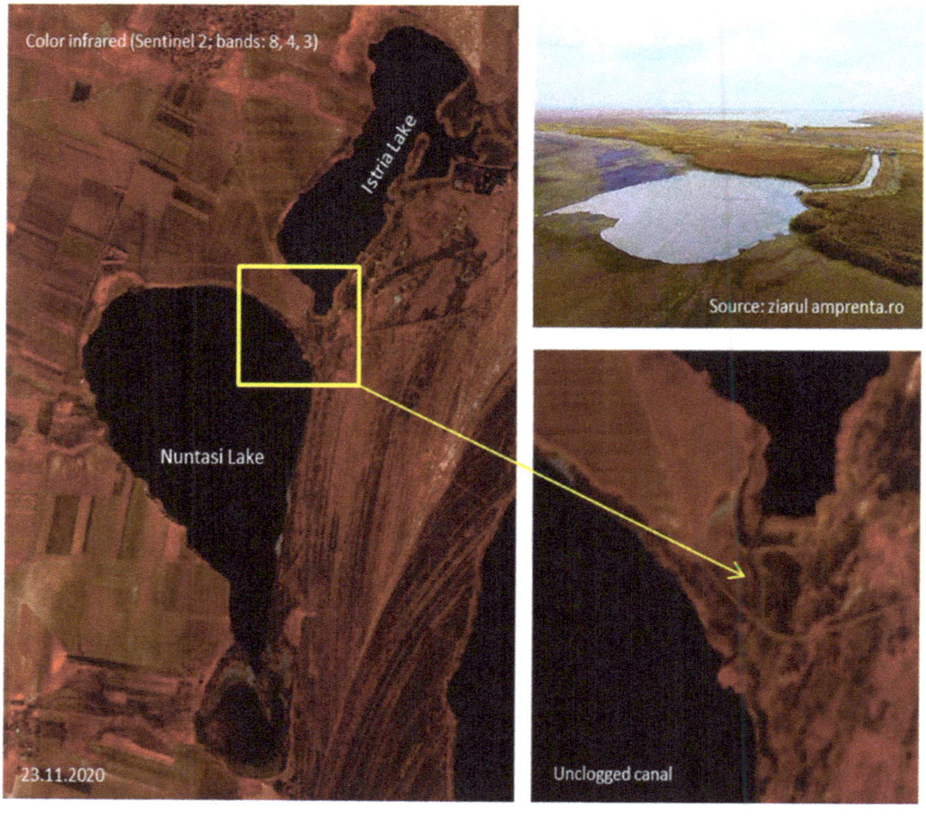

Figure 8. Nuntași and Tuzla Lakes after refilling process.

Although our initial analyses of 2013 and 2020 led us to think that in dry years the number of birds in our study area decreased, a simple analysis of the average number of birds over a period of 10 years (2010–2020) revealed that, in the years when the aquatic surface of the lakes decreased, the size of the flocks of some groups of birds increased compared to the years when the WSs of the lakes were not affected by drought (Figure 9). Even in this situation, pelicans were the most affected and only small numbers of them were identified in 2013 and 2020 (Figure 9).

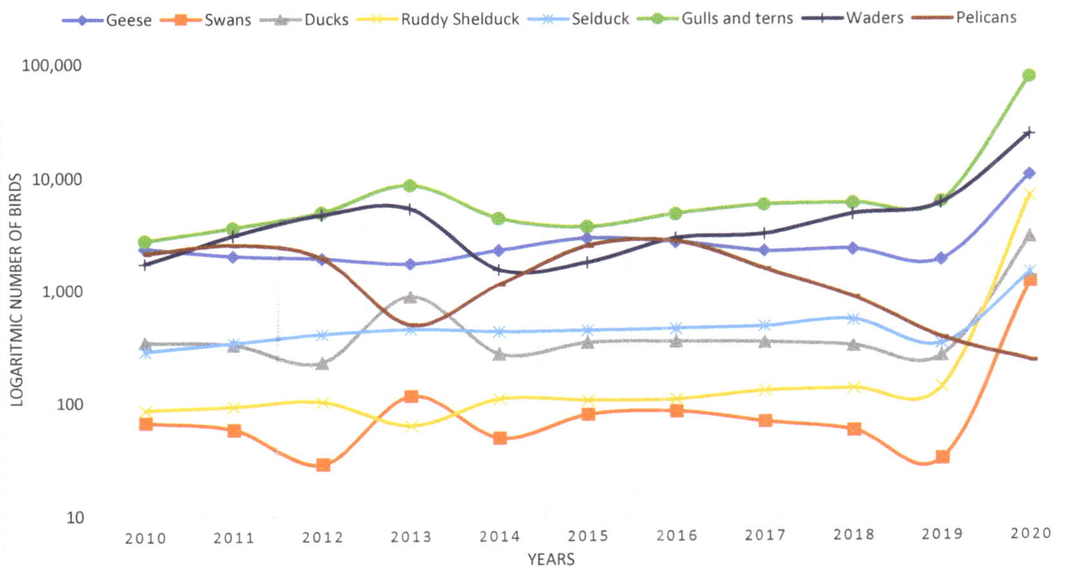

Figure 9. Average number of bird assemblages from 2010–2020.

Furthermore, a Kruskal–Wallis test showed that between 2020 and some of the other years (2010, 2014 and 2019) there is a statistically significant difference ($p < 0.005$) concerning the bird assemblage (Table 3) with regard to the WS [47].

Table 3. Kruskal–Wallis test regarding birds assemblages among years 2010–2020.

p-Values:	2010	2011	2012	2013	2014	2015	2016	2017	2018	2019	2020
2010	1	0.807	0.822	0.611	0.969	0.725	0.551	0.653	0.674	0.883	0.031
2011	0.807	1	0.984	0.792	0.777	0.914	0.725	0.837	0.860	0.922	0.056
2012	0.822	0.984	1	0.777	0.792	0.899	0.710	0.822	0.845	0.938	0.054
2013	0.611	0.792	0.777	1	0.584	0.876	0.930	0.953	0.930	0.717	0.100
2014	0.969	0.777	0.792	0.584	1	0.695	0.525	0.625	0.646	0.853	0.028
2015	0.725	0.914	0.899	0.876	0.695	1	0.807	0.922	0.945	0.837	0.072
2016	0.551	0.725	0.710	0.930	0.525	0.807	1	0.883	0.860	0.653	0.120
2017	0.653	0.837	0.822	0.953	0.625	0.922	0.883	1	0.977	0.762	0.089
2018	0.674	0.860	0.845	0.930	0.646	0.945	0.860	0.977	1	0.784	0.083
2019	0.883	0.922	0.938	0.717	0.853	0.837	0.653	0.762	0.784	1	0.045
2020	0.031	0.056	0.054	0.100	0.028	0.072	0.120	0.089	0.083	0.045	1

For 2013 and 2020, the Poisson regression showed that all of the bird assemblages were generally influenced by the meteorological data (the frequency of precipitation, temperature mean, maximum temperature and minimum temperature) and the WS variations (Table 3). Moreover, the changes in the weather parameters and fluctuation of the WS influenced the birds' presence on the Nuntași–Tuzla lakes. This finding achieved $p < 0.05$, which means that there is a significant influence of the intercept among the meteorological variables on bird assemblages on the one hand and the WS on the birds' presence on other hand.

5. Discussion

Although the use of satellite imagery to track the evolution of water bodies is an already widely used method, either we focus on large areas over long periods of time [54,55] or we are talking about smaller areas [56,57], as in our case. However, we must recognize that the involvement of satellite remote sensing in the dynamics of avifauna as correlated with field data and meteorological data is a category of complex studies that require numerous resources, both in terms of the staff involved and the financing of the undertaking [58,59]. Regarding what we just stated, it could be considered an advantage to approach such a subject with the help of satellite images because, in the absence of data from specific institutions, they allow us to analyze a phenomenon that occurred in the past, without having to involve a large number of staff or a large financial effort. On the other hand, the frequent monitoring of the birds in an area such as the studied one involves both a large number of people and a considerable budget, which would otherwise be limiting factors. However, field monitoring also reveals other observations that can be particularly important in certain circumstances.

Regarding the bird assemblages in the Nuntași–Tuzla area, their presence is determined by their feeding and roosting possibilities, which depend on the meteorological factors or the WS area [60]. In this sense, waterfowl are influenced by weather factors of course, but it is important to emphasize that each group of birds reacts differently depending on their ecology. Climate fluctuations seem to be manifested by characteristics that tend to extremes, with a higher frequency of heatwaves [61], which is certainly reflected in the abundance and distribution of the groups of birds that were targeted in this study.

Moreover, knowing the dynamic pattern of the numerical changes and dispersals of the birds leads to an understanding of the time periods and weather conditions that can induce changes in the presence of the birds and, consequently, can establish ways of satellite monitoring combined with field monitoring and active conservation measures in order to keep a high diversity index. A high value of the diversity index (which captures the number of ecological groups and number of birds) can ensure the premises for long-term biodiversity conservation in accordance with European strategies [62], but also for the purpose of designating the Danube Delta Biosphere Reserve [63].

The fact that, in the case study, all of the meteorological factors generally influenced the presence of the birds induces the idea that the birds in the area are strictly dependent on the atmospheric conditions. However, the birds may be versatile and may find food and roost in adjacent areas regardless of the climatic parameters. On a large scale, previous studies [64] have stated that, in the context of climate change, long-distance migratory bird species will suffer less than sedentary species, precisely because of their mobility. It should also be noted that sedentary bird species are much more numerous than migratory ones, which highlights the vulnerability of birds to climate change [65,66], but also the need for case studies in smaller areas.

The high mobility of the birds also induces a certain independence of them from the WS area in the Nuntași and Tuzla lakes, considering that large lakes are found nearby that are also part of the Danube Delta Biosphere Reserve. On the other hand, the Nuntași and Tuzla lakes are prone to strong fluctuations in their water levels and this is a major limiting factor for most submerged plants as shallow sites may dry out. A temporary reduction of the water level is frequently used to control submerged vegetation in reservoirs [67]. Obviously, water level fluctuations are particularly problematic for the vegetation of these lakes, where marginal and shallow water-covered places are the only suitable places for growing vegetation. If the water level is kept relatively constant, we may expect that submerged vegetation will survive in the previously mentioned locations. The lack/disappearance of submerged vegetation leads to an increase in the availability of nutrients in the water and to the intensification of floating plant layers. Severe reduction of the WS or even the complete drying of the lakes, like in the 2020 case, results in the loss of fish and most plant species. Also, the increased frequency of droughts is associated with significant decreases in invertebrate diversity [68]. Consequently, the number/presence of herbivorous and piscivorous bird species are affected in the short term by the reduction of the aquatic surface or a complete dry out in the extremely dry years. In contrast, the bird species that feed on aquatic and benthic fauna are favored by these temporarily larger feeding habitats (the presence of mudflats) and higher prey density (a greater concentration of aquatic fauna).

On the other hand, if the lake has a lot of water, it will be an excellent roost place because it is close to the feeding areas that are represented by the excellent cereal fields for geese and winter swans.

Moreover, swans and pelicans are the most sensitive and are strictly dependent on water, as the results show. In case of a lack of water, swans and pelicans leave the area. The other groups can use the areas in the middle of the lakes without water or with very little water as a place of rest, considering that they can spot terrestrial predators from a distance. In general, the reduction of the water surface induces a change in the habitats and negatively affects the bird populations [69,70], but not in this particular case. The number of birds can decline due to the lack of adequate management measures over a longer period of time [71,72].

The lack of water in 2020 lasted for months, until the application of some hydrological management measures. This also indicates the need for rapid and immediate intervention for the better conservation of wetland areas in order to ensure the living conditions of waterfowl [70].

Under the conditions of water restriction, a very rich foraging resource area for many bird species (waders, shelducks, ducks, gulls and terns) is created as the lake dries, which explains the birds' large numbers in the case of the droughts in the studied years. At the same time, seagulls and terns take advantage of the small water surface area because their food is concentrated in a small volume of water. So, given the fact that the restriction of the water surface or even the drying up in one season of the studied lakes causes an increase in the number of birds, it makes us wonder if the same would happen if the lakes were dried up for a longer period of time. Would the birds leave the area due to a lack of food in the mudflats or would they continue to use it, but only for resting? How birds behave or respond to restrictive conditions may be a future direction of scientific investigation based on this idea. Hence the need for the water level to be controlled in the sense of reducing the water level when it is too high to be favorable for bird species [73]. Even if the water luster is missing but there is enough wet mud, it is an excellent substrate for the development of the many invertebrate species that determine the presence of large numbers of birds. At the same time, the partial or complete drying of the lake must be avoided for long periods of time, in order to reduce the disappearance of the birds from this area and their dispersion over areas that are not so favorable, given the unique characteristics of the lakes complex.

As the frequency of droughts increases, there is an increasing impact on wetlands and, consequently, for birds, as is shown in the results. Thus, we find it necessary to analyze satellite images in conjunction with the monitoring of bird groups so that, through proper management, the Nuntași–Tuzla area remains a permanent hotspot for birds in one of the most valuable wetlands in the world, the Danube Delta Biosphere Reserve.

6. Conclusions

The present study reveals the characteristics of the current context, a context that is dominated by both climate change and habitat change, either as a result of a natural course or as a result of anthropogenic influence. In order to keep up with the dynamics of these changes, the inclusion of satellite images and field data (climate factors and biodiversity data) helps a lot in following the phenomenon and in observing the repercussions.

Our study outlines the suitability of long-term waterbird observations as an efficient tool to validate the qualitative and quantitative hydrological and biological transformations of wetlands, eventually supporting studies that are aimed at investigating climate change scenarios while creating synergies, as much as possible, between climatic and biological data. Corroborating ornithological and hydrological observations may eventually enhance the development of integrated monitoring programs that can offer more comprehensive information regarding climate change's impact on wetlands.

This paper considered the presentation of 2 scenarios, one from 2013 in which a partial drying was analyzed and the other one from 2020 in which the total drying of the lakes for a period longer than 2 months was analyzed. The fluctuations of the water surface in the Nuntași and Tuzla lakes show an impact both on the submerged vegetation and on the vegetation near the lakes and, in the scenario of a complete drought, the impact that extended to fish and crustaceans, the potential food sources for the analyzed bird groups. Swans were the most affected by a total drought, leaving the area, which proves their dependence on the presence of water. A similar situation was encountered in the case of pelicans. Conversely, the other groups of birds tended to use the center of the dry lake as a resting area, having the opportunity to observe predators from a distance.

Given similar scenarios in the future, this paper will represent a starting point for analyzing the impact of such phenomena on birds and it will be easier to follow their behavior and their habitat preferences.

Given the fact that the area near the Nuntași and Tuzla lakes is suitable for birds (to the north, west and south there is agricultural land for feeding and to the east there is the Istria and Sinoie lakes) and that the saltier water delays the freezing of the lakes, these factors make this location a biodiversity hotspot. It is the case studies from 2013 and 2020 that highlight the fragility of this ecosystem in the context of changing natural factors and, at the same time, the need to involve human society in habitat conservation.

Author Contributions: Conceptualization, V.J., P.I. and C.I.; methodology, V.J., P.I. and C.I.; software, V.J., P.I. and C.I.; validation: V.J. and A.D.; formal analysis, V.J., P.I. and C.I.; investigation, V.J. and A.D.; resources, P.I., A.D. and C.I.; data curation, V.J., P.I., A.D. and C.I.; writing—V.J., C.I. and A.D.; writing—review and editing, V.J., P.I. and A.D.; visualization, P.I. and V.J.; supervision, V.J.,C.I. and P.I. All authors contributed equally and have equal rights to this research paper. All authors have read and agreed to the published version of the manuscript.

Funding: This research received no external funding.

Institutional Review Board Statement: Not applicable.

Informed Consent Statement: Not applicable.

Data Availability Statement: The data presented in this study are available on request from the correspondence author.

Acknowledgments: This project received technical support from the Department of Geography, Faculty of Geography and Geology, Alexandru Ioan Cuza University of Iasi, Romania, which offered us full access to the remote sensing and GIS laboratories. Vasile Jitariu is supported by the Romanian Young Academy, which is funded by Stiftung Mercator and the Alexander von Humboldt Foundation for the period 2021–2022. Pavel Ichim is supported by the Romanian Young Academy, which is funded by Stiftung Mercator and the Alexander von Humboldt Foundation for the period 2020–2022. Acknowledgment is also given by C.I. to the infrastructure support from the Operational Program Competitiveness 2014–2020, Axis 1, under POC/448/1/1 research infrastructure projects for public R&D institutions/Sections F 2018, through the Research Center with Integrated Techniques for Atmospheric Aerosol Investigation in Romania (RECENT AIR) project, under grant agreement MySMIS no. 127324.

Conflicts of Interest: The authors declare no conflict of interest.

References

1. Dai, A. Drought under global warming: A review. *WIREs Clim. Chang.* **2011**, *2*, 45–65. [CrossRef]
2. Wilhite, D.A.; Glantz, M.H. Understanding: The Drought Phenomenon: The Role of Definitions. *Water Int.* **1985**, *10*, 111–120. [CrossRef]

3. Wang, R.; Peng, W.; Liu, X.; Wu, W.; Chen, X.; Zhang, S. Responses of Water Level in China's Largest Freshwater Lake to the Meteorological Drought Index (SPEI) in the Past Five Decades. *Water* **2018**, *10*, 137. [CrossRef]
4. Hao, Z.; Singh, V.P. Drought characterization from a multivariate perspective: A review. *J. Hydrol.* **2015**, *527*, 668–678. [CrossRef]
5. Zhao, H.; Gao, G.; An, W.; Zou, X.; Li, H.; Hou, M. Timescale differences between SC-PDSI and SPEI for drought monitoring in China. *Phys. Chem. Earth Parts A/B/C* **2017**, *102*, 48–58. [CrossRef]
6. IPCC. *Climate Change 2013: The Physical Science Basis: Summary for Policymakers, a Report of Working Group I of the IPCC: Technical Summary, a Report Accepted by Working Group I of the IPCC but Not Approved in Detail: And Frequently Asked Questions: Part of the Working Group I Contribution to the Fifth Assessment Report of the Intergovernmental Panel on Climate Change*; Stocker, T., Qin, D., Eds.; WMO: Geneva, Switzerland; UNEP: Nairobi, Kenya, 2013; ISBN 978-92-9169-138-8.
7. Dobri, R.-V.; Sfîcă, L.; Amihăesei, V.-A.; Apostol, L.; Țîmpu, S. Drought Extent and Severity on Arable Lands in Romania Derived from Normalized Difference Drought Index (2001–2020). *Remote Sens.* **2021**, *13*, 1478. [CrossRef]
8. Minea, I.; Iosub, M.; Boicu, D. Multi-scale approach for different type of drought in temperate climatic conditions. *Nat. Hazards* **2022**, *110*, 1153–1177. [CrossRef]
9. Pedro-Monzonís, M.; Solera, A.; Ferrer, J.; Estrela, T.; Paredes-Arquiola, J. A review of water scarcity and drought indexes in water resources planning and management. *J. Hydrol.* **2015**, *527*, 482–493. [CrossRef]
10. Bolger, D.T.; Patten, M.A.; Bostock, D.C. Avian reproductive failure in response to an extreme climatic event. *Oecologia* **2005**, *142*, 398–406. [CrossRef]
11. McCreedy, C.; Van Riper, C. Drought-caused delay in nesting of Sonoran Desert birds and its facilitation of parasite- and predator-mediated variation in reproductive success. *Auk* **2015**, *132*, 235–247. [CrossRef]
12. Magoulick, D.D.; Kobza, R.M. The role of refugia for fishes during drought: A review and synthesis: Refugia for Fishes during Drought. *Freshw. Biol.* **2003**, *48*, 1186–1198. [CrossRef]
13. Batanero, G.L.; León-Palmero, E.; Li, L.; Green, A.J.; Rendón-Martos, M.; Suttle, C.A.; Reche, I. Flamingos and drought as drivers of nutrients and microbial dynamics in a saline lake. *Sci. Rep.* **2017**, *7*, 12173. [CrossRef] [PubMed]
14. Cady, S.M.; O'Connell, T.J.; Loss, S.R.; Jaffe, N.E.; Davis, C.A. Species-specific and temporal scale-dependent responses of birds to drought. *Glob. Chang. Biol.* **2019**, *25*, 2691–2702. [CrossRef] [PubMed]
15. Albright, T.P.; Pidgeon, A.M.; Rittenhouse, C.D.; Clayton, M.K.; Wardlow, B.D.; Flather, C.H.; Culbert, P.D.; Radeloff, V.C. Combined effects of heat waves and droughts on avian communities across the conterminous United States. *Ecosphere* **2010**, *1*, 1–22. [CrossRef]
16. Rajpar, M.N.; Zakaria, M. Bird Species Abundance and Their Correlationship with Microclimate and Habitat Variables at Natural Wetland Reserve, Peninsular Malaysia. *Int. J. Zool.* **2011**, *2011*, 1–17. [CrossRef]
17. Baschuk, M.S.; Koper, N.; Wrubleski, D.A.; Goldsborough, G. Effects of Water Depth, Cover and Food Resources on Habitat use of Marsh Birds and Waterfowl in Boreal Wetlands of Manitoba, Canada. *Waterbirds* **2012**, *35*, 44–55. [CrossRef]
18. Liu, W.T.; Kogan, F.N. Monitoring regional drought using the Vegetation Condition Index. *Int. J. Remote Sens.* **1996**, *17*, 2761–2782. [CrossRef]
19. Huovinen, P.; Ramírez, J.; Caputo, L.; Gómez, I. Mapping of spatial and temporal variation of water characteristics through satellite remote sensing in Lake Panguipulli, Chile. *Sci. Total Environ.* **2019**, *679*, 196–208. [CrossRef]
20. Croitoru, A.E.; Piticar, A.; Sfîcă, L.; Roșca, C.-F.; Tudose, T.; Horvath, C.; Ionuț, M.; Ciupertea, A.-F.; Scripcă, S.; Harpa, G. *Extreme Temperature and Precipitation Events in Romania, Editura Academiei Române*; The Romanian Academy: Bucharest, Romania, 2018; p. 359. ISBN 978-973-27-2833-8.
21. Șerban, C.; Maftei, C.; Dobrică, G. Surface Water Change Detection via Water Indices and Predictive Modeling Using Remote Sensing Imagery: A Case Study of Nuntasi-Tuzla Lake, Romania. *Water* **2022**, *14*, 556. [CrossRef]
22. Available online: https://ebird.org/region/RO-CT/hotspots?yr=all&m= (accessed on 15 December 2021).
23. Vespremeanu-Stroe, A.; Preoteasa, L.; Hanganu, D.; Brown, A.G.; Bîrzescu, I.; Toms, P.; Timar-Gabor, A. The impact of the Late Holocene coastal changes on the rise and decay of the ancient city of Histria (southern Danube delta). *Quat. Int.* **2013**, *293*, 245–256. [CrossRef]
24. Maftei, C.; Dobrica, G.; Cerneaga, C.; Buzgaru, N. Drought Land Degradation and Desertification—Case Study of Nuntasi-Tuzla Lake in Romania. In *Water Safety, Security and Sustainability: Threat Detection and Mitigation*; Vaseashta, A., Maftei, C., Eds.; Springer International Publishing: Cham, Switzerland, 2021; pp. 583–597.
25. Breier, A. *Lacurile de pe Litoralul Romanesc al Marii Negre*; Studiu Hidrogeografic; Editura Academiei Republicii Socialiste România: Bucuresti, Romania, 1976.
26. Jeppesen, E.; Peder Jensen, J.; Søndergaard, M.; Lauridsen, T.; Junge Pedersen, L.; Jensen, L. Top-Down Control in Freshwater Lakes: The Role of Nutrient State, Submerged Macrophytes and Water Depth. In *Shallow Lakes '95*; Kufel, L., Prejs, A., Rybak, J.I., Eds.; Springer: Dordrecht, The Netherlands, 1997; pp. 151–164. ISBN 978-94-010-6382-1.
27. Heredia, B.; Rose, L.; Painter, M. *Globally Threatened Birds in Europe: Action Plans*; Council of Europe: Strasbourg, France, 1996; ISBN 92-871-3066-3.
28. Available online: https://europe.wetlands.org/wpcontent/uploads/sites/3/2016/08/Protocol_for_waterbird_counting_En_.pdf (accessed on 10 November 2021).
29. Dumitrescu, A.; Birsan, M.-V. ROCADA: A gridded daily climatic dataset over Romania (1961–2013) for nine meteorological variables. *Nat. Hazards* **2015**, *78*, 1045–1063. [CrossRef]

30. McFeeters, S.K. Using the Normalized Difference Water Index (NDWI) within a Geographic Information System to Detect Swimming Pools for Mosquito Abatement: A Practical Approach. *Remote Sens.* **2013**, *5*, 3544–3561. [CrossRef]
31. Environmental Protection Agency; Queensland Government. *Wetland Mapping and Classification Methodology: Overall Framework; A Method to Provide Baseline Mapping and Classification for Wetlands in Queensland*; Environmental Protection Agency; Queensland Government: Brisbane, QL, Australia, 2005; ISBN 978-0-9757344-6-9.
32. Murray, N.J.; Phinn, S.R.; Clemens, R.S.; Roelfsema, C.M.; Fuller, R.A. Continental Scale Mapping of Tidal Flats across East Asia Using the Landsat Archive. *Remote Sens.* **2012**, *4*, 3417–3426. [CrossRef]
33. Rokni, K.; Ahmad, A.; Selamat, A.; Hazini, S. Water Feature Extraction and Change Detection Using Multitemporal Landsat Imagery. *Remote Sens.* **2014**, *6*, 4173–4189. [CrossRef]
34. Fathizad, H.; Tazeh, M.; Kalantari, S.; Shojaei, S. The investigation of spatiotemporal variations of land surface temperature based on land use changes using NDVI in southwest of Iran. *J. Afr. Earth Sci.* **2017**, *134*, 249–256. [CrossRef]
35. Muttitanon, W.; Tripathi, N.K. Land use/land cover changes in the coastal zone of Ban Don Bay, Thailand using Landsat 5 TM data. *Int. J. Remote Sens.* **2005**, *26*, 2311–2323. [CrossRef]
36. Kinyanjui, M.J. NDVI-based vegetation monitoring in Mau forest complex, Kenya. *Afr. J. Ecol.* **2011**, *49*, 165–174. [CrossRef]
37. Pei, F.; Wu, C.; Liu, X.; Li, X.; Yang, K.; Zhou, Y.; Wang, K.; Xu, L.; Xia, G. Monitoring the vegetation activity in China using vegetation health indices. *Agric. For. Meteorol.* **2018**, *248*, 215–227. [CrossRef]
38. Sruthi, S.; Aslam, M.M. Agricultural Drought Analysis Using the NDVI and Land Surface Temperature Data; a Case Study of Raichur District. *Aquat. Procedia* **2015**, *7*, 1258–1264. [CrossRef]
39. Jitariu, V.; Vasiliniuc, I.; Rusu, C.; Rosca, B. The Use of Sentinel 2 Images for Drought Phenomenon Monitoring in Apple Orchards. In Proceedings of the International Multidisciplinary Scientific GeoConference SGEM, Albena, Bulgaria, 28 June–6 July 2019; Volume 19, pp. 671–678.
40. Gao, Y.; Quevedo, A.; Szantoi, Z.; Skutsch, M. Monitoring Forest Disturbance Using Time-Series MODIS NDVI in Michoacán, Mexico. *Geocarto Int.* **2021**, *36*, 1768–1784. [CrossRef]
41. Schultz, M.; Clevers, J.G.P.W.; Carter, S.; Verbesselt, J.; Avitabile, V.; Quang, H.V.; Herold, M. Performance of vegetation indices from Landsat time series in deforestation monitoring. *Int. J. Appl. Earth Obs. Geoinf.* **2016**, *52*, 318–327. [CrossRef]
42. Dong, Z.; Wang, Z.; Liu, D.; Song, K.; Li, L.; Jia, M.; Ding, Z. Mapping Wetland Areas Using Landsat-Derived NDVI and LSWI: A Case Study of West Songnen Plain, Northeast China. *J. Indian Soc. Remote Sens.* **2014**, *42*, 569–576. [CrossRef]
43. Omute, P.; Corner, R.; Awange, J.L. The use of NDVI and its Derivatives for Monitoring Lake Victoria's Water Level and Drought Conditions. *Water Resour. Manag.* **2012**, *26*, 1591–1613. [CrossRef]
44. Han, Q.; Niu, Z. Construction of the Long-Term Global Surface Water Extent Dataset Based on Water-NDVI Spatio-Temporal Parameter Set. *Remote Sens.* **2020**, *12*, 2675. [CrossRef]
45. Zhai, K.; Wu, X.; Qin, Y.; Du, P. Comparison of surface water extraction performances of different classic water indices using OLI and TM imageries in different situations. *Geo Spat. Inf. Sci.* **2015**, *18*, 32–42. [CrossRef]
46. Rouse, J.W., Jr.; Haas, R.H.; Schell, J.A.; Deering, D.W. Paper a 20. In Proceedings of the Third Earth Resources Technology Satellite-1 Symposium: Section AB. Technical presentations; Scientific and Technical Information Office, National Aeronautics and Space, Greenbelt, MD, USA, 10–14 December 1973; Volume 1, p. 309.
47. Marusteri, M.; Bacarea, V. Comparing Groups for Statistical Di Erences: How to Choose the Right Statistical Test? *Indian J. Ophthalmol.* **2011**, *59*, 85–86.
48. Van Strien, A.; Pannekoek, J.; Hagemeijer, W.; Verstrael, T. A Loglinear Poisson Regression Method to Analyse Bird Monitoring. *Data. Bird* **2004**, *482*, 33–39.
49. Azizoğlu, E.; Adizel, O.; Kara, R. A statistical approach on seasonal population changes and habitat preferences on coastal and waterfowl species around Ekşisu Reeds (Erzincan-Turkey): Using negative binomial regression. *Appl. Ecol. Environ. Res.* **2021**, *19*, 653–665. [CrossRef]
50. García-Núñez, R.M.; Romero-Díaz, C.; Ugalde-Lezama, S.; Tinoco-Rueda, J.A. Vegetation and habitat structure that determines the diet of insectivorous birds in agroforestry systems. *Rev. Mex. Cienc. Agríc.* **2020**, *11*, 4.
51. *Encyclopædia Britannica*; Encyclopædia Britannica, Inc.: Chicago, IL, USA, 2016.
52. Rousi, E.; Rust, H.W.; Ulbrich, U.; Anagnostopoulou, C. Implications of Winter NAO Flavors on Present and Future European Climate. *Climate* **2020**, *8*, 13. [CrossRef]
53. Lindsey, R.; Dahlman, L. Climate Variability: North Atlantic Oscillation. Climate. gov. Last modified 30 August 2009. 2009. Available online: https://www.climate.gov/news-features/understanding-climate/climate-variability-north-atlantic-oscillation (accessed on 15 February 2022).
54. Wan, W.; Xiao, P.; Feng, X.; Li, H.; Ma, R.; Duan, H.; Zhao, L. Monitoring lake changes of Qinghai-Tibetan Plateau over the past 30 years using satellite remote sensing data. *Chin. Sci. Bull.* **2014**, *59*, 1021–1035. [CrossRef]
55. Nitze, I.; Grosse, G.; Jones, B.M.; Arp, C.D.; Ulrich, M.; Fedorov, A.; Veremeeva, A. Landsat-Based Trend Analysis of Lake Dynamics across Northern Permafrost Regions. *Remote Sens.* **2017**, *9*, 640. [CrossRef]
56. Deus, D.; Gloaguen, R. Remote Sensing Analysis of Lake Dynamics in Semi-Arid Regions: Implication for Water Resource Management. Lake Manyara, East African Rift, Northern Tanzania. *Water* **2013**, *5*, 698–727. [CrossRef]
57. Policelli, F.; Hubbard, A.; Jung, H.C.; Zaitchik, B.; Ichoku, C. Lake Chad Total Surface Water Area as Derived from Land Surface Temperature and Radar Remote Sensing Data. *Remote Sens.* **2018**, *10*, 252. [CrossRef]

Besnard, A.G.; Davranche, A.; Maugenest, S.; Bouzillé, J.-B.; Vian, A.; Secondi, J. Vegetation Maps Based on Remote Sensing Are Informative Predictors of Habitat Selection of Grassland Birds across a Wetness Gradient. *Ecol. Indic.* **2015**, *58*, 47–54. [CrossRef]

Koma, Z.; Seijmonsbergen, A.C.; Grootes, M.W.; Nattino, F.; Groot, J.; Sierdsema, H.; Foppen, R.P.B.; Kissling, W.D. Better together? Assessing different remote sensing products for predicting habitat suitability of wetland birds. *Divers. Distrib.* **2022**, *28*, 685–699. [CrossRef]

Constantin, I.; Ştefan, B.E.; Ursu, A.; Lucian, S.; Elena, I.A.; Constantin, S.C. Relationship between Environmental Features and Bird Assemblages in the Wetlands of Eastern Romania. *Present Environ. Sustain. Dev.* **2019**, *13*, 265–290. [CrossRef]

Sfîcă, L.; Croitoru, A.-E.; Iordache, I.; Ciupertea, A.-F. Synoptic Conditions Generating Heat Waves and Warm Spells in Romania. *Atmosphere* **2017**, *8*, 50. [CrossRef]

Available online: https://ec.europa.eu/environment/nature/biodiversity/strategy/index_en.htm (accessed on 10 November 2021).

Available online: http://sgglegis.gov.ro/legislativ/docs/2014/01/w2nphs85kcdr0b1zgj3f.pdf (accessed on 10 November 2021).

Sekercioglu, C.H.; Schneider, S.H.; Fay, J.P.; Loarie, S.R. Climate Change, Elevational Range Shifts, and Bird Extinctions. *Conserv. Biol.* **2008**, *22*, 140–150. [CrossRef]

Sekercioglu, C.H. Conservation Ecology: Area Trumps Mobility in Fragment Bird Extinctions. *Curr. Biol.* **2007**, *17*, R283–R286. [CrossRef]

Cox, G.W. *Bird Migration and Global Change*; Island Press: Washington, DC, USA, 2010; ISBN 1-59726-969-7.

Cooke, G.D.; Welch, E.B.; Peterson, S.A.; Newroth, P.R. *Restoration and Management of Lakes and Reservoirs*; Lewis Publishers: Boca Raton, FL, USA, 1993.

Atkinson, S.T.; Cale, D.; Pinder, A.; Chambers, J.M.; Halse, S.A.; Robson, B.J. Substantial long-term loss of alpha and gamma diversity of lake invertebrates in a landscape exposed to a drying climate. *Glob. Chang. Biol.* **2021**, *27*, 6263–6279. [CrossRef] [PubMed]

MARINOV, M.; HULEA, D.; CONDAC, M. Dinamica avifaunei din RBDD în corelatia cu principalii factori de mediu. *Anal. Ştiintifice Inst. Delta Dunării* **1996**, *5*, 211–226.

Gonzalez, A.L.; Sepúlveda, P.V.; Schlatter, R. Waterbird Assemblages and Habitat Characteristics in Wetlands: Influence of Temporal Variability on Species-Habitat Relationships. *Waterbirds* **2009**, *32*, 225–233. [CrossRef]

Marinov, M.; Doroșencu, A.; Alexe, V.; Bolboacă, L.-E.; Kiss, J.B. Current status of bird species in annex 1 of the birds directive for which special protection areas have been designated on the territory of the danube delta biosphere reserve. *Rom. J. Biol. Zool.* **2020**, *65*, 61–70.

Ma, Z.; Cai, Y.; Li, B.; Chen, J. Managing Wetland Habitats for Waterbirds: An International Perspective. *Wetlands* **2010**, *30*, 15–27. [CrossRef]

Sebastián-González, E.; Green, A.J. Habitat Use by Waterbirds in Relation to Pond Size, Water Depth, and Isolation: Lessons from a Restoration in Southern Spain. *Restor. Ecol.* **2014**, *22*, 311–318. [CrossRef]

land

Article

How Do the New Residential Areas in Bucharest Satisfy Population Demands, and Where Do They Fall Short?

Mihaela Preda [1], Iuliana Vijulie [1,*], Ana-Irina Lequeux-Dincă [1], Marta Jurchescu [2], Alina Mareci [1] and Alexandru Preda [3]

1 Faculty of Geography, University of Bucharest, 1. Blv. Nicolae Bălcescu, 010041 Bucharest, Romania; mihaela.preda@geo.unibuc.ro (M.P.); ana.dinca@geo.unibuc.ro (A.-I.L.-D.); alina.mareci@unibuc.ro (A.M.)
2 Institute of Geography, Romanian Academy, 12 Dimitrie Racoviță, 023993 Bucharest, Romania; marta_jurchescu@yahoo.com
3 National Institute of Hydrology and Water Management (INHGA), 97E București-Ploiești, 013686 Bucharest, Romania; alexandru.preda@hidro.ro
* Correspondence: iuliana.vijulie@g.unibuc.ro; Tel.: +40-721-548-408

Abstract: In recent years, Bucharest's residential dynamics have thrived, fueled by growing demand and an insufficient housing fund. This study aimed to analyze the residential satisfaction of those living in newly built dwellings. Its objectives were to identify the characteristics of three new residential areas and analyze the satisfaction level among residents regarding both their dwellings and neighborhoods. The investigation employed direct observations during the fieldwork phase (through observation sheets and mapping methods) and surveys (through questionnaires with residents and interviews with developers). Its results highlighted spaces that exhibit an increase in residential constructions, with a tendency to expand toward suburban areas, without necessarily meeting legislative requirements. When measuring the population's residential satisfaction level, the study observed a general satisfaction regarding dwellings' modernity and price but noticeable differences within the sample residential nuclei. The solutions proposed by residents mainly target authorities, who were held responsible for developing the urban infrastructure prior to granting building permits, as well as for vetting developers better and requiring them to respect the legislation. Hence, scientists, local authorities, real-estate developers and the local population represent the beneficiaries of the current study's results.

Keywords: residential satisfaction; new urban nuclei; infrastructure characteristics; legislation; Bucharest; Romania

1. Introduction

Cities are complex systems whose economic and social functionality depends on implementing adequate housing development policies. Housing quality in urban areas is an ongoing concern for all involved actors, because, on the one hand, it represents a direct response to issues connected to urban systematization, e.g., traffic, pollution, lack of green areas, waste management, etc.; and, on the other hand, as population preferences regarding modern housing standards are rapidly changing, studying them is one way of keeping up with market demands [1]. Residential satisfaction is a heterogeneous and multidimensional concept that underlines that the evaluation of the living place and neighborhood is a balance between expectations and benefits. It stems from housing characteristics, amenities, and quality of the environment associated with expenditure [2] and refers to a perception of the place of residence from both physical and social perspectives [3].

Certain studies emphasize the importance of social environment as a third element generating residential satisfaction, indicating as frequent variables: "safety, upkeep, social ties, quietness, housing quality, greenery, age and income level" (p. 114, Reference [4]). Some authors, such as Berköz and Kellekçi (p. 41, Reference [5]), name "the characteristics of residential units, characteristics of users, managerial, environmental and locational factors" as variables that influence residential satisfaction, while Cho, Park, and Echevarria-Cruz [6] conclude that neighborhoods have experienced a reduction in their role as traditional communities due to decades of rapid industrialization and urbanization. Still, understanding how residents rate neighborhoods remains an essential element of urban planning policies, as neighborhood satisfaction is a component of quality of life which can form a basis for public policy feedback [7,8]. Neighborhood satisfaction is perceived as a multifaceted construct and includes relationships with neighbors, housing characteristics, demographic factors, security, presence of and access to services (amenities), aesthetics (i.e., appearance), and air quality/pollution [9,10]. Nowadays, many studies move away from the somewhat-limited past approach to neighborhood satisfaction (that analyzes only one or just more than one level of the environment) [11] and attempt to evaluate inhabitants' preferences about attributes of the residential environment as one of the multiple variables of neighborhood satisfaction [12–15]. The current evolution of cities should consider the population's needs regarding housing quality, while also maintaining the principles that govern neighborhoods' economic and social functionality. Housing satisfaction is evaluated based on components from living conditions to the environment, physical infrastructure, public service, governance, and culture, as these represent urban development pillars in most countries [16]. Nonetheless, it depends to a large extent on successful housing policies [17], which do not necessarily consider the constantly evolving demographic needs and rely more on the developers' point of view [15]. Unfortunately, urbanization does not always respect legislative demands, and many large cities develop chaotically and uncontrollably, frequently leading to traffic, pollution, lack of green areas, and poorly developed infrastructure [18].

Nowadays, many large cities display a competitive city profile that considers both investors' and residents' choices. They have to incorporate an extensive range of amenities and services to satisfy people's complex network of needs for the place in which they decide to live [19].

A wide range of studies recognize the subjective nature of personal expectations for new dwellings and the subjective attributes of the residential environment as an indivisible part of "the systemic model of residential satisfaction" (p. 2, Reference [12]) [19,20].

Undoubtedly, residential satisfaction is a complex process that displays different perspectives because urban regions' socio-spatial patterns differ significantly from one geographical region to another and even from one neighborhood to another [4,13]. Therefore, an obstacle in researching the relationships between various dimensions of the residential environment is that, in general, predicting satisfaction in an urban community context is a challenging task.

1.1. Residential Attractiveness in Central and Eastern Europe (CEE) Countries

The post-industrialization period and the political changes after the fall of the communist regimes meant that the urban areas of CEE countries saw "the formation of a new socio-economic framework of cities" (p. 23, Reference [21]). These areas embraced old and completely new functions and generated population relocations toward new residential projects. The different paradigms on housing quality brought forward by the open-market economy implemented after 1990 influence the intention to stay or move out in former communist countries [22]. One consequential result was that the central authorities no longer decided on the dwellings' dimensions, functionality, compartmentalization, endowments, or design [23–26]. A new housing culture was necessary due to the numerous problems of housing built during communism which were only aggravated by insufficient investments and inadequate management, the physical degradation of the buildings, and their numerous technical issues. The lifespan of these buildings was long overdue, the investment capital oriented for their rehabilitation was scarce, the government programs limited, and the financial power of their inhabitants virtually inexistent [27].

In this context, in states such as Poland, Estonia, Hungary, Eastern Germany, Bulgaria, the Czech Republic, Slovakia, and Romania, the population started preferring newer dwellings, and suburban sprawl became a dominant phenomenon caused by a lack of locally implemented territorial planning programs [28]. In fact, the specialized literature recognizes that suburbanization has been particularly important in transforming the metropolitan regions of CEE countries such as Romania in the last decades. Because it was artificially limited during communism, after 1990, some countries even experienced a "chaotic suburbanisation", which decreased the quality of the residential environment (p. 160, Reference [29]); (p. 46, References [30,31]). This is a direct consequence of a neoliberal housing market that has not always found integrated and coordinated development solutions applicable throughout all the newly appearing residential areas [32–37]. The "differentiation" or, rather, "(re)-differentiation" process of "reordering the scale and location of urban land uses" had a higher intensity in the cities of the post-communist countries than in the "classic western European ones" (p. 23, Reference [21]).

Neighborhood satisfaction is also linked to the principles of sustainable urban development [38,39]. However, although integrated strategies for sustainable urban development that address economic, environmental, climate, social, and demographic challenges exist in many Central and Eastern European cities, they are only superficially implemented. Difficulties arise from the fact that financial resources must be concentrated in an integrated way, especially for areas facing specific urban challenges [40]. Moreover, local authorities are the ones responsible for implementing the integrated urban development strategies, respectively, selecting the projects that ensure sustainable development, which only slowed these processes down or stopped them completely [41].

1.2. Residential Dynamics in Bucharest during the Post-Communist Period

In post-communist Romania, sustainable housing policies have been applied to a limited degree and many times with only a locally oriented vision, without considering urban regeneration or durable systematization, and this trend has had negative repercussions, especially in terms of residential satisfaction [42].

The capital city of Bucharest registered a solid territorial dynamic after 1990. Its residential area extended toward its periphery in an urban spill and was dictated more by land prices and less by local development or systematization policies [43]. The emergence of a housing market that did not exist before 1990 was caused mainly by Law No. 18/1991, which returned land plots to their former owners and allowed for changes in their function of agricultural and forested plots, which, in time, saw many of them used for residential, commercial, or industrial purposes [44]. In this context, residential satisfaction was a secondary preoccupation for Bucharest's urban development.

The city developed in stages, especially in the post-communist period; its evolution focused on its historical center and extended outwardly [45–47]. According to Suditu et al. [48], when analyzing Bucharest's residential areas, three relatively homogenous divisions stand out: (i) the central area corresponding to the historical center, which developed spontaneously and lacked a controlled systematization process; (ii) an intermediary area represented by communist workers' neighborhoods, totally systematized and in the same time functional because it incorporates services (e.g., health and education), as well as parks and recreational areas; and (iii) an exterior area that contains new residential nuclei alongside rural patches.

Before 1990, authorities of the communist states decided on a city's development rhythm by controlling its spread, as well as the characteristics and number of dwellings built. However, today, the large capitals of these countries, including Bucharest, undergo an urban-sprawl-like suburbanization process [49]. In the post-communist period, Romania's large cities' urban landscape registered tendencies of rapidly extending their residential and services areas, especially storage and retail, particularly inside their area of influence [50], and simultaneously de- and relocating some of their other functions (especially the industrial one) toward the periphery (Figure 1). Against continuous deindustrialization, the market resorted to extending residential areas. The removal of the industrial structures built during communism came after successive unsuccessful restructuring attempts, which led to their bankruptcy and demolition and, in the end, replacement with residential nuclei [51–53]. As a result, after the 2000s, Bucharest's territorial development was primarily aimed at its peri-urban area due to these recently vacant plots, with previous agricultural areas around the city converted into residential nuclei [54]. In actuality, the city extended not according to legislative measures or by following a development strategy established by authorities but according to demand and supply, with developers strongly influencing this process. The emergence of these residential nuclei resulted from the changes the city itself went through, namely the migration of certain inhabitants belonging to a "matured" middle class gaining higher incomes [46] and the economic relocation of certain functions (e.g., industrial production, storage and retail, other services, etc.), including the residential function toward the periphery (Figure 1). Apart from people looking for housing in different areas, the development of new residential nuclei came down to Bucharest's increased number of inhabitants. This was fueled by large waves of newcomers from around the country, encouraged by the abrogation of an urbanism communist law that controlled urban growth and defined the so-called "closed cities". During the communist period, the Decree No. 68 of 17 March 1976 controlled the evolution of the urban population, thereby allowing provincial-born citizens to live in large cities (Bucharest, Iași, Cluj-Napoca, Timișoara, Constanta, and Brașov) only if they could prove a familial connection or an economic interest through a contract of employment or a state dispensation [55]. The city continues to be seen as an economic and innovation national pole, always needing a specialized workforce in sectors such as business, services, and newly emerging industries [53]. Its area of influence stretches primarily toward the north (Băneasa-Otopeni), west (Chiajna commune), and, in later years, south (Berceni commune), the development of the city translating into an extension to the outer rural area, interspersing in all directions urban spaces with areas with rural characteristics [56].

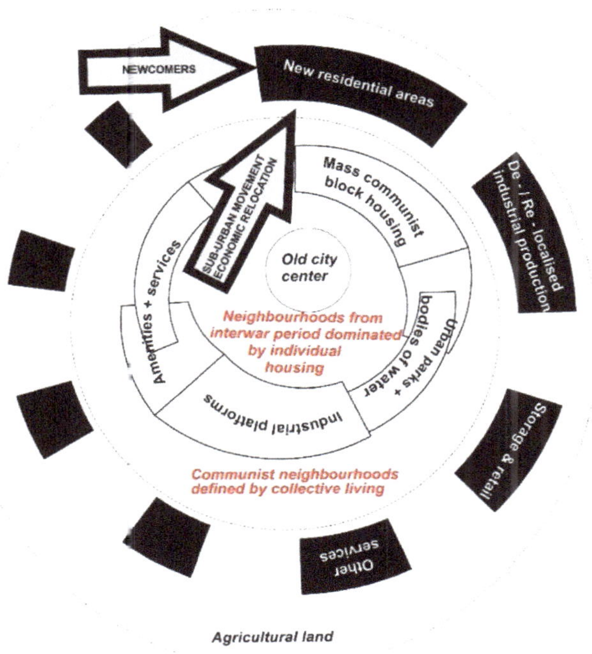

Figure 1. Conceptual framework of Bucharest's urban structure and dynamics.

Young families also fed the high demand for new housing, because, in Romanian society, having a home is an important criterion when getting married and starting a new family [57]. In order to meet this social constraint, a governmental program was introduced immediately after the financial crisis of 2008. It was called "The First Home" and later renamed "The New Home" and offered new families the possibility to buy a dwelling with the help of a state-guaranteed low-interest loan (for old homes, the state guaranteed 40% of the amount, and for new buildings, 50%). The program aimed to revitalize the construction sector severely affected by the 2008 crisis and offer new young families the possibility of owning a home [58]. Real-estate developers kept in mind that their buyers would have to access a bank loan according to their limited incomes; as such, the newly built dwellings had to fit into a budget as small as possible. One way to ensure this was to use cheap land plots outside Bucharest.

Some aspects of the Eastern European regional socio-spatial pattern, such as "the dominance of owner-occupiers of the existing housing" or "the lack of affordable housing and underdeveloped mortgage markets" (p. 61, Reference [13]), emphasized the new residential nuclei as viable, attractive solutions for both newcomers and locals. Similarly, the local and regional patterns of "certain historically specific housing forms and space consumption and its affordability" could not be neglected because Romania is, unfortunately, one of the most imbalanced countries considering its over-dominant 98% homeownership rate and least mobile and financialized housing market within the EU (p. 1056, Reference [59]). After the year 1990 in Romania, similar to other former communist countries, the living preferences in terms of dwelling type changed, and parts of the population gravitated toward either unifamilial villa-type houses or large apartments, both of which were located inevitably in the city's periphery or its peri-urban area [37,60–62]. According to Soaita and Dewilde [59], there were also constraints brought by the old housing stock built during communism (75%) that offers minimal options for developing more spacious and comfortable homes, which higher-income dwellers might require.

Accordingly, the new residential nuclei should meet modern standards in terms of quality of living, both from a functional and aesthetic point of view, because the new consumers perceived them to be in opposition to the residential areas built during communism with their numerous problems and shortcomings. Nonetheless, specialists and numerous journalistic inquiries decried the lack of superior living standards in these residential areas, identifying and presenting problems and inconsistencies in their functional zoning and urban endowments [63–66].

1.3. The Aim of the Study

This exploratory study aims to approach, through applied research within Bucharest, Romania, a subject of high interest for administrative authorities, planners interested in local development strategies or housing policies, and their direct beneficiaries: the local population. Its primary focus is to evaluate whether the analyzed new residential nuclei correspond to the needs and wishes of their residents in terms of amenities, comfort, utilities, location, price, etc. These residential areas' chaotic development is a subject debated arduously by the civil society, press, and administrative structures, all emphasizing residential satisfaction as an influential element for their future evolution.

In the above context, our study has a twofold contribution, firstly to bring more in-depth insights from the new residential nuclei built in the last two decades in the suburban area of Bucharest, and secondly to analyze the inhabitants' residential satisfaction in these areas.

To achieve this, we distinguished the following main research objectives:

- O1. Identifying the characteristics of the residential nuclei inside the study area;
- O2. Analyzing the satisfaction level of the inhabitants concerning the residential area they live in;
- O3. Analyzing the satisfaction level of the inhabitants concerning their dwelling (individual or collective).

2. Methodology

The objectives set for this study required a threefold methodological approach, as illustrated in Figure 2.

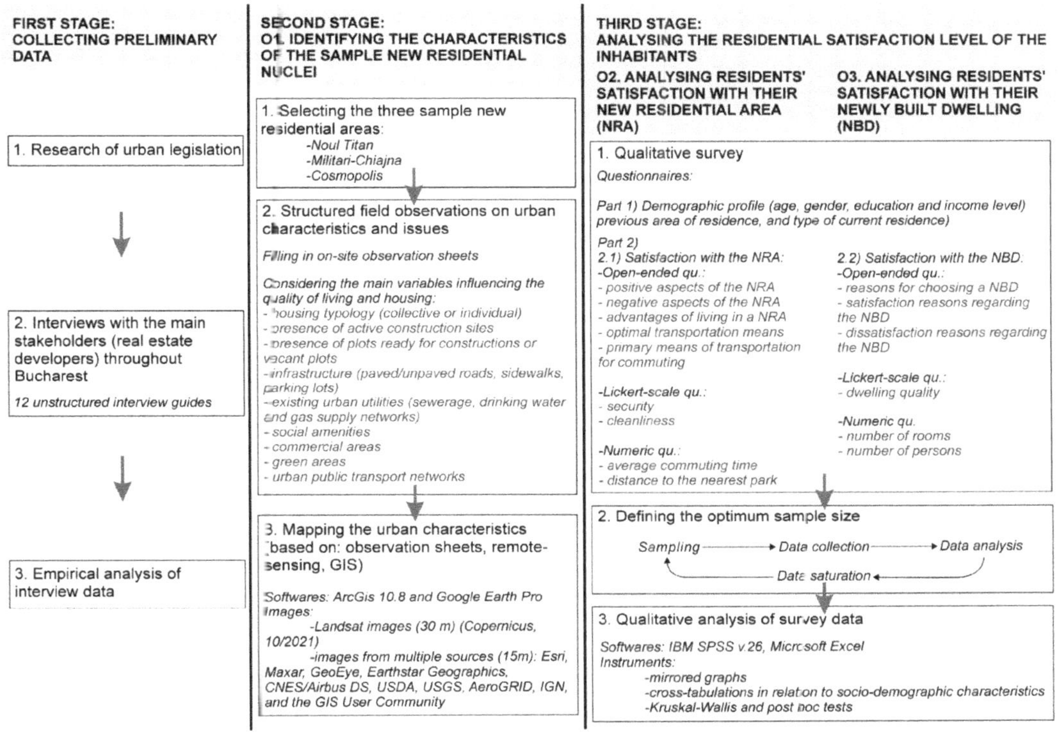

Figure 2. Methodological framework of the study.

The first stage was a preliminary information-gathering one. It required studying urban planning legislation, and it was followed by interviewing the actors who directly shape the residential space: real-estate developers. Real-estate developers were interviewed between March and April 2021 because of their overall importance, participation with real-estate projects in the study area, as well as their seniority and experience in the construction sector in Bucharest, Romania. Furthermore, the analysis of these interviews helped shape the observation sheets' content and select the variables for the questionnaire applied later to residents. The tool used in this stage was an unstructured interview guide (12 such interview guides were conducted) with developers throughout Bucharest, regardless of the shape and amplitude of their real-estate projects. These interviews probed the difficulties in developing real-estate projects; the factors they consider when choosing the dimensions of a dwelling; how the existing infrastructure and the technical and municipal utilities condition the execution of a project; the interconnection between the final price and quality of the construction materials; and the difficulties created by the existing legislation.

The second methodological stage consisted of filling in on-site observation sheets during May 2021. This stage aimed to map the characteristics and issues of the newly developed residential nuclei, while also considering the interviews conducted with the real-estate developers. The criterion used in selecting the three sample residential areas was their geographical location in relation to the capital city. The first nucleus is Noul Titan, located in the eastern part of Bucharest, inside the administrative city limit. The second nucleus is

Militari–Chiajna, located in close proximity to the administrative city limit. Finally, the third studied residential nucleus is Cosmopolis, located 15 km outside Bucharest on its road belt (DNCB) (Figure 3). The three areas were selected to compare the difference in residential features, causes determining these differences, and disparities in how residents perceive housing conditions in the three newly built areas and the neighborhoods themselves.

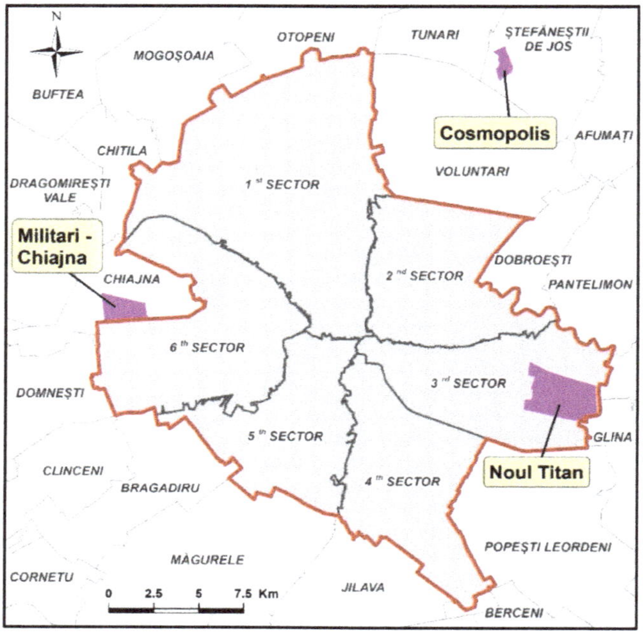

Figure 3. Position of the sample new residential areas in relation to Bucharest.

The first one, **Noul Titan**, started developing after the year 2000 and is located inside the city and extends up to its administrative limit. This residential nucleus, in the eastern part of the city, is actually an extension of the Titan neighborhood (i.e., Noul Titan translates into The *New* Titan), which was built between 1950 and 1970 to serve as a dormitory area for the workers of the large industrial plant located in its vicinity. In the post-communist period, the development of this new residential space happened in stages. It was an inhomogeneous process, with single-family homes, villas, block-type residential buildings, or residential complexes slowly being built on vacant plots, former agricultural areas, or plots newly available after the demolition of former industrial units (the Policolor factory, Bucharest, Romania). Today, Noul Titan is still growing, with former agricultural lands gradually transforming into residential areas.

The second study area, **Militari–Chiajna**, is located in the western part of the city in the immediate vicinity of the administrative limit of Bucharest's 6th sector. The sample area is peri-urban, administratively belonging to the Chiajna commune, Ilfov County. The authors chose it to emphasize the amplification of the urban overflow phenomenon. The low prices of these plots and the area's location (i.e., the first commercial platform appeared here after 1989, along Highway 1) aided the emergence of this new residential nucleus, as its interconnectivity with Bucharest represented one of its marketing advantages.

The third chosen residential nucleus, **Cosmopolis**, is located 15 km to the northeast of Bucharest, placed in its urban area but administratively belonging to the commune Ştefăneştii de Jos, Ilfov County, Romania. It was built to incorporate many facilities. Real-estate developers included from the very beginning a public pool, an urban beach, green areas, playgrounds, commercial units, and gyms in the construction project.

The structured field observation process considered variables that influence the quality of living and housing, taking into account the authors' expertise, as well as the elements that resulted from the interviews with the real-estate developers (Figure 2). At this stage, the work methodology also included observing, analyzing, and mapping the urban characteristics of the sample residential nuclei. The maps were constructed by correlating the information from the observation sheets with remote sensing and GIS techniques to identify the distribution of different land uses (residential, commercial, green areas, etc.). The mapping was performed during the second half of 2021, using Landsat satellite images (Copernicus, 25 October 2021) in Google Earth Pro, with a 30 m resolution [67]. This information was refined with images from multiple sources: Esri, Maxar, GeoEye, Earthstar Geographics, CNES/Airbus DS, USDA, USGS, AeroGRID, IGN, and the GIS User Community [68]. The final cartographic material was obtained by using ArcGIS 10.8 and Google Earth Pro.

In the third stage of the research process, during July and August of 2021, questionnaires were conducted to support objectives O2 and O3 of the research. The authors used the qualitative survey method suitable for the exploratory design of this study and its research objectives [69], namely residents' perceptions regarding their neighborhoods (residential nuclei) and dwellings. The research considered rather "the logic of the study than statistical probability" (p. 2, Reference [70]) and used the data previously obtained from observation sheets and interviews to define the variables in the survey. Expressly, qualitative research is intensely employed in themed studies on living satisfaction, with many researchers also incorporating several quantitative approaches, such as cross-tabulation of their data [70]. The target population referred to residents (both renters and owners) from the three residential nuclei, and the optimum sample size was obtained by considering the parameters of the phenomenon studied and the research objectives [69,71]. According to different authors, the sample size in qualitative research should not be ample, because this would impede observing and analyzing the phenomenon [72,73]. This stage aimed to gather a large variety of opinions through mainly open-ended questions and fewer semi-opened/Likert scale questions.

The coding and primary analysis of data allowed us to use data saturation techniques [69]; adjust the optimal sample size; and combine sampling, data collection, and data analysis, rather than treating each of them separately in a linear order [74]. The theoretical saturation is the moment when no new information is offered that would allow the researcher to identify additional features of the studied phenomenon [69,75] and is often combined in practice with data saturation [73]. Starting from this rationale, the authors considered the final number of respondents as optimal [76–78] while also fulfilling the representativeness level of the sampling for each residential area and avoiding biased responses because each respondent was a unique representative of a household of the sample areas. The random sampling technique was used to identify respondents, and the sample stratification for each residential area was rather a result in this case. The simple random sampling, suitable for generalization in exploratory studies [72,78], was used multiple times to evaluate residential satisfaction in urban environments [4,13,79]. Although the authors did not aim for a respondents' stratification based on sociodemographic variables, to ensure the sample representativeness and accuracy and to cover the diversity of situations that may occur in terms of living conditions, the study sample was "a sizeable proportion" (p. 81, Reference [68]) of the target group (10% of all households living in the selected areas) and represented a balanced coverage of both types of housing (i.e., individual and collective). Because the survey was carried out during the COVID-19 pandemic crisis, the qualitative questionnaires were applied online through owners' associations' groups on Facebook or WhatsApp. The questionnaires filled incorrectly or incompletely were eliminated. As a result, the final valid number of respondents was 126 (34 from Noul Titan, 52 from Militari–Chiajna and 40 from Cosmopolis).

The first part of the questionnaire contained queries relating to the respondents' demographic profile (i.e., age, gender, education, and income level) (Table 1), as well as variables important to the regional specificity of the area and the case study (previous area of residence, i.e., Bucharest or the rest of the country; and type of current residence, i.e., apartment in a block of flats within a residential area, apartment in a block of flats outside a residential area, house within a residential area, or house outside a residential area).

Table 1. Respondents' sociodemographic profile in the three sample residential nuclei.

Age		Noul Titan	Militari–Chiajna	Cosmopolis	Total
18–29	% within the sample area	15.0%	19.2%	47.1%	-
	% of total	4.8%	7.9%	12.7%	25.4%
30–39	% within the sample area	55.0%	44.2%	26.5%	-
	% of total	17.5%	18.3%	7.1%	42.9%
40–49	% within the sample area	17.5%	28.8%	20.6%	-
	% of total	5.6%	11.9%	5.6%	23.0%
50–65	% within the sample area	10.0%	7.7%	5.9%	-
	% of total	3.2%	3.2%	1.6%	7.9%
>65	% within the sample area	2.5%	0.0%	0.0%	-
	% of total	0.8%	0.0%	0.0%	0.8%
Gender		**Noul Titan**	**Militari–Chiajna**	**Cosmopolis**	**Total**
male	% within the sample area	22.5%	50.0%	50.0%	-
	% of total	7.1%	20.6%	13.5%	41.3%
female	% within the sample area	77.5%	50.0%	50.0%	-
	% of total	24.6%	20.6%	13.5%	58.7%
Education		**Noul Titan**	**Militari–Chiajna**	**Cosmopolis**	**Total**
High school	% within the sample area	2.5%	17.3%	17.6%	-
	% of total	0.8%	7.1%	4.8%	12.7%
University	% within the sample area	97.5%	82.7%	82.4%	-
	% of total	31.0%	34.1%	22.2%	87.3%
Monthly income/person (EUR)		**Noul Titan**	**Militari–Chiajna**	**Cosmopolis**	**Total**
<607	% within the sample area	17.5%	21.2%	14.7%	-
	% of total	5.6%	8.7%	4.0%	18.3%
607–1214	% within the sample area	42.5%	65.4%	47.1%	-
	% of total	13.5%	27.0%	12.7%	53.2%
>1214	% within the sample area	40.0%	13.5%	38.2%	-
	% of total	12.7%	5.6%	10.3%	28.6%

The income level differs among residents from the three residential nuclei, but in Cosmopolis and Noul Titan, most residents split relatively evenly between earning between 607 and 1214 EUR (3000 and 6000 RON) and above 1214 EUR (6000 RON); the Militari–Chiajna area sees most residents earning between 607 and 1214 EUR (3000 and 6000 RON) (Table 1). The threshold of 3000 RON (607 EUR) was chosen instead of the national minimum wage of 458 EUR on 1 January 2021 [80], as the first represents, according to the latest studies [81], the special minimum gross salary for a series of domains in Romania. This better corresponds to the occupational profile of our respondents, as those working in Bucharest generally have higher incomes than people in other regions of the country. The age-group distribution sees a predominance of young adults and adults, as these age categories are primarily eligible for accessing "The First Home" program (Table 1).

The second part of the questionnaire was built around the two research objectives. Understanding the residential satisfaction and mobility patterns was essential in the complex sociocultural context of this study, and using open-ended questions (e.g., positive vs. negative aspects) or Likert scales in order to characterize more in-depth certain key variables (e.g., security and cleanliness) was the best way to obtain genuine, unbiased answers.

The questions measuring neighborhood satisfaction targeted the positive and negative aspects of these areas as perceived by the respondents and were then cross-checked with an additional question on their perceived benefits of living in a new residential nucleus.

Another set of questions focused on the satisfaction vs. dissatisfaction reasons regarding the respondents' new dwellings and the reasons that prompted them to buy one in order to identify the advantages and deficiencies that owners discovered in their newly built dwellings. Moreover, questions on the number of rooms and the number of residents had the role of statistically checking the comfort level of their chosen new residential nuclei.

For the bivariate viewing and analysis of the questionnaire data (performed mainly through the IBM SPSS Statistics v.26 software, Armonk, NY, USA), we used mirrored graphs to portray the satisfaction with new residential nuclei and dwellings, while also distinguishing among the three sample neighborhoods. We further used cross-tabulations between two categorical variables, which display the data in the two-dimensional space, given by rows and columns, thus examining possible relationships between row and column variables. In these tables, the relative frequencies of respondents' perceptions of the new residential nuclei that they are inhabiting and their newly built dwellings were arranged in relation to several of their descriptive sociodemographic characteristics (e.g., age, education, income, their previous area of residence, or type of building they currently live in). Because the non-parametric Kruskal–Wallis test does not assume normal distributions of the data, it was used to assess if there are significant differences in the rank or numerical type of responses provided by the interviewees as a function of the residential nuclei they inhabit. The null hypothesis states that the distributions of all groups are equal, while the alternative hypothesis is that at least one group's distribution is different from that of another group. A significant Kruskal–Wallis test result indicates that the null hypothesis is rejected and that, thus, at least one sample stochastically dominates one other sample. In cases where significant differences among the groups were observed, the analysis was further completed by a post hoc test to explore where and for how many pairs of groups the stochastic dominance occurs [82]. The post hoc analysis was conducted by using the pairwise Dunn's test with the Bonferroni correction.

Word clouds were used as visual tools to present phrases frequently mentioned by residents to describe their recommendations on improving the overall quality of living based on their recent experiences. This graphic instrument uses an algorithm based on basic linear, power, and logarithmic representation of font sizes, thus summarizing the texts [83].

3. Results

3.1. Urban Planning Rules and Real-Estate Developers

Analyzing the urban planning legislation allowed us to identify the notable documents that regulate territorial planning. This stage aimed to identify both the rules that govern this process and their coherence, as well as the existing dysfunctionalities that might result in inadequate urban development.

The "National Housing Strategy for 2017–2030" recognizes several problems that the Romanian housing market is experiencing. They number an ageing housing fund, a small amount of public housing, and an increasingly unofficial and unregulated renting market. As the need for housing manifested stronger in larger cities, uncontrollably meeting this demand has led to several imbalances in the housing market, with many of them impacting housing quality [84]. That is why the most poignant problems of the new residential nuclei relate to insufficient or missing means of transportation, urban utilities, or socio-cultural infrastructure of the lack of green areas [85].

The rules that govern territorial and urban planning are set by Law No. 350 from 2001, which corrects the previous legislation and details the process of authorizing building permits and some measures relating to dwellings [86]. This law went through numerous changes that resulted in ambiguities and contradictions regarding its interpretation and applicability.

The central public administrations (i.e., the Ministry for Development, Public Works and Administration; and the Ministry of Investments and European Projects) and the county and local councils are responsible for implementing development strategies, constructing and updating urban planning documents, as well as supervising the execution of the General Urban Regulations [87].

Decision No. 525 from 1996 on approving the General Urban Regulations is meant to prevent dysfunctionalities and ensure living standards per the existing legislation. There are, however, numerous ways in which it fails to achieve its purpose. The field observation proved many instances where the following stipulations were not fully respected: fire-extinguishing vehicles must have access to the construction through public roads; pedestrian access (sidewalks, pedestrian streets, or squares) must exist according to the importance and destination of the building; and new residents must have the possibility to connect to existing drinking-water, sewerage, gas, and electrical infrastructures. The legislation is also very convoluted. Article 28, for example, stipulates that enhancing the capacity of public utility networks falls, partially or fully, into the responsibility of the real-estate developer or beneficiary depending on a series of conditions which can, in fact, deprive any of them of this responsibility. Real-estate developers face copious obligations, as they are responsible for conserving existing green areas and creating new ones offering a minimum of 2 sq. m per inhabitant [88], they must also construct parking lots according to the type of construction and surface area, serving as much as 60% of the residents (i.e., for individual housing) and 100% (i.e., for collective housing) [89].

The numerous exceptions that allowed for interpreting the current legislation have convinced governmental and local authorities to initiate changes to relevant laws, thus affecting the activities of real-estate developers directly. For example, a proposal by the Ministry of Environment, Water, and Forests to ban individual/apartment gas boilers is currently in public debate and is contested by many real-estate developers (i.e., who complain that the final construction costs would increase) and by consumers alike, with the latter preferring an independent heating solution given the deficient performance of the existing centralized heating system [90,91]. Another debate is ongoing in Bucharest, the capital city, over a proposal that real-estate developers should build kindergartens, schools, roads, and parks as part of the residential nuclei, not the local authorities who previously assumed this task [92].

The interview guides we conducted with real-estate developers showed that their most important criticism revolves around the current state of the infrastructure generated, in their opinion, by the lack of local development strategies, "*City Hall has its own responsibilities. Authorities should invest in road, water, and sewerage networks and pave the existing roads*" (developer, Bucharest's fifth sector). They described the existing legislation as "*cumbersome and often unclear*" (developer, Bucharest's first sector).

All of them pointed out that, when it comes to housing nowadays, most buyers rely on bank loans to purchase a dwelling, with those loans being a reflection of their income level, so they often "*chose to decrease the housing area so that people can afford the price*" (developer, Bucharest's second sector). If they were to comply fully with modern living standards in terms of housing area, facilities, quality of materials, and interconnectivity among parts of the residential area, as well as between the area itself and the city, the final prices would increase beyond the affordability threshold of the average buyer, as "*it would increase the price of an apartment to 4000 euros per built sq. m., and no one would buy them*" (developer, Bucharest's fourth sector).

3.2. *Identifying the Residential Typologies of the Three Study Cases*

The territorial evolution of the new residential areas in Bucharest differs depending on their location. The emergence of new nuclei depends on many factors. The first one is the availability of empty land plots within the city's administrative limits. After the year 1990, these were occupied immediately by constructions; thus, especially after the year 2000, the possibility of new residential nuclei appearing inside the city limits decreased exponentially. A second factor was the reconversion of industrial plants or agricultural plots located outside the city's administrative limits into residential areas.

The information extracted from the observation sheets cumulated with GIS analysis shows that the residential pattern of the Noul Titan nucleus is a mix of individual housing in the form of houses or villas, small duplex villas complexes, P + 4 block-type residential buildings, or residential areas with collective housing of either P + 4 or P + 10 levels (Figure 4).

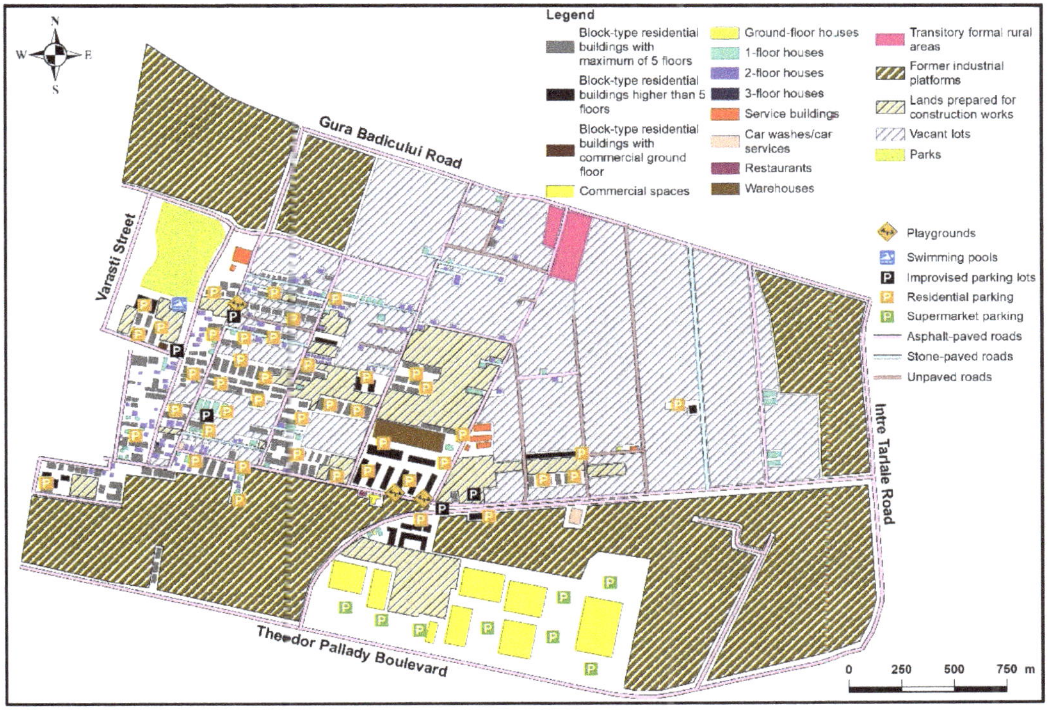

Figure 4. Residential features of the Noul Titan area.

There is no systematization regarding the buildings' height regime or designation, with built-up areas interposing with vacant lots or areas prepared for construction works. In addition, there are still transitory former rural areas containing old houses with rustic features constructed before the communist age on what was then agricultural land.

Centralized sewerage and drinking-water networks cover the entire residential area, and developers connect the new constructions to the already existing infrastructure. In addition, the interconnectivity and accessibility of the public transport network are ensured by numerous subway, tram, and bus stations located very close to the residential area. The primary and secondary roads are asphalt-paved and have sidewalks.

From a residential pattern point of view, Militari–Chiajna includes individual (i.e., P + 2 villas) and collective housing that see a wider variety: P + 4 block-type residential buildings located near the villas and P + 10 block-type residential buildings, and residential complexes that, in addition to residential buildings, also include leisure, commercial, finance/banking, education, and other units (Figure 5).

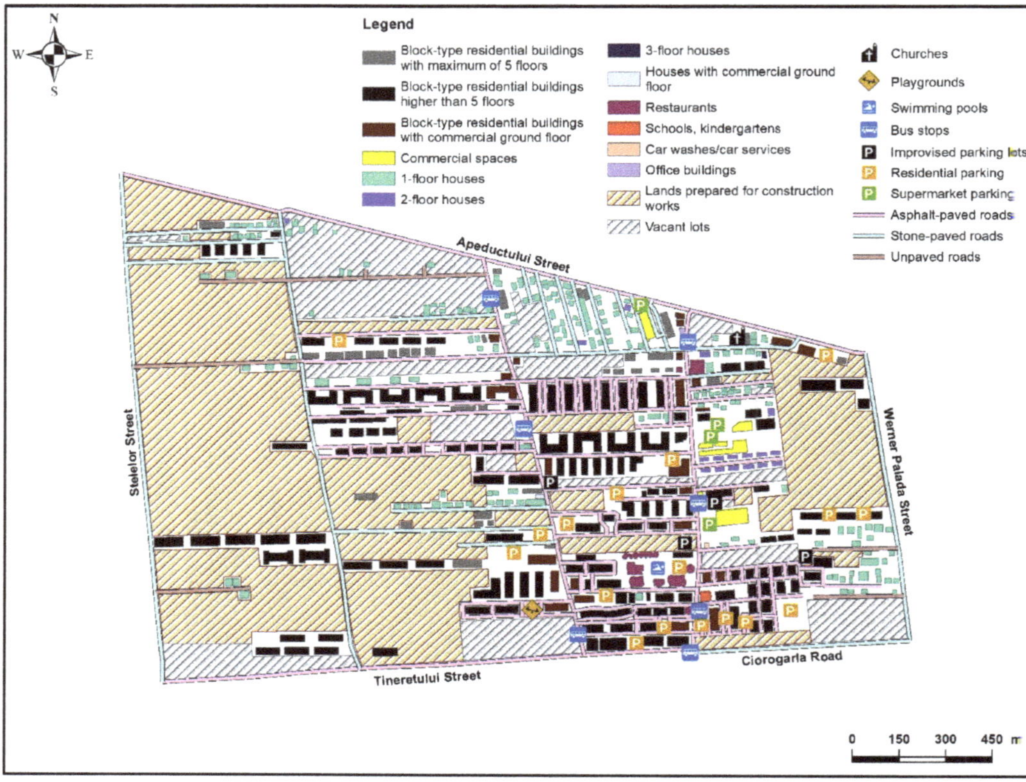

Figure 5. Residential features of the Militari–Chiajna area.

This nucleus extended without a systematization plan implemented by local authorities, so utilities such as centralized sewerage or drinking-water networks are absent, and developers connected the new buildings to septic tanks and surface wells. Accessibility is a significant issue, as, on account of its peripheral location, the transport interconnectivity with public services is supplied by a single bus line with a private transport line also available. The primary roads are asphalt-paved, and some of the secondary roads are stone-paved but are not wide enough to allow access to large intervention vehicles. The sidewalks are discontinuous since neither the developers nor the local authorities included them in the construction projects.

The Cosmopolis residential nucleus partially meets the "gated community" criteria, as it proposes a lifestyle specific to a community with above-average income levels in the form of a "prestige project", focused on image, intimacy, a few shared facilities, amenities, and security inside an area where access is limited and always controlled [93,94].

This nucleus' residential pattern comprises collective housing, such as P + 5, P + 6, P + 10, and P + 12 block-type residential buildings and individual P + 1, P + 2, and P + 3 houses; and P + 1 duplex villas (Figure 6).

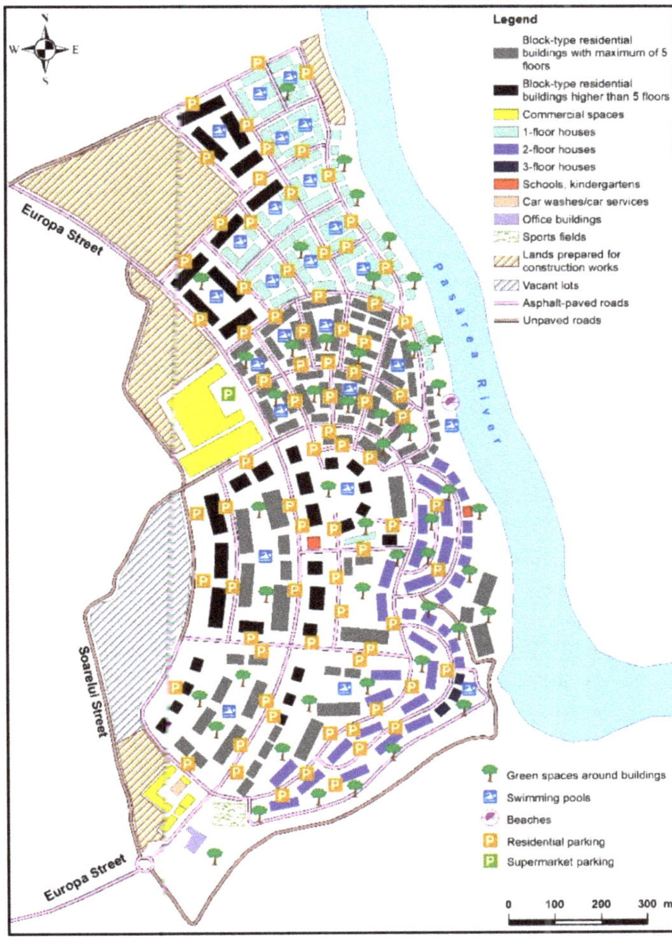

Figure 6. Residential features of the Cosmopolis area.

In terms of technical-urban utilities, the information extracted from the observation sheets determined that no centralized sewerage or drinking-water networks exist. Therefore, the developer set up a system strictly for the residential nucleus, using water supplied from surface wells and septic tanks for sewage evacuation. Cosmopolis is not connected to Bucharest's public transport system, so residents need to use personal cars or private transport that eventually connects them to areas covered by the public means of transportation in the northern part of the city. The primary roads (i.e., Bucharest's ring road) and the secondary roads (i.e., certain roads inside the complex) are asphalt-paved, with the latter also having sidewalks. The need for such a development was heightened by its isolated position and the fact that the local administration of the commune Ștefăneștii de Jos offered limited public amenities related to security and urban or social utilities. This residential nucleus was built on agricultural lands that were subsequently reclassified as urban, so, in contrast to the other study areas, its design benefited from functional zoning suitable for housing. Today, Cosmopolis continues to expand, as high demand has motivated the developer to build new housing.

3.3. *Analyzing the Neighborhood Satisfaction in the New Residential Nuclei*

Residents' evaluation of their housing depends on their needs, preferences, and expectations. Dissatisfaction appears when the housing conditions (i.e., as they perceive them) do not match their expectations [4,95]. Expectations are defined according to their needs, which can differ from family dimension to financial, social, or cultural needs but are conditioned, in the end, by income level, which dictates housing typology and features. Considering these restraints, the future residents mold their expectations to include the benefits that a new residential nucleus, their future neighborhood, might offer. The survey showed that the most often expressed expectation was *new dwellings with modern amenities and comfort* (43.7% of all answers). A second criterion for "choosing to live in a new residential nucleus" was *security*, followed by *affordable price*, with shares of 12.7% and 11.9%, respectively, of all answers. The first criterion is the most often cited regardless of the income bracket, but it is worth mentioning that those who earn less than 607 EUR place the *affordable price* in the second most important place, while those earning between 607 and 1214 EUR monthly and more than 1214 EUR are more interested in *security, private parking place* and *recreational facilities* (with *security* being cited as second most important by the latter group) (Table 2).

Table 2. Advantages of living in a new residential area by income brackets cross-tabulation.

Monthly Income/Person (EUR)		Advantages of Living in a New Residential Area							Total
		Affordable Price	Easy Access to Commer-cial Services	New Dwellings with Modern Amenities and Comfort	Private Parking Place	Recreational Facilities	Security	No Advantage	
<607	% within income	17.4%	0.0%	43.5%	8.7%	4.3%	8.7%	17.4%	100.0%
	% of total	3.2%	0.0%	7.9%	1.6%	0.8%	1.6%	3.2%	18.3%
607–1214	% within income	10.4%	6.0%	38.8%	13.4%	10.4%	11.9%	9.0%	100.0%
	% of total	5.6%	3.2%	20.6%	7.1%	5.6%	6.3%	4.8%	53.2%
>1214	% within income	11.1%	0.0%	52.8%	5.6%	5.6%	16.7%	8.3%	100.0%
	% of total	3.2%	0.0%	15.1%	1.6%	1.6%	4.8%	2.4%	28.6%
Total	% of total	11.9%	3.2%	43.7%	10.3%	7.9%	12.7%	10.3%	100.0%

The analysis of the age group distribution showed that, when choosing a new residential area, respondents aged 18 to 49 mainly favored their *modern amenities and comfort* (40% to 52%). Specifically, modernity was viewed as the main benefit of a new residential nucleus to its largest share (43.6%) by those aged between 30 and 39. On the other hand, although to a lesser degree, *security* mattered more for those aged over 50 (20%). Conversely, *recreational facilities* were more consequential in selecting a new residential area for the younger segment of respondents, namely those aged between 18 and 29 and 30 to 39. However, when considering the overall survey, *recreational facilities* were only relevant for as much as 7.9%. Another point worth mentioning is that 1 out of 10 respondents in general and 3 out of 10 respondents aged over 50 declared there are *no advantages* to living in a new residential nucleus. This signals the shortcomings these areas experience due to insufficient planning in incorporating beneficiaries' needs. In terms of the three chosen sample areas, more than three-quarters of the residents from Chiajna–Militari in this age segment declared the lack of any advantage, clearly showing the area's deficient housing standards, especially concerning age-friendly housing for the elderly.

In terms of their origin, respondents who had previously lived in Bucharest were more interested in *new dwellings with modern amenities and comfort* than those who came to the capital city from elsewhere in the country (43.7% to 37%, respectively). On the other hand, they were less interested in a *private parking place* (8.3% against 18.5% within their category), explained by the latter group's higher need for a car as a means of transportation within the city and especially outside it, to connect with family and friends from their previous residence.

The comparative analysis of the "negative vs. positive aspects of the new residential nuclei" shows that the highest number of respondents identified *security* as a positive aspect, while mostly complaining about *amenities*, with the latter being difficult to interpret since the same appears to be also a broadly mentioned positive aspect. On the other hand, the "negative aspects" that seem to bother residents the least are *security* and *cleanliness*. However, considering the three residential nuclei individually, there are some differences. For example, *asphalted roads*, *amenities* (sidewalks, public lighting, playgrounds, and gyms), *air quality*, and *sewerage* infrastructure are weak points in Militari–Chiajna, while *security*, *price*, and *neighbors* were viewed as assets by residents when considering this area as their future home. Cosmopolis's major disadvantages relate to *transport* (i.e., it lacks public transport) and *neighbors*, while its *security*, *amenities*, and *air quality* are seen as assets. For the Noul Titan nucleus, respondents have not mentioned a large number of disadvantages, while its advantages (*transport* and *amenities*) stem from its intra-urban location (Figure 7).

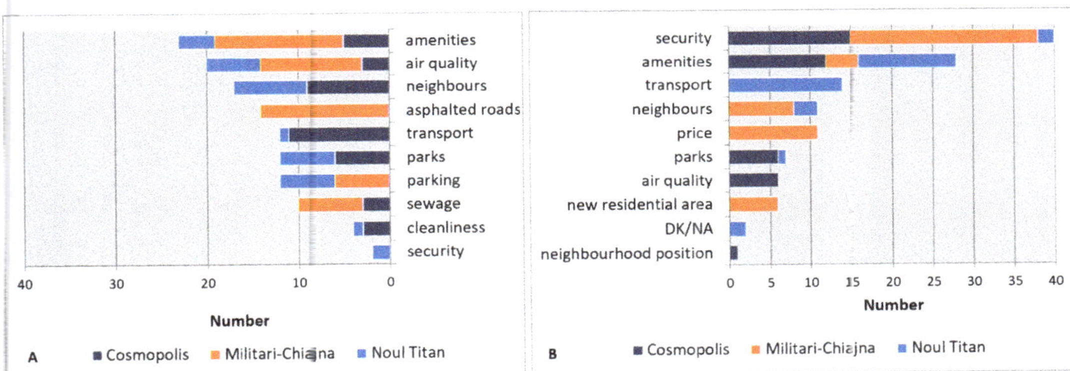

Figure 7. Negative (**A**) and positive aspects (**B**) of the new residential nuclei as per residents' opinions.

Table 3 displays the Kruskal–Wallis test results of the questions with answers expressed by variables of ordinal and numerical type. The analysis was performed to check if there are differences among the three sample residential nuclei in terms of elements defining everyday comfort and living standards, such as "security", "cleanliness", "average commuting time", and "distance to the nearest park". According to the significance level ($p < 0.05$), the null hypothesis is rejected in the case of "security", "average commuting time", and "distance to the nearest park". Therefore, for these three elements, at least one of the neighborhoods is different from the others, with the most significant differences related to the "average commuting time". In the case of "cleanliness", identified above as one of the least bothering negative aspects, the null hypothesis is not rejected, pointing to the relative similarities in the interviewees' relative satisfaction in all neighborhoods.

Table 3. The non-parametric-test results for the questions' items with Likert scale responses.

Question	Independent-Samples Kruskal–Wallis Test Summary		
	Test Statistic	Degree of Freedom	Asymptotic Sig. (2-Sided Test)
Security	16.164	2	0.000
Cleanliness	0.017 [a,b]	2	0.992
Average commuting time	50.885 [a]	2	0.000
Distance to the nearest park	43.110 [a]	2	0.000

[a] The test statistic is adjusted for ties. [b] Multiple comparisons are not performed because the overall test does not show significant differences across samples.

The post hoc analysis results from the Dunn's tests, represented in Table 4, reveal that Militari–Chiajna neighborhood displays strong differences in terms of "security". These results are confirmed by the outstandingly large number of answers from Militari–Chiajna' residents identifying security as a positive aspect of their neighborhood (Figure 7B). For the "average commuting time", meaning the time spent traveling from home to the city, the pairwise comparison shows significant differences, suggesting that each of the three residential areas displays an individual pattern in terms of accessibility and interconnectivity with the city. The commuting time values are lowest for Cosmopolis and highest for Militari–Chiajna. Although Cosmopolis is the farthest away from the city, its location next to Bucharest's ring road is advantageous when analyzing the "average commuting time". The Militari–Chiajna nucleus, while located close to the city, has the most problematic interconnectivity due to its proximity to an extensive service and commercial area that only overburdens an insufficiently developed road infrastructure. In the case of the "distance to the nearest park", the test shows contrasting situations between Militari–Chiajna and the other two nuclei, which could be explained by its scarcity of parks, in contrast Noul Titan, which incorporates a park, and Cosmopolis, where green spaces and gardens have been integrated into the initial project and are located frequently around buildings (Figures 4–6).

Table 4. Post hoc analysis results of pairwise Dunn's tests for the questions' items with Likert scale responses, which yielded significant Kruskal–Wallis test results.

Question	Pairwise Comparisons of Name					
	Sample 1–Sample 2	Test Statistic	Std. Error	Std. Test Statistic	Sig.	Adj. Sig. [a]
Security	Noul Titan–Cosmopolis	11.890	8.195	1.451	0.147	0.441
	Noul Titan–Militari–Chiajna	30.265	7.749	3.906	0.000	0.000
	Cosmopolis–Militari–Chiajna	−18.375	7.389	−2.487	0.013	0.039
Average commuting time	Cosmopolis–Noul Titan	−22.012	8.423	−2.613	0.009	0.027
	Cosmopolis-Militari–Chiajna	−53.484	7.594	−7.043	0.000	0.000
	Noul Titan–Militari–Chiajna	31.471	7.964	3.952	0.000	0.000
Distance to the nearest park	Noul Titan–Cosmopolis	2.596	8.486	0.306	0.760	1.000
	Noul Titan–Militari–Chiajna	44.576	8.023	5.556	0.000	0.000
	Cosmopolis–Militari–Chiajna	−41.981	7.651	−5.487	0.000	0.000

Each row tests the null hypothesis that the Sample 1 and Sample 2 distributions are the same. Asymptotic significances (2-sided tests) are displayed. The significance level is 0.05. [a] Significance values have been adjusted by the Bonferroni correction for multiple tests.

The three sample areas register substantial differences when analyzing their public transport accessibility and coverage. The authors again correlate these with the distance to Bucharest's center and the type of residential project each nucleus was part of (e.g., inside or outside the city's administrative limit). For example, in Noul Titan, two-thirds of respondents declared the residential nucleus to be sufficiently covered by *public transport* (which was also chosen as the "primary means of transportation for commuting"), thus explaining the lack of *private transport* options among residents. Secondly, in Militari–Chiajna, more than three-quarters of respondents declared the *low frequency of public transportation means* and the choice among using a *personal car, public,* and *private means of transportation* is the most balanced among the three nuclei. On the other hand, 90% of the respondents living in Cosmopolis chose *private transport* when asked about the neighborhood's "optimal transportation means".

Due to the long distance between their home and the city, workplace, or previous residence (i.e., respondents that previously resided in Bucharest vs. outside it) and the issues of the public transport system (i.e., units' low frequency or insufficient number to cover some residential nuclei) meant most residents need a personal car. Consequently, the "primary means of transportation" was designated to be the *personal car* by 75% of the respondents residing in Cosmopolis, 41.18% of those residing in Noul Titan, and 32.69% of those residing in Militari–Chiajna.

3.4. Analyzing the Housing Satisfaction in the Newly Built Dwellings of the Sample Residential Nuclei

Defined as "a function of household, dwelling and neighborhood characteristics", residential satisfaction is also a result of the interaction between residential attributes and moving intention (p. 154, Reference [96]), as both dissatisfactions with previous residence and attractive elements in new residential areas may convince people to move.

Choosing a newly built dwelling to the detriment of an old one directly results from people's expectations regarding the differences in living conditions. The "old" blocks of flats built in Bucharest during the communist regime that today dominate the city's landscape, and the houses built in the inter- and post-war periods are both inaccessible due to their prices and inadequate in terms of current housing standards.

Similar to the reasons for choosing to live in a new residential nucleus, the main "reason for choosing a newly built dwelling" was the need for *modernity and comfort* (almost 59%), regardless of income level (over 50% within each category of income), age, or type of current residence (individual or collective). The second most-often cited reason, but at a relatively large distance, was the *affordable price* (17.5%).

The prevalence of opting for *modernity and comfort* increases with the age of the respondents: 50% within 18–29 years, 55.6% within 30–39 years, reaching 69% within 40–49 years, and even 80% within the age group 50–65. Conversely, the *affordable price* is an important variable for the younger age groups of respondents (25% for those aged 18 to 29, 18.5% for the age group 30 to 39, against only 10.3% of those aged 40 to 49). This shows the considerable financial limitations of younger people in Bucharest and the advantageous offer that the new residential nuclei have brought to the city. The opportunity for a *private parking place* does not rank very high for any age group except those aged 50 to 65. It is nonetheless worth mentioning that, even if it was mentioned to a lesser extent, the *private parking place* was still one of the reasons listed when considering buying a newly built dwelling to the detriment of an older one. The reasons for this include the peripheral locations of the new residential nuclei and because it is a piece of established information that new residential nuclei incorporate private parking right from their project design phase (Table 5).

Education level shows slight differences between respondents' choices, with *neighborhood position* and *security* being more favored when purchasing a newly built dwelling by those with higher education. These dwellings' *increased value with time* is more important for those with a university-level education than for those with a high-school-level education. In terms of the previous residence, an *affordable price* proved more important for those coming from outside Bucharest than those that had already lived in the city, pointing to the first group being more vulnerable to the capital's expensive real-estate market.

Table 5. Reasons for choosing a newly built dwelling by age cross-tabulation.

Age		Reasons for Choosing a Newly Built Dwelling							
		Affordable Price	Owning a House	Increased Value with Time	Modernity and comfort	Neighborhood Position	Private Parking Place	Security	Total
18–29	% within age group	25.0%	0.0%	3.1%	50.0%	3.1%	9.4%	9.4%	100.0%
	% of total	6.3%	0.0%	0.8%	12.7%	0.8%	2.4%	2.4%	25.4%
30–39	% within age group	18.5%	1.9%	5.6%	55.6%	7.4%	7.4%	3.7%	100.0%
	% of total	7.9%	0.8%	2.4%	23.8%	3.2%	3.2%	1.6%	42.9%
40–49	% within age group	10.3%	3.4%	0.0%	69.0%	6.9%	3.4%	6.9%	100.0%
	% of total	2.4%	0.8%	0.0%	15.9%	1.6%	0.8%	1.6%	23.0%
50–65	% within age group	0.0%	0.0%	0.0%	80.0%	0.0%	20.0%	0.0%	100.0%
	% of total	0.0%	0.0%	0.0%	6.3%	0.0%	1.6%	0.0%	7.9%
>65	% within age group	100.0%	0.0%	0.0%	0.0%	0.0%	0.0%	0.0%	100.0%
	% of total	0.8%	0.0%	0.0%	0.0%	0.0%	0.0%	0.0%	0.8%
Total	% of total	17.5%	1.6%	3.2%	58.7%	5.6%	7.9%	5.6%	100.0%

The correlation between the "reasons for choosing a newly built dwelling" and the type of current residence (Table 6) proved that, regardless of the typology of their current housing, all categories prioritized the *modern and comfortable* aspect, which is seen as lacking in communist built housing. Furthermore, respondents living in blocks of flats were also influenced by the *affordable price*, in contrast to those living in houses, which can offer a glimpse into how demand shapes the evolution of the market toward continuing to build block-type structures. Another important reason for those living in blocks of flats is the possibility to benefit from a *private parking place*. As the city-owned residential parking offer is limited, buying a parking place is nearly impossible for those living in older dwellings in the capital city. Finally, the *neighborhood position* was also identified as a reason for choosing a newly built dwelling, as emphasized by respondents from Cosmopolis who declared that they looked forward to escaping Bucharest's congestion and air pollution (Table 6).

When analyzing the "reasons for satisfaction with the newly built dwellings" in the three sample nuclei, relatively equal shares in the residents' answers are split between *well-balanced rooms* and *living space dimensions*. Another reason for residents' satisfaction is the opportunity for individual water and heating *consumption metering* that allows consumers to manage their costs and budgets better and more accurately. On the other hand, the main "dissatisfaction reasons" were poor *finishes* and *living-space dimensions* for residents from Cosmopolis, where expectations mirrored the specificities of this gated-community-type residential nucleus. In addition to the above-stated, Militari–Chiajna and Noul Titan residents brought up issues regarding *soundproofing*, *residential density*, and *distance from the city center* (Figure 8).

The interior index of density directly depends on income level; as such, the Cosmopolis residential nucleus registers a value of 0.87 inhabitants per room, and Noul Titan registers a value of 0.91; meanwhile, in Militari–Chiajna, families opted for a lower number of rooms per inhabitant due to financial limitations, with the index capping at 1.29 inhabitants per room. This ranking of the interior index of density is mainly the result of housing pricing, with developers having to meet the high demand for lower-priced dwellings. In the end, they choose to build with fewer rooms or lower living surfaces. Within the three sample nuclei, the majority of dwellings in collective housing are two-bedroom flats, while for individual housing, we most often find three bedrooms in duplex houses, so their living areas and, hence, their costs are as low as possible.

Table 6. Reasons for choosing a newly built dwelling by the type of current residence cross-tabulation.

Type of Current Residence		Affordable Price	Owning a House	Increased Value with Time	Modernity and Comfort	Neighborhood Position	Private Parking Place	Security	Total
		Reasons for Choosing a Newly Built Dwelling							
Block of flats in a residential area	% within type of current residence	20.2%	0.0%	3.4%	56.2%	6.7%	6.7%	6.7%	100.0%
	% of total	14.3%	0.0%	2.4%	39.7%	4.8%	4.8%	4.8%	70.6%
Block of flats outside a residential area	% within type of current residence	18.2%	0.0%	0.0%	63.6%	0.0%	18.2%	0.0%	100.0%
	% of total	3.2%	0.0%	0.0%	11.1%	0.0%	3.2%	0.0%	17.5%
House within a residential area	% within type of current residence	0.0%	18.2%	9.1%	54.5%	9.1%	0.0%	9.1%	100.0%
	% of total	0.0%	1.6%	0.8%	4.8%	0.8%	0.0%	0.8%	8.7%
House outside of a residential area	% within type of current residence	0.0%	0.0%	0.0%	100.0%	0.0%	0.0%	0.0%	100.0%
	% of total	0.0%	0.0%	0.0%	3.2%	0.0%	0.0%	0.0%	3.2%
Total	% of total	17.5%	1.6%	3.2%	58.7%	5.6%	7.9%	5.6%	100.0%

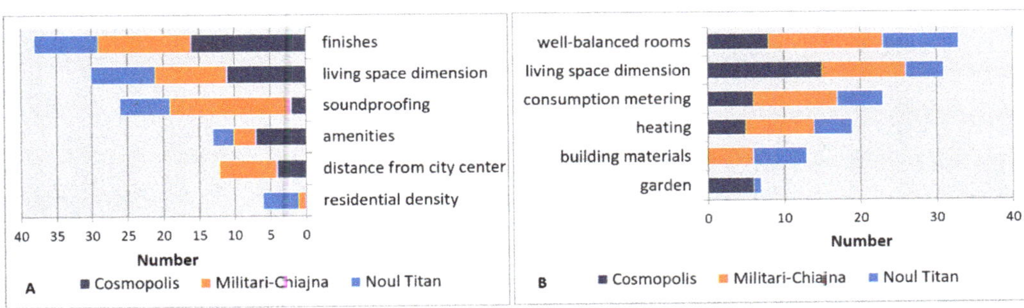

Figure 8. Residents' dissatisfaction (**A**) and satisfaction (**B**) reasons regarding their newly built dwellings.

4. Discussion

A lack of living space evidently marks Bucharest. A significant barrier, which only grew during the last decades, is the saturation of its available interior plots, as identified by Soaita and Dewilde [59] in their study of realignment between socioeconomic and housing stratification. The need for new housing has led to the development of construction entrepreneurship, with real-estate developers focusing on building primarily outside the city. Demand is fueled by the low number of self-built housing (with low-quality construction materials) and the rising preference for another category of residences, such as more modern villas. Both of these are a result of an old or very old housing fund that lacks the comfort elements necessary for current living standards and needs [59]. The very high number of private property homes and, hence, the low mobility of the population from a renter position to a homeowner position are characteristic of other CEE capital cities, as well (e.g., Prague by Špačková, Dvořáková and Tobrmanová [25]), and further prove the need for and importance of new residential nuclei in Bucharest. Apart from the fact that the newly built dwellings are more modern and bring added comfort, they also have the advantage of an affordable price in a market where the housing fund is limited, and everything built before 1990 is expensive. The new houses are in complete antithesis to the flats built during the communist regime, where compartmentalization and size were meant to meet basic necessities rather than design choices. The age distribution of respondents opting for modernity and comfort as the main reason for choosing a newly built dwelling emphasizes the particular requirements of the older-adult segment, who usually have higher-than-average income levels and for whom housing stratification and meeting housing expectations are important factors determining their mobility [59].

Residents' perception of local dysfunctionalities strongly influences neighborhood satisfaction within the three sample areas, and local authorities are considered to be the ones capable of preventing or fixing them. A large number of recommendations and a wide variety of problems identified by respondents prove that the urban planning policies meant to ensure better living standards are inconsequentially implemented, thus resulting in situations where housing standards in new residential nuclei are sometimes lower than in the blocks of flats built during communism [85]. For example, in Noul Titan, residents declared the need to modernize the road system, build more parking places and green areas, and improve urban lighting and overall cleanliness (Figure 9A).

For the residential nuclei located outside Bucharest's administrative city limits, recommendations refer to the development and improvement of paved road infrastructure, sewerage and drinking-water central networks, connectivity with the city (i.e., lack of public means of transportation), green areas, and air-quality issues (e.g., in the case of Militari–Chiajna, as a result of its proximity to a landfill) (Figure 9B,C).

The comparison of the urban utilities in the three residential nuclei shows a direct connection between the amplitude of identified dysfunctionalities and increased distance from the city center. Figure 10 summarizes, using the data gathered from the observation sheets and the survey, the dysfunctionalities characterizing the urban infrastructure and heating systems within the three sample residential nuclei.

Figure 9. (**A**) Noul Titan, (**B**) Militari–Chiajna, and (**C**) Cosmopolis: residents' recommendations to local authorities for improving housing standards, based on survey-derived information.

The observation sheets filled in during the fieldwork stage showed significant differences between the three new residential nuclei, which can be correlated with their position and distance from the city center. The construction density (i.e., the larger number of taller buildings) is higher outside the city's administrative limits. For example, Noul Titan, where urban planning rules impose a height regime, sees fewer tall buildings than the other two sample areas. Urban utilities, such as the water supply and sewerage system, are more deficient outside the city (e.g., Militari–Chiajna and Cosmopolis) than in its interior (Noul Titan), and the solutions implemented by the developers (septic tanks and wells) have to be employed, in most cases, for each housing unit. Noul Titan's intra-urban position also gives it the advantage of a well-developed road system and good interconnectivity with the rest of the city.

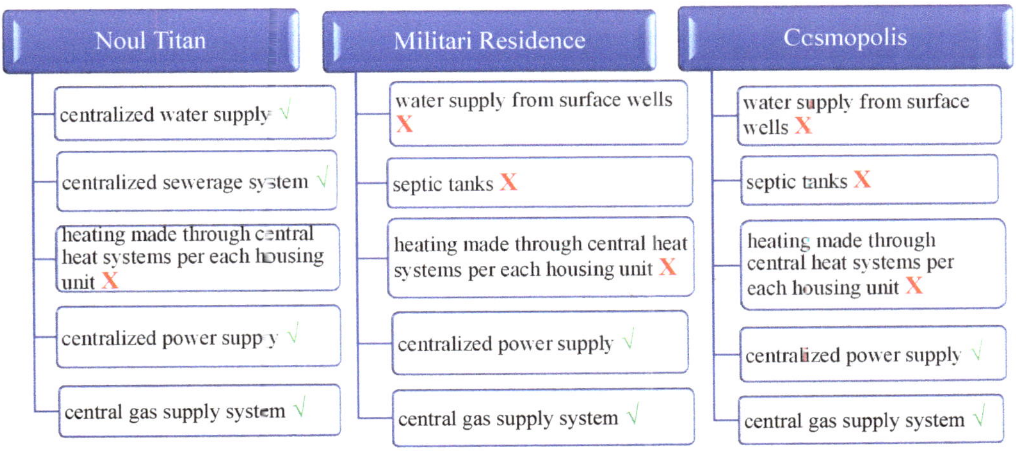

Figure 10. Comparative summary of the urban utilities in the three chosen residential nuclei (assets—√ vs. dysfunctionalities—X) based on data gathered from the observation sheets and the survey.

All sample nuclei benefit from access to the centralized gas and electricity network, but only Noul Titan is connected to the city's central sewerage and drinking-water network. A common flaw of all sample nuclei that also increases pollution levels is the type of heating system currently in use; that is, each dwelling uses individual gas boilers [97,98]. This is seen as an asset by residents, as they do not depend on the city's central heating system, which was installed in the 1960s and is a source of often poor and high-priced services [90,99]. However, the latest proposals by the Ministry of Environment, Water and Forests of replacing individual gas boilers with less polluting block or condominium heating boilers as per the European Green Deal policies might make residents rethink this element that is essential to their housing comfort [100].

Authorities have submitted two bills to public debate, namely the National Housing Strategy and the Code of Spatial Planning, Urbanism, and Constructions, precisely to eliminate the problems that the newly built residential nuclei are currently facing [101,102].

Bucharest's residential space's morphologic and spatial evolution is closely connected to housing prices. In addition, the prices from the housing market inside Bucharest also pressure the three sample residential nuclei. The housing market in Romania's large cities, and especially in Bucharest, is obviously getting more expensive, particularly for newly built dwellings, due to "more and more people looking to buy new flats or houses", increasing prices for construction materials, and a labor-force crisis in the construction sector, as many people immigrate to other EU countries [103].

In many cases, developers exploit authorities' PR strategies for future investment HUBs and development projects and increase housing prices for those areas. For example, Bucharest's first and third sectors register the highest prices per built square meter due to the Pipera area, Bucharest, Romania, which hosts the headquarters of many multinational firms (first sector) or famous commercial units such as Ikea (third sector) and, in both cases, the possibility of future extension of the transport infrastructure [104]. In other cases, the price is influenced by the time of sale, the commissions of real-estate agencies, the fees of notary offices, etc. [105].

This study also identified an imbalance between the conceptual idea of a new residential nucleus with its intrinsic benefits and the financial responsibility this would bring. Respondents mentioned financial reasons (lower prices) as a primary reason for choosing a dwelling in a new residential nucleus, but they also enumerated problems that are a direct result of the budget they are willing to spend: quality of construction materials, living space dimensions, or even issues with their neighbors which they directly correlate with education and income levels ("the price of the house shows the quality of the people"—Cosmopolis resident).

5. Conclusions

Bucharest's sample new residential nuclei differ in terms of utilities and interconnectivity, with the one located inside the city administrative limits (Noul Titan) having advantages such as access to the city's existing central utilities and public transport system and the ones outside the city limits (Militari–Chiajna and Cosmopolis) having to find alternative solutions.

In terms of residents' expectations and their neighborhood satisfaction, it is worth mentioning that they preferred new residential nuclei because they were looking for modern living standards, including private parking and opportunities for recreational activities. Some of these expectations were met, for example, those relating to security, but many were not, for example, those relating to urban systematization.

In terms of housing satisfaction in the newly built dwellings of residential nuclei, respondents were content primarily with their modernity and price. The main reasons buyers chose a newly built dwelling at the expense of an older one relate to their shift in preferences for housing standards. Prices were also an important reason, and it is crucial to bear in mind that prices for dwellings in the new residential nuclei are, in general, lower than for the rest of Bucharest. While many respondents decry the living-space dimensions, much praise was given to the well-balanced rooms. Residents were also dissatisfied with finishes and soundproofing, while praising the possibility of water consumption and heating metering directly affecting their ability to balance their finances. In the opinion of the respondents participating in this survey, many of the existing dysfunctionalities can be avoided in the future if developers respect the existing legislation and if local authorities develop the urban infrastructure prior to granting construction permits.

Residential satisfaction depends on how successfully local authorities implement urban systematization policies. Numerous existing derogations and legislative gaps lead to the opportunities for interpreting urban planning rules to benefit different parties and create fundamental urban dysfunctions, which ultimately affect the population that has chosen to live in new residential nuclei.

The issues surrounding new residential nuclei in Bucharest have been highly debated in the Romanian press in recent years, as buyers are often forced to pay high amounts of money for homes with poorer amenities than what developers promised or even not benefiting from prepaid apartments due to real-estate projects being closed. All of these aspects resulted in a decrease in housing quality, which was rarely analyzed in specialized studies of the last years. In this context, we consider it necessary to continue the study of these new residential nuclei and possibly more profound research for Bucharest or a continuation in other large cities in Romania, where the housing market is equally challenging.

The study faced inevitable limitations imposed by its development during the Covid-pandemic context, limiting face-to-face interactions. This resulted in the survey either being distributed online on Facebook or the WhatsApp groups of the owners' associations in each residential nucleus or filled in by phone.

Similar to other exploratory research [106], our study raises several further research questions, particularly referring to how existing residents perceive new neighborhoods/ residential nuclei and how neighborhoods and dwellings need to look to satisfy comfort needs and reach the current quality-of-life standards but also consider future urbanites. Follow-up research may tackle these areas from a more in-depth stratified consumer perspective in order to obtain more precise responses for one or more elements identified in our study as development priorities (e.g., transport solutions, parks and green areas, etc.). Therefore, this exploratory study may help both scientists and professionals project further studies on residential satisfaction and territorial planning. Decision-makers and politicians may also find this study an informative source when making decisions and shaping policies to formulate urban planning measures.

Author Contributions: The authors contributed equally to this work. Conceptualization, M.P. and A.-I.L.-D.; methodology, M.P., M.J., A.M., I.V., A.P. and A.-I.L.-D.; software, M.P., M.J., A.M., I.V., A.P. and A.-I.L.-D.; validation, M.P., I.V., M.J., A.M. and A.-I.L.-D.; formal analysis M.P., A.-I.L.-D. and M.J.; investigation, M.P.; resources, A.M., A.P., A.-I.L.-D. and I.V.; data curation, M.P.; writing—original draft preparation, M.P. and A.-I.L.-D.; writing—review and editing, I.V., A.M. and M.J.; visualization, A.P. and I.V.; supervision, I.V., A.M. and M.P.; project administration, M.P.; funding acquisition, M.P. and A.M. All authors have read and agreed to the published version of the manuscript.

Funding: The publication of this research has partially been funded by the University of Bucharest, Bucharest, Romania.

Institutional Review Board Statement: Ethical review and approval were waived for this study because the interview guides did not collect personal data that could lead to the identification of research subjects. The information was collected and encoded in an anonymized database.

Informed Consent Statement: Informed consent was obtained from all subjects involved in the study.

Data Availability Statement: The data presented in this study are available upon request from the corresponding author.

Acknowledgments: The authors would like to thank all the presidents of the Owners Associations, as well as the respondents for accepting to participate in our study. The authors would also like to thank the editors and anonymous reviewers for their constructive comments and helpful suggestions.

Conflicts of Interest: The authors declare no conflict of interest.

References

1. Ziółkowska-Weiss, K. Satisfaction with Selected Indicators of the Quality of Urban Space by Polonia in the Greater Toronto Area. *Land* **2021**, *10*, 778. [CrossRef]
2. Borgoni, R.; Michelangeli, A.; Pirola, F. Handling Heterogeneity in Assessing Residential Satisfaction. *Geogr. Anal.* **2021**, *53*, 447–466. [CrossRef]
3. Biswas, B.; Sultana, Z.; Priovashini, C.; Ahsan, M.N.; Mallick, B. The emergence of residential satisfaction studies in social research: A bibliometric analysis. *Habitat Int.* **2021**, *109*, 102336. [CrossRef]
4. Türkoğlu, H.; Terzi, F.; Salihoğlu, T.; Bölen, F.; Okumuş, G. Residential satisfaction in formal and informal neighborhoods: The case of Istanbul, Turkey. *Int. J. Archit. Res.* **2019**, *13*, 112–132. [CrossRef]
5. Berköz, L.; Kellekçi, Ö.L. Mass Housing: Residents Satisfaction with Their Housing and Environment. *Open House Int.* **2007**, *32*, 41–49. [CrossRef]
6. Cho, Y.; Park, G.; Echevarria-Cruz, S. Perceived neighborhood characteristics and the health of adult Koreans. *Soc. Sci. Med.* **2005**, *60*, 1285–1297. [CrossRef]
7. Greenberg, M.; Crossney, K. Perceived neighborhood quality in the United States: Measuring outdoor, housing and jurisdictional influences. *Socio-Econ. Plan. Sci.* **2007**, *41*, 181–194. [CrossRef]
8. Salleh, A. Neighbourhood factors in private low-cost housing in Malaysia. *Habitat Int.* **2008**, *32*, 485–493. [CrossRef]
9. Bonaiuto, M.; Fornara, F.; Bonnes, M. Perceived residential environment quality in middle and low extension Italian cities. *Eur. Rev. Appl. Psychol.* **2006**, *56*, 23–34. [CrossRef]
10. Rioux, L.; Werner, C. Residential satisfaction among aging people living in place. *J. Environ. Psychol.* **2011**, *31*, 158–169. [CrossRef]
11. Amole, D. Residential satisfaction in students' housing. *J. Environ. Psychol.* **2009**, *29*, 76–85. [CrossRef]
12. Amerigo, M.; Aragones, J.I. A theoretical and methodological approach to the study of residential satisfaction. *J. Environ. Psychol.* **1997**, *17*, 47–57. [CrossRef]
13. Herfert, G.; Neugebauer, C.S.; Smigiel, C. Living in residential satisfaction? Insights from large-scale housing estates in Central and Eastern Europe. *Tijdschr. Econ. Soc. Geogr.* **2013**, *104*, 57–74. [CrossRef]
14. Kwon, H.J.; Beamish, J.O. Older Adults in Multifamily Housing: Residential Satisfaction and Intention to Move. *Fam. Consum. Sci. Res. J.* **2013**, *42*, 40–54. [CrossRef]
15. Caplan, A.J.; Akhundjanov, S.B.; Toll, K. Measuring heterogeneous preferences for residential amenities. *Reg. Sci. Urban. Econ.* **2021**, *87*, 103646. [CrossRef]
16. Wang, P.; Qin, X.; Li, Y. Satisfaction Evaluation of Rural Human Settlements in Northwest China: Method and Application. *Land* **2021**, *10*, 813. [CrossRef]
17. Lu, M. Determinants of residential satisfaction: Ordered logit vs. regression models. *Growth Change* **1999**, *30*, 264–287. [CrossRef]

18. Yilmaz, H.; Surat, H.; Özhanci, E.; Yeșil, P.; Yeșil, M. Urban Living Area Satisfaction and Public Preference. *J. For. Fac.* 20 *15*, 319–329. [CrossRef]
19. Potapov, D.; Shafranskaya, I.; Bozhya-Volya, A. Happiness and the city. An empirical study of the interaction between subje well-being and city satisfaction. *J. Place Manag. Dev.* **2016**, *9*, 313–330. [CrossRef]
20. Dekker, K.; De Vos, S.; Musterd, S.; Van Kempen, R. Residential Satisfaction in Housing Estates in European Cities: A Multi-le Research Approach. *Hous. Stud.* **2011**, *26*, 479–499. [CrossRef]
21. Földi, Z. *Neighbourhood Dynamics in Inner-Budapest—A Realist Approach*; Netherlands Geographical Studies 350; Urban a Regional Research Centre Utrecht, Faculty of Geosciences, Utrecht University: Utrecht, The Netherlands, 2006; Available onli https://dspace.library.uu.nl/handle/1874/13840 (accessed on 10 March 2022).
22. Barreira, A.P.; Nunes, L.C.; Guimarães, M.H.; Panagopoulos, T. Satisfied but thinking about leaving: The reasons beh residential satisfaction and residential attractiveness in shrinking Portuguese cities. *Int. J. Urban Sci.* **2019**, *23*, 67–87. [CrossRe
23. Brade, I.; Smigiel, C.; Kovács, Z. Suburban Residential Development in Post-socialist Urban Regions: The Case of Moscow, So and Budapest. In *German Annual of Spatial Research and Policy*; Kilper, H., Ed.; Springer: Berlin/Heidelberg, Germany, 20 pp. 79–104. [CrossRef]
24. Kährik, A.; Tammaru, T. Soviet prefabricated panel housing estates: Areas of continued social mix or decline? The case of Tal *Hous. Stud.* **2010**, *25*, 201–219. [CrossRef]
25. Špačková, P.; Dvořáková, N.; Tobrmanová, M. Residential Satisfaction and Intention to Move: The Case of Prague's Ne Suburbanites. *Geogr. Ann. Ser. B Hum. Geogr.* **2016**, *98*, 331–348. [CrossRef]
26. Maleszyk, P.; Kędra., A. Intention to move and residential satisfaction: Evidence from Poland. *Equilib. Q. J. Econ. Econ. Pol* **2020**, *15*, 341–360. [CrossRef]
27. Nedučin, D.; Škorić, M.; Krklješ, M. Post-socialist Development and Rehabilitation of Large Housing Estates in Central an Eastern Europe: A Review. *Teh. Vjesn.* **2019**, *26*, 1853–1860. [CrossRef]
28. Szafrańska, E. Transformations of large housing estates in Central and Eastern Europe after the collapse of communism. *Geog Pol.* **2015**, *88*, 621–648. [CrossRef]
29. Szmytkie, R. Suburbanisation processes within and outside the city: The development of intra-urban suburbs in Wrocław, Pola *Morav. Geogr. Rep.* **2021**, *29*, 149–165. [CrossRef]
30. Bertaud, A. The spatial structures of Central and Eastern European cities. In *The Urban Mosaic of Post-Socialist EUROPE: Spa Institutions and Polic*; Tsenkova, S., Nedovic-Budic, Z., Eds.; Springer: Berlin/Heidelberg, Germany, 2006; pp. 91–110.
31. Dumitrache, L.; Zamfir, D.; Nae, M.; Simion, G.; Stoica, I.V. The Urban Nexus: Contradictions and Dilemmas of (Post)Communis (Sub)Urbanization in Romania. *Hum. Geogr.* **2016**, *10*, 39–58. [CrossRef]
32. Kährik, A.; Leetmaa, K.; Tammaru, T. Residential decision-making and satisfaction among new suburbanites in the Tallinn urba region, Estonia. *Cities* **2012**, *29*, 49–58. [CrossRef]
33. Kubeš, J. Analysis of regulation of residential suburbanisation in hinterland of post-socialist 'one hundred thousands' city Česke Budějovice. *Bull. Geogr. Socio-Econ. Ser.* **2015**, *27*, 109–131. [CrossRef]
34. Leetmaa, K.; Tammaru, T. Suburbanisation in countries in transition: Destinations of suburbanisers in the Tallinn metropolita area. *Geogr. Ann. Ser. B Hum. Geogr.* **2007**, *89*, 127–146. [CrossRef]
35. Ruoppila, S. Establishing a market-oriented urban planning system after state socialism: The case of Tallinn. *Eur. Plan. Stud.* 20 *15*, 405–427. [CrossRef]
36. Timár, J.; Váradi, D. The uneven development of suburbanisation during transition in Hungary. *Eur. Urban Reg. Stud.* 20 *18*, 349–360. [CrossRef]
37. Wießner, R. Urban development in East Germany—Specific features of urban transformation processes. *GeoJournal* **1999**, *49*, 43–5 [CrossRef]
38. Morano, P.; Guarini, M.R.; Tajani, F.; Anelli, D. Sustainable redevelopment: The cost-revenue analysis to support the urba planning decisions. In Proceedings of the International Conference on Computational Science and Its Applications, Cagliari, Ital 1–4 July 2020; pp. 968–980. [CrossRef]
39. Turkoglu, H. Sustainable development and quality of urban life. *Procedia Soc. Behav. Sci.* **2015**, *202*, 10–14. [CrossRef]
40. Ministerul Dezvoltării, Lucrărilor Publice și Administrației (MDLPA) [Ministry for Development, Public Works and Administra tion]. Dimensiunea Teritoriala 2014–2020 [Territorial Dimension 2014–2020]. Available online: https://www.mdlpa.ro/pages dimensiuneateritoriala20142020 (accessed on 15 May 2019).
41. The World Bank. Ghid Pentru Promovarea Activităților de Dezvoltare Urbană Durabilă, 2015 [Guide for Promoting Sustainabl Urban Development Activities, 2015]. Available online: https://www.fonduri-ue.ro/images/files/studiianalize/48101/11. Ghid_promovare_activ_dezv_urbana_durabila_ro.pdf (accessed on 15 May 2019).
42. Merciu, F.C.; Merciu, G.L. GIS solutions adapted to the needs of local public administration in the urban regeneration of the olc centre of Petrosani municipality (Hunedoara county). *J. Appl. Eng. Sci.* **2014**, *2*, 60–66.
43. Grigorescu, I.; Dumitrică, C.; Dumitrașcu, M.; Mitrică, B.; Dumitrașcu, C. Urban Development and the (Re)use of the Communist Built Industrial and Agricultural Sites after 1990. The Showcase of Bucharest–Ilfov Development Region. *Land* **2021**, *10*, 104 [CrossRef]
44. Mitrică, B.; Şerban, P.; Mocanu, I.; Grigorescu, I.; Damian, N.; Dumitraşcu, M. Social development and regional disparities in the rural areas of Romania: Focus on the social disadvantaged areas. *Soc. Indic. Res.* **2020**, *152*, 67–89. [CrossRef]

45. Nae, M. Facteurs limitatifs de l'habitat urbain bucarestois lors de l'époque (post)transitionnelle à l impact visible sur la qualité de vie'. *An. Univ. Din Bucureşti. Ser. Geogr.* **2008**, *1*, 63–74.
46. Ianoş, I.; Sîrdoev, I.; Pascariu, G.; Henebry, G. Divergent patterns of built-up urban space growth following post-socialist changes. *Urban Stud.* **2015**, *53*, 3172–3188. [CrossRef]
47. Stoica, I.V.; Vîrghileanu, M.; Zamfir, D.; Mihai, B.A.; Săvulescu, I. Comparative Assessment of the Built-Up Area Expansion Based on Corine Land Cover and Landsat Datasets: A Case Study of a Post-Socialist City. *Remote Sens.* **2020**, *12*, 2137. [CrossRef]
48. Suditu, B.; Dumitrache, L.; Vîrdol, D.; Vâlceanu, D.G. New trajectories of post-socialist residential mobility in Bucharest. *Hum. Geogr.* **2014**, *8*, 75–82. [CrossRef]
49. Tammaru, T.; Leetmaa, K.; Silm, S.; Ahas, R. Temporal and spatial dynamics of the new residential areas around Tallinn. *Eur. Plan. Stud.* **2009**, *17*, 423–439. [CrossRef]
50. Merciu, C.; Mavru, I.; Iliescu, C.B.; Merciu, G.L. Delineation of the urban influence area using the multi-criteria assessment method. The case study of Focşani city, Romania. *J. Settl. Spat. Plan.* **2019**, *10*, 13–28.
51. Nae, M.; Turnock, D. The new Bucharest: Two decades of restructuring. *Cities* **2011**, *28*, 206–219 [CrossRef]
52. Stoica, I.V.; Zamfir, D.; Vîrghileanu, M. Evaluating the Territorial Impact of Built-Up Area Expansion in the Surroundings of Bucharest (Romania) through a Multilevel Approach Based on Landsat Satellite Imagery. *Remote Sens.* **2021**, *13*, 3969. [CrossRef]
53. Popescu, C. From command to market-driven economy: The changing role of manufacturing industries in Romania. *Hum. Geogr.* **2021**, *15*, 127–146. [CrossRef]
54. Simion, G.; Nistor, C. Spatial structure changes inside post-communist capital city of Bucharest. *Hum. Geogr.* **2012**, *6*, 79–89. [CrossRef]
55. Portal Legislativ [Legislative Portal]. Decizia nr. 68, 17 Martie 1976, privind schimbarea domiciliului din alte localităţi în oraşe declarate, potrivit legii, oraşe mari [Decision no. 68, from 17 March 1976, Regarding the Change of Domicile from Other Localities to Cities Declared, by Law, to be Big Cities]. Published in the Official Bulletin no. 24, from 20 March 1976. Available online: https://legislatie.just.ro/Public/DetaliiDocumentAfis/411 (accessed on 18 May 2022).
56. Ion, F.; Pirvu, D. Territorial discrepancies in the rural-urban fringe zone of Bucharest. *Sci. Bull.-Econ. Sci. Univ. Pitesti* **2015**, *14*, 56–63.
57. Preda, M.; Mareci, A.; Tudoricu, A.; Taloș, A.M.; Bogan, E.; Lequeux-Dincă, A.I.; Vijulie, I. Defining the Concept of Family through the Lens of Fertile Aged Women in Bucharest, Romania—Between Traditionalism and Inclusion. *Sustainability* **2020**, *12*, 2691. [CrossRef]
58. Legislative Portal. Decision no. 717, from 17 June 2009, on the Approval of the Implementation Rules for "The First Home" program. Published in the Official Monitor no. 418, from 18 June 2009. Available online: http://legislatie.just.ro/Public/DetaliiDocumentAfis/107177 (accessed on 27 December 2021).
59. Soaita, A.M.; Dewilde, C. Housing stratification in Romania: Mapping a decade of change. *J. Hous. Built. Environ.* **2021**, *36*, 1055–1076. [CrossRef]
60. Krišjāne, Z.; Bērziņš, M. Post-socialist urban trends: New patterns and motivations for migration in the suburban areas of Riga, Latvia. *Urban Stud.* **2012**, *49*, 289–306. [CrossRef]
61. Krišjāne, Z.; Bērziņš, M.; Ivlevs, A.; Bauls, A. Who are the typical commuters in the post-socialist metropolis? The case of Riga, Latvia. *Cities* **2012**, *29*, 334–340. [CrossRef]
62. Kubeš, J. European post-socialist cities and their near hinterland in intra-urban geography literature. *Bull. Geogr. Socio-Econ. Ser.* **2013**, *19*, 19–43. [CrossRef]
63. The Romanian Order of Architects. Raportul Pentru București 2018 [Report for Bucharest 2018]. Publication of the Bucharest Order of Architects, Made through the Architectural Stamp of the Romanian Order of Architects, 2018. PIM Publishing, Iași. Available online: https://www.oar-bucuresti.ro/documente/Raportul-pentru-Bucuresti_2018.pdf (accessed on 16 December 2021).
64. Digi24. Blocuri noi în București, Cu Parcări Fictive. Legislația Impune un loc de Parcare Pentru Fiecare Apartament sub 100 de mp [New Blocks of Flats in Bucharest have Fictitious Parking Lots. The Legislation Requires a Parking Space for Each Apartment under 100 sqm]. 2021. Available online: https://www.digi24.ro/stiri/actualitate/social/blocuri-noi-in-bucuresti-cu-parcari-fictive-legislatia-impune-un-loc-de-parcare-pentru-fiecare-apartament-sub-100-de-mp-1720781 (accessed on 2 November 2021).
65. Ziarul Financiar. Blocurile noi, un Pericol Pentru Cumpărători. "Este un Grad Mare de Inconștiență" [New Blocks of Flats, a Danger to Buyers. "There is a High Degree of Unconsciousness"]. 2021. Available online: https://www.zf.ro/constructii-imobiliare/blocurile-noi-un-pericol-pentru-cumparatori-este-un-grad-mare-de-20034657 (accessed on 28 October 2019).
66. Știrile Pro Tv [Pro Tv News]. Coșmarul Celor Care și-au Luat Case Lângă BUCUREȘTI: "Se Ridică Blocurile ca Nişte Închisori" [The Nightmare of Those Who Bought Their Houses near Bucharest: "The Blocks of Flats Are Rising like Prisons"]. 2019. Available online: https://stirileprotv.ro/stiri/inspectorul-pro/cosmarul-celor-care-si-au-luat-case-langa-bucuresti-se-ridica-blocurile-ca-niste-inchisori.html (accessed on 8 October 2019).
67. Google Earth Pro. București 44°25′36.36″ N, 26°06′09.14″ E, Eye Alt. 400 m, Imagery Date 25 October 2021. Available online: https://earth.google.com/web/search/Bucuresti (accessed on 25 October 2021).
68. Esri. "Imagery" [Basemap]. Scale Not Given. "World Imagery". Imagery Date: 22 November 2020. Available online: https://www.arcgis.com/home/item.html?id=10df2279f9684e4a9f6a7f08febac2a9 (accessed on 25 September 2021).
69. Marshall, B.; Cardon, P.; Poddar, A.; Fontenot, R. Does sample size matter in qualitative research?: A review of qualitative interviews in is research. *J. Comput. Inf. Syst.* **2013**, *54*, 11–22. [CrossRef]

70. Alasuutari, P.; Bickman, L.; Brannen, J. Social Research in Changing Social Conditions. In *The Sage Handbook of Social Research Methods*; Alasuutari, P., Bickman, L., Brannen, J., Eds.; Sage: London, UK, 2008; pp. 1–8.
71. Farugia, B. WASP (Write a Scientific Paper): Sampling in qualitative research. *Early Hum. Dev.* **2019**, *133*, 69–71. [CrossRef]
72. Onwuegbuzie, A.J.; Leech, N.L. Sampling designs in qualitative research: Making the sampling process more public. *Qual. Rep.* **2007**, *12*, 238–254. [CrossRef]
73. Onwuegbuzie, A.J.; Leech, N.L. The role of sampling in qualitative research. *Acad. Exch. Q.* **2005**, *9*, 280–284.
74. Saunders, B.; Sim, J.; Kingstone, T.; Baker, S.; Waterfield, J.; Bartlam, B.; Burroughs, H.; Jinks, C. Saturation in qualitative research: Exploring its conceptualization and operationalization. *Qual. Quant.* **2018**, *52*, 1893–1907. [CrossRef]
75. Glaser, B.; Strauss, A. *The Discovery of Grounded Theory: Strategies for Qualitative Research*; Aldine Publishing Company: New York, NY, USA, 1967.
76. Guest, G.; Bunce, A.; Johnson, L. How Many Interviews are Enough?: An Experiment with Data Saturation and Variability. *Field Methods.* **2006**, *18*, 59–82. [CrossRef]
77. Flick, U. *An Introduction to Qualitative Research: Theory, Method, and Applications*; Sage: London, UK, 1998.
78. Strauss, A.; Corbin, J. *Basics of Qualitative Research: Grounded Theory Procedures and Techniques*; Sage: Washington, DC, USA, 1990.
79. Firestone, W.A. Alternative arguments for generalizing from data, as applied to qualitative research. *Educ. Res.* **1993**, *22*, 16–23. [CrossRef]
80. EUROSTAT. Disparities in Minimum Wages across the EU. 2021. Available online: https://ec.europa.eu/eurostat/web/products-eurostat-news/-/ddn-20210205-1 (accessed on 15 May 2022).
81. Guga, A. Salariul Minim și Traiul Minim Decent. De la Mituri la Oportunităţi [Minimum Wage and Decent Minimum Living from Myths to Opportunities]. SYNDEX Knowledge for Action, 2021 Friedrich Ebert Stiftung. Available online: http://library.fes.de/pdf-files/bueros/bukarest/17786.pdf (accessed on 15 May 2022).
82. Agresti, A. *Categorical Data Analysis*, 3rd ed.; John Wiley & Sons, Inc.: Hoboken, NJ, USA, 2013.
83. Heimerl, F.; Lohmann, S.; Lange, S.; Ertl, T. Word Cloud Explorer: Text Analytics Based on Word Clouds. In Proceedings of the 47th Hawaii International Conference on System Sciences, Waikoloa, HI, USA, 6–9 January 2014; pp. 1833–1842. [CrossRef]
84. The Ministry of Environment, Water and Forests, National Housing Strategy. 2017. Available online: http://www.mmediu.ro/app/webroot/uploads/files/2017_11_09_Strategia_Nationala_a_Locuirii_2017-2030.pdf (accessed on 17 January 2022).
85. Toșa, C.; Mitrea, A.; Sato, H.; Miwa, T.; Morikawa, T. Economic growth and urban metamorphosis: A quarter century of transformations within the metropolitan area of Bucharest. *J. Transp. Land Use* **2018**, *11*, 273–295. [CrossRef]
86. Chamber of Deputies. LAW No. 350 from 6 July 2001 on Spatial Planning and Urbanism. Published in the Official Monitor No. 373 from 10 July 2001. 2001. Available online: http://www.cdep.ro/pls/legis/legis_pck.htp_act?ida=30200 (accessed on 5 January 2022).
87. Legislative Portal. Law no. 350, from 6 June 2001, on Spatial Planning and Urbanism. Published in the Official Monitor no. 373 from 10 July 2001. Available online: http://legislatie.just.ro/Public/DetaliiDocument/29453 (accessed on 20 December 2021).
88. Chamber of Deputies. DECISION No. 525 from 27 June 1996 for the Approval of the General Urban Planning Regulations. Re-Published in the Official Monitor No. 856 from 27 November 2002. 1996. Available online: http://www.cdep.ro/pls/legis/legis_pck.htp_act_text?idt=38952 (accessed on 15 January 2022).
89. Legislative Portal. Decision no. 525, from 27 June 1996 (Republished and Updated) for the Approval of the General Urban Planning Regulations. Official Monitor no. 856, from 27 November 2002. Available online: https://legislatie.just.ro/Public/DetaliiDocument/9499 (accessed on 20 December 2021).
90. Digi24. Administratorul Elcen: Sistemul de Încălzire al Bucureştiului, în Colaps. E ca în Bancul cu Petrică şi Lupul [Elcen's Administrator: Bucharest' Heating System is Collapsing. It's like the Story of the Boy who Cried Wolf]. 2020. Available online: https://www.digi24.ro/stiri/economie/administratorul-elcen-sistemul-de-incalzire-al-bucurestiului-in-colaps-e-ca-in-bancul-cu-petrica-si-lupul-1380998 (accessed on 5 September 2020).
91. Ivanov, C. Dezbatere Publică. Este Interzicerea Centralelor de Apartament la Blocurile noi o Idee Bună? Ce Spun Primarii, Activiștii de Mediu și Dezvoltatorii Imobiliari [Public Debate. Is Banning Apartment Gas Boilers in Newly-Built Blocks of Flats a Good Idea? What Mayors, Environmental Activists and Real Estate Developers Say]. HotNews. 2021. Available online: https://www.hotnews.ro/stiri-administratie_locala-25196097-dezbatere-publica-este-interzicerea-centralelor-apartament-blocurile-noi-idee-buna-spun-primarii-activistii-mediu-dezvoltatorii-imobiliari.htm (accessed on 30 December 2021).
92. Milea, A. Noi Reguli Pentru Dezvoltatorii Imobiliari în București. Obligatoriu Şcoli, Creşe, Drumuri și Parcuri, în Cartierele Rezidențiale [New Rules for Real Estate Developers in Bucharest. It Becomes Mandatory for Residential Neighborhoods to Have Schools, Nurseries, Roads and Parks]. Observator. 2021. Available online: https://observatornews.ro/eveniment/noi-reguli-pentru-dezvoltatorii-imobiliari-in-bucuresti-obligatoriu-scoli-crese-drumuri-si-parcuri-in-cartierele-rezidentiale-450222.html (accessed on 14 December 2021).
93. Blakely, E.J.; Snyder, M.G. *Fortress America: Gating Communities in the United States*; Brookings Institution Press and Lincoln Institute of Land Policy: Washington, DC, USA, 1997.
94. Grant, J.; Mittelsteadt, L. Types of Gated Communities. *Environ. Plan B Urban Anal. City Sci.* **2004**, *31*, 913–930. [CrossRef]
95. Kahana, E.; Lovegreen, L.; Kahana, B.; Kahana, M. Person, environment and person environment fit as influences on residential satisfaction of elders; *Environ. Behav.* **2003**, *35*, 434–453.

Olfindo, R. Transport accessibility, residential satisfaction, and moving intention in a context of limited travel mode choice. *Transp. Res. Part A Policy Pract.* **2021**, *145*, 153–166. [CrossRef]

Bianchi, A.M.; Marinescu, M.; Hera, D.; Dimitriu, S.; Ivan, G.; Ionescu, M.; Băltăreţu, F.L. Sisteme de Alimentare Centralizată cu Energie Termică în România [Centralised Thermal Energy Supply Systems in Romania]. In Proceedings of the Conferinţa Naţională de Termotehnică cu Participare Internaţională CNT 17 [National Thermotechnical Conference with International Participation CNT 17], Brașov, Romania, 21–22 May 2009. Available online: http://193.254.231.99:8080/jspui/handle/123456789/1240 (accessed on 30 November 2021).

Profit.ro. ELCEN Atacă Centralele de Apartament: E Nedrept că nu Plătesc Taxă de Poluare [ELCEN Attacks Apartment Gas heating Boilers: It's Unfair that They don't Incur a Pollution Tax]. 2019. Available online: https://www.profit.ro/povesti-cu-profit/energie/elcen-ataca-centralele-de-apartament-e-nedrept-ca-nu-platesc-taxa-de-poluare-19149588 (accessed on 29 December 2021).

Icleanu, E. Analiză: Sistemul de Termificare din București Este în Prag de Colaps. Agenda Construcțiilor [Analysis: The Heating System in Bucharest is on the Verge of Collapse. Construction Agenda]. 2019. Available online: https://www.agendaconstructiilor.ro/files/actualitatea-interna/analiza-sistemul-de-termificare-din-bucuresti-este-in-prag-de-colaps.html (accessed on 24 November 2021).

10. Observatornews. Centralele Termice pe gaz ar Putea fi Interzise din 2025. Care sunt Paşii Cheie Pentru a Salva Planeta? [Gas-fuelled Power Plants Could be Banned from 2025. What are the Key Steps to Save the Planet?]. 2021. Available online: https://observatornews.ro/economic/centralele-termice-pe-gaz-ar-putea-fi-interzise-din-2025-pentru-a-ajunge-la-tinta-de-emisiizero-420930.html (accessed on 5 January 2022).

11. Ministerul Dezvoltării, Lucrărilor Publice și Administrației (MDLPA) [Ministry for Development, Public Works and Administration]. Proiect de Hotarâre de Guvern pentru aprobarea Strategiei Naționale a Locuirii [Draft Government Decision for the approval of the National Housing Strategy]. Available online: https://www.mdlpa.ro/pages/proiecthgaprobatesnl21042022 (accessed on 15 May 2022).

12. Ministerul Dezvoltării, Lucrărilor Publice și Administrației (MDLPA) [Ministry for Development, Public Works and Administration]. Lege pentru aprobarea codului amenajării teritoriului, urbanismului si construcțiilor [Law for the approval of the Code for Territorial Planning, Urbanism and Constructions]. Available online: https://mdlpa.ro/pages/proiectlegeaprobarecodamenajareteritoriuurbanismconstructii (accessed on 15 May 2022).

13. Digi24. Orașul în Care Apartamentele s-au Scumpit Cel Mai Mult. Ce Se va Întâmpla Anul Viitor pe Piața Imobiliară din Romania [The City with the Most Expensive Apartments. What will Happen Next Year to the Romanian Real Estate Market]. 2022. Available online: https://www.digi24.ro/imobiliare/orasul-in-care-apartamentele-s-au-scumpit-cel-mai-mult-ce-se-va-intampla-anul-viitor-pe-piata-imobiliara-din-romania-1774345 (accessed on 14 January 2022).

14. Ziare.com. Cele Mai Scumpe Apartamente din Bucuresti—Topul pe Sectoare [The Most Expensive Apartments in Bucharest—the Toplist by City Sectors]. 2015. Available online: https://ziare.com/casa/preturi-apartamente/cele-mai-scumpe-apartamente-din-bucuresti-topul-pe-sectoare-1349514 (accessed on 25 October 2019).

15. Munteanu, G.M.; Stan, A.M. The marketing perspective on the risks of unqualified real estate market. *Bull. Transilv. Univ. Bras.* **2015**, *8*, 95–104.

16. Schwarz, L.; Keler, A.; Jukka, A.; Krisp, M. Improving urban bicycle infrastructure-an exploratory study based on the effects from the COVID-19 Lockdown. *J. Urban Mobil.* **2022**, *2*, 100013. [CrossRef]

land

MDPI

Article

Could Lavender Farming Go from a Niche Crop to a Suitable Solution for Romanian Small Farms?

Iuliana Vijulie, Ana-Irina Lequeux-Dincă *, Mihaela Preda, Alina Mareci and Elena Matei

Faculty of Geography, University of Bucharest, 1. Blv. Nicolae Balcescu, 010041 Bucharest, Romania; iuliana.vijulie@g.unibuc.ro (I.V.); mihaela.preda@geo.unibuc.ro (M.P.); alina.mareci@unibuc.ro (A.M.); elena.matei@g.unibuc.ro (E.M.)

* Correspondence: ana.dinca@geo.unibuc.ro; Tel.: +40-726-691-272

Abstract: Lavender crops have had an impressive continuous development in recent years, being currently a suitable alternative to other traditional crops because they can yield a high profit per hectare. This can be especially useful in Romania, with its high prevalence of subsistence and semi-subsistence farms. This study aims to analyse the issue of small emergent lavender farms in the context of the current Romanian agricultural background, including the framework mechanisms for implementing the Common Agricultural Policy at a national level. The research uses the qualitative survey method to provide broad, synthetic, analytical insights into small lavender farms/businesses in Romania, considering the perspective of the following two target groups: farm owners and civil servants with agricultural expertise. The main results show that both sample groups agree that lavender farms can be successful and satisfactory solutions. Increasing participation in information and training sessions may improve farmers' access to financing mechanisms, but both small farmers and civil servants with agricultural expertise identify a series of problems, mainly regarding the absence of a dedicated market for lavender-based products and a lack of labour force, both essential for maintaining the farming–processing–commercialising chain. The authors also conclude that a more flexible and future harmonisation between Romania's agricultural realities, the Common Agricultural Policy, and the National Rural Development Programme would improve lavender farming's social and economic impact. Follow-up research may envisage more in-depth market analyses for this emerging sector in Romania, facing obvious competition, but which could also benefit from good practice exchanges in the region.

Keywords: small farmers; lavender farms; barriers; civil servants with agricultural expertise; perception; tourism; Romania

1. Introduction

Lavender is an aromatic plant from the *Lamiaceae* family with approx. 40 species and numerous varieties [1–3] all over the world, and is native to the Mediterranean area, where it has been cultivated since ancient times, starting with the Greeks and the Romans [4–6]. In recent years, it acclimatised to continental climates [7], but its spread continues to be hindered by temperature limitations [5,8]. It is used for its therapeutical properties in curing a series of illnesses [9–12]. Its chemical qualities differ depending on the variety [13,14] and soil [15], and it is also an essential resource for the cosmetics and perfume industry [16,17] and even the food industry [18]. In addition, lavender is valued for its landscaping attributes in urban planning [19,20] and as a tourism attractiveness element [4,21–23], thus playing a significant role in helping small communities by championing investments in rural areas, creating new jobs, and ensuring an increase in residents' income level, ultimately bringing local economic benefits [17,24,25].

Globally, lavender farming has evolved two-fold. At first, Mediterranean states developed it strictly for its economic value in the cosmetics and perfume industry. However, this crop's applications later diversified, with the focus shifting towards its landscaping value, which was capitalised through tourism. As a result, iconic lavender fields in Provence (France), Isparta (Turkey), Tuscany (Italy), and Croatia became part of rural tourism, promoted through festivals or a new segment—aroma tourism [26].

Presently, the biggest European lavender cultivators are Bulgaria and France [27–31]. Worldwide, the regions or countries renowned for essential oil production are Bulgaria [32,33], France [4], Italy [34], Spain [35,36], Turkey [37,38], Kashmir [39,40], South Africa [41], and regions in Northern Africa [15].

Lavender farming is examined by a whole range of studies belonging to different disciplines that are mainly interested in the lavender oil market [38,39], the ecological sustainability of crops from different points of view [36,39,42] or its economic sustainability for rural and local development [32,37,40,43], including the added value for the tourism sector [44]. Existing research approaches regions with important lavender production results and with tradition in cultivating this aromatic plant [33,37,38,45,46]. To the best of our knowledge, lavender farming in Romania, and especially the stakeholders' perspective on it, has not been approached by scientific studies, despite the growing interest of subsistence small-scale family farms in the economic advantages of this crop.

Lavender farming became important in Romania around 2010, encouraged by the Common Agricultural Policy (CAP) incentive measures and the conditions created by climate change, which constituted a favourable factor for acclimatising aromatic plants typical for Mediterranean Europe. However, compared with countries where lavender has an established tradition and is extensively cultivated, this crop covers mostly small or very small agricultural areas in Romania. This is down to the following two aspects: firstly, it is a new crop introduced relatively late, and secondly, agricultural properties in this country have been transformed and fragmented by historical factors, even more so in recent years. This is proven by the fact that several studies dedicated to small-scale farming and its typology in Europe [47] or particularly to its efficiency and sustainability for Central and Eastern Europe (CEE) that have undergone critical socio-economic transformations during the post-communist period [48], inevitably also included Romania. In fact, according to the National Institute of Statistics, out of the total number of agricultural holdings in Romania, those smaller than 1 ha make up 53%, and those between 1 and 5 ha make up 38.6%, but only cover 28.7% of the utilised agricultural surface of the country [49]. This study aimed to evaluate, from an empirical perspective, small-scale lavender farming as a suitable economic and environmental solution in the light of the European Union's Common Agricultural Policy. In this respect, the perspectives of both farmers and civil servants with agricultural expertise were surveyed through semi-structured interviews.

The research gap in this specific type of analysis stems from the fact that studies on lavender farming in Romania have focused on technical information about planting/cultivation. Most of the published works are, in fact, reports detailing how lavender should be grown [50–52], climate and soil requirements [50,51,53], details on lavender varieties [54,55], maintenance works [52,53,56], production [54], harvesting [52,54], and on the manufacturing and use of lavender essential oil [53]. However, to date, there are no articles that qualitatively investigate the perspective of farmers and decision-makers regarding this niche crop, despite the growing interest of subsistence and semi-subsistence family farms in the economic benefits of lavender farming. Thus, the economic and social problems that lavender farmers face, their perception of the success of their business, and the effects of this crop on local communities have not yet been addressed by specialised studies.

The layout of the manuscript mirrors its exploratory approach, which sought to capture the complexity of lavender farming on small plots in the local agricultural landscape as an alternative to other traditional crops because they can produce a considerable profit per hectare.

The article was structured into the following five parts: introduction, methodology, results, discussions, and limitations of the study, ending with the chapter on conclusions. Thus, the study opens with an introductory chapter dedicated to the literature review, which starts with a general view of the topic and then focuses on a particular aspect of its relationship with the study area. This is followed by a presentation of the geographical and environmental considerations of lavender culture in Romania as well as EU agricultural policies that support its local development through a series of measures and sub-measures to support small farmers with the help of specialised institutions in the country.

The next chapter describes the research methodology, which used the qualitative survey method to provide broad analytical perspectives on small lavender farms in Romania, taking into account the views of the following two target groups: small lavender farm owners and civil servants with agricultural expertise. The article continues with the main results and discussions, including the study's limitations, followed by the conclusions of the research.

1.1. Background of Lavender Farming in Romania

1.1.1. Geographic and Environmental Considerations

Out of all aromatic plants, lavender is the most well known in Romania, even if it is a relatively recently introduced crop. The first bushes were recorded to have been planted in the 1950s, around Bucharest, probably by Bulgarian gardeners [56]. Afterwards, this crop extended south and west. In 1990, over 3000 ha were planted with lavender. However, between 1990 and 2005, this coverage was drastically cut to approx. 420 ha, with most farms located in southern Romania [57–59].

The year 2010 saw a gradual increase in lavender farming in almost all country regions, but the crop was mainly cultivated on small farms. According to recent data from the Agency for Payments and Intervention in Agriculture—APIA (Agenţia de Plăţi şi Intervenţii pentru Agricultură—APIA), lavender farms with an area of between 0.03 and 2 ha make up 86.18% of the total farm number and are supported by very small households practising subsistence farming. The next type of plots cultivated with lavender is farms between 2 and 5 ha, which constitute 10.9% of all farms number, being small semi-subsistence households. In contrast, the most extensive farms, covering between 5 and 25 ha, are owned by companies specialising in lavender farming but only make up 2.92% of their total number (Table 1) [60].

Table 1. Lavender farms in Romania.

No.	Area Cultivated with Lavender per Locality (ha)	Total Coverage (%)	No. of Localities with Lavender Farms per Category
1	<2 ha	86.18%	237
2	2 ha–5 ha	10.9%	30
3	> 5 ha	2.92%	8

Source: APIA data, 2019 [60].

Lavender is cultivated in many areas of Romania with a high insolation degree due to the plants' ecological adaptive and drought-resistant capabilities and because it has no particular demands on types of soil—though it prefers well-drained ones [53].

Most Romanian lavender farms are centred on very small and small agricultural plots (<5 ha), which make out 97.08% of the total farms' number. Their national distribution is unequal, but the most crucial region for lavender farming is the Transylvania Macroregion (MR1), with its high concentration of farms accounting for 42.69% of their total number. The second most important macroregion is Moldova-Dobrogea—MR2 (26.23%), followed by Banat-Oltenia—MR4 (16.85%) and Muntenia—MR3 (14.23%).

Conversely, all larger farms (5–25 ha) are located in just a few localities (e.g., Mihail Kogălniceanu, Corbii Mari, Tecuci, Bărăganul, Poiana, Nicolae Bălcescu, Simeria, and Pischia) in southern Romania (in Moldova-Dobrogea—MR2, Muntenia—MR3, and Banat-Oltenia—MR4 Macroregions). The latter area's territorial distribution and the associated sizeable lavender production farms are explained by the optimal pedoclimatic conditions in the southern parts of the country that resemble lavender' Mediterranean regions of origin (Figure 1).

Figure 1. Areas covered with lavender farms per locality in Romania (2019). Source: AFIA data, 2019.

1.1.2. Farming and Lavender Farming in Romania

Lavender farming in Romania is mainly limited to small farms, which is a dominant feature of the Romanian agricultural landscape. Subject to profound transformations induced by the transition from a socialist to an open market economic system, family farming and small-scale agriculture are dominant in the CEE countries, among which Romania distinguishes itself with the largest number of small farms [48].

These types of farms are excessively numerous in our country and represent 33% of the total European Union's (EU) farms of approx. 9.9 million [61,62]. Called family farms in the EU, in Romania they could be defined as both subsistence farms (smaller than 2 ha with a less than 2000 euro/year standard economic value) as well as semi-subsistence farms (between 2 and 5 ha with an economic value of 8000 euro/year) [63]. Subsistence farms mainly produce for the household's consumption and are managed and worked by family members, often the elderly. As far as 50% or sometimes more family members manage semi-subsistence farms to ensure household consumption and commercialise a small part of their agricultural production [62,64,65]. According to the international/European size hierarchy, these farms can be categorised into very small and small farms [66]. Many of them face obstacles such as poverty and a dire lack of labour, leading to reduced economic competitiveness and productivity [67]. Reasons for this include farm fragmentation (i.e., nowadays, a Romanian agricultural holding is formed of multiple small plots) [68], existing demographics, the commercial and technological context, and severe depopulation of rural areas. Moreover, in many cases, they might disappear due to larger farms absorbing them or because they are straightforwardly abandoned [69].

The multitude of small farms in the eastern EU—a phenomenon called "rural subsistence" by the specialised literature [70]—is a challenge for rural development. Moreover, it will continue to be a challenge as their presence will persist due to historical, political, and financial factors and the mentality of former communist states [71]. Many Romanian farmers rejected the idea of associations after the 1990 land re-allotment government measures [72,73], as they still lament the communist collectivisation process (1949–1962) when they were forced to surrender their properties to the Agricultural Production Cooperatives [74–76]. Furthermore, the Romanian landowners who nowadays form associations usually receive insignificant financial remuneration, with most economic benefits going towards those who organised the association [77]. As such, landowners of small farms further lose motivation to associate, even despite successful examples in their vicinity (e.g., the Moldova Lavender Association, which recently received member status in the European Federation of Essential Oils—EFEO) [78]. The Moldova Lavender Association showcases an excellent example of how niche crops such as lavender can be "profitable". They can be a lucrative option that would allow farmers or landowners of subsistence or semi-subsistence farms to continue their activities in their current form when faced with the emergence in their vicinity of more extensive agricultural holdings [79–81].

Even though they face some obstacles and limitations, lavender farms are expanding in Romania. Many specialised producers and firms are creating new plantations, processing and/or exporting raw products or essential oils [59]. Furthermore, the following characteristics of lavender farms facilitate their success: the plant requires minimal work and low maintenance costs, and the finite products obtained through only primary processing bring significant added value.

1.1.3. EU Agricultural Policies

After Romania's EU accession, the Common Agricultural Policy (CAP) implementation has supported rural development and sustainable agriculture [82]. Created in 1962, CAP is dynamic, adaptable, constructive, and committed to answering the UN Sustainable Development Goals (SDG) 2030. Although CAP is preparing for the 2023–2027 exercise, it continues the actions of the 2014–2020 exercise regulations by following its three domains as follows: direct support [83], marketing measures [84], and rural development [85].

CAP values small farms considering they generate the following: picturesque rural landscapes, biodiversity, local products, and they maintain traditions and ethnocultural specificities. Moreover, small farms substantially increase rural areas' attractiveness for business, rural residency, or leisure and relaxation activities for tourists [86].

In transposing CAP at a national level, the Romanian Ministry of Agriculture and Rural Development (MARD) had to support farmers by activating a series of instruments to consolidate agricultural competitiveness using programmes and measures that energise agricultural activities through subsidies and loans or guarantees. The Ministry implements European policies via its 42 county agricultural directorates (CAD), 8 regional centres, and 42 county offices of the Agency for Rural Financing and Investments—ARFI (Agenţia pentru Finanțarea Investițiilor Rurale—AFIR), and the 42 county centres of the Agency for Payments and Intervention in Agriculture—APIA (Agenţia de Plăţi şi Intervenţii pentru Agricultură—APIA) [87–90].

Among the financial measures (M) and sub-measures (sM) that aim to support farms (including lavender) as well as other agricultural activities in Romania within the National Rural Development Programme (PNDR) implemented through the ARFI relevant for this study are *sM 6.1. Setting up young farmers* (112 measure from the previous exercise); *sM 4.1. Investments in agricultural holdings; sM 4.2. Component for agricultural processing; sM 17.1. Mutual funds: helping farmers with insurance premiums, covering agricultural risks, economic losses due to climate hazards, plant diseases, pest infestation, or environmental accidents* [86].

Within the PNDR 2014–2020, APIA finances *M 10. Environment and climate* and *M 11. Ecological agriculture,* and farmers can apply to both these ongoing programmes for yearly subsidies [91–93]. Similar to any other agricultural entrepreneur, lavender farmers can, if eligible, receive financial support from multiple measures and sub-measures. For example, farmers who meet the eligibility criteria can apply to *M 10.4 Agro-environment—green crops*—and receive a yearly subsidy of 130 euro/ha [94]. Among the eligibility criteria, the most important one is that the minimum surface of the cultivated plot is 0.3 ha. For farms smaller than 1 ha, this is the main impediment to applying for APIA subsidies (Table 2).

Table 2. Main types of lavender farming in Romania and associated elements.

Type of Farm	Very Small Farms/ Subsistence Farms	Small Farms/ Semi-Subsistence Farms	Larger Farms
Dimension	<2 ha	2–5 ha	>5 ha
Type of owner	Farmer	Farmer/Company	Company
Access to EU financing mechanisms	Conditioned	√	√
Business financing	Private savings/Bank loan/SF (structural funds)	Private savings/Bank loan/SF (structural funds)	Private savings/Bank loan/SF (structural funds)
Degree of using /accessing SF (structural funds)	Very weak	Weak	Moderate/good
Labour force structure	Mainly family members	Family members + employees	Mainly employees
Distribution	X	√	√
Production's orientation	Other purposes + oil	Oil + other purposes	Oil + other purposes
Other associated activities	√	√	√

Source: Authors composition and APIA data, 2019.

The farmers who are certified as practising ecological agriculture can benefit from M 11., and receive support from the more suitable of the two sub-measures available, either *sM 11.1. Support for converting to ecological agriculture practices and methods* for which they receive 365 euro/ha/year, or *sM 11.2. Support for maintaining ecological agriculture practices and methods*, for which they receive 350 euro/ha/year [88]. In addition to these financial mechanisms, APIA offers compensatory payments to lavender farmers in the unique payment-per-plot land scheme (SAPS) with a total of 80 euro/ha. and a national transition support (ANT 1) regardless of product with a total of 20 euro/ha, which means that a farmer can theoretically receive subsidies of up to 600 euros per year because the M and sM payments are cumulative for the same plot if the farmers are deemed eligible [94].

One important aspect worth mentioning is that despite this crop's opportunities in terms of agricultural profitability, there are no political measures or stipulations particularly adapted for marketing lavender products. On the contrary, the market integration of agricultural holdings is generally emphasised as a significant problem for small-scale farms in Romania [95] In terms of creating market access for food products from small farms in Romania, studies have shown few successful attempts. These products contribute to a large extent to the 'informal food exchanges,' partially leading to the so-called 'unseen food' or 'self-provisioning' products [96]. As a cautionary tale, when national measures to support various crops (e.g., the Tomata Programme, the Garlic Programme) are implemented, farmers cannot capitalise on the excess of their yearly production because many of them lack training and essential experience for basic business tools and operations. The excess production cannot be absorbed by the local consumer market or supermarket supply chains, as these regularly appeal to imported products to the detriment of local farmers [97]. Moreover, the same distribution problems are even more prevalent in the case of small or very small farms (Table 2), which represent the target of the current study and, especially for lavender as a new type of crop, need a different market than the one for food products.

In this context, it is even more surprising that those who capitalised first on the opportunities brought forward by this niche crop were precisely the owners of small land plots (i.e., confronted most often with self-production agricultural activities and income insecurity) and not political actors or large corporate entities. Over time, business actors such as banks started offering designated loans for profitable agricultural endeavours. Transilvania Bank is worth mentioning, as it created a series of financing measures dedicated to lavender farming in Romania (i.e., it publishes and promotes a Complete Guide for Lavender Farming in Romania) [98] and even defined in business terms the "small farmer" and encouraged start-up lavender businesses with dedicated personal loans [99].

2. Data and Methodology

In order to meet the aim of the study, of analysing the situation of the small emergent lavender farms within the present-day agricultural Romanian background, this research was based on qualitative surveys to provide broad analytical insights into small lavender farms from the perspective of the two target groups.

The owners of the small and very small lavender farms were interviewed first to obtain an overview of their perception of lavender-based businesses following a series of variables relevant in the particular regional context of Eastern Europe.

This study focused on two types of farms defined by the EU and adapted to the Romanian context according to multiple factors (e.g., the size of cultivated area, associated economic benefits, etc.) (Table 2). The first refers to subsistence (very small) farms and the second to semi-subsistence (small) farms. These two types were considered because of their high number and socio-economic importance in Romania. Farms smaller than 2 ha and those between 2 and 5 ha make up more than 97.08% of the total number of lavender farms (Table 1). This study focused on these specific types of farms because of the complexity of problems their owners face and the overall acute need to stimulate agricultural entrepreneurship (Table 2). In order to simplify their description, the owners of small and very small farms interviewed by the authors will be referred to as "small farmers".

The second set of interviews was conducted with civil servants with agricultural expertise because their opinions on the present situation are relevant to the future perspectives of developing lavender farms in Romania in the framework of applying CAP.

2.1. Data Source and Methods

The secondary data were mined from the Romanian National Institute of Statistics and APIA (total area cultivated with lavender in 2019, number of farms, farm surface, data on CAP and MARD operational structures) for each macroregion of the country (MR1, MR2, MR3, MR4). Data from APIA regarding the location of lavender farms in the country were mapped using the proportional symbols cartographic method in ArcGIS Pro 2.8 (ESRI, Redlands, CA). On this map, Romania's four macroregions were identified to more clearly represent the level of development of lavender farming according to its territorial spread (Figure 1).

Primary data in our research came from two in-depth qualitative surveys interviewing both small farmers and civil servants with agricultural expertise. The latter category included people working in the following local agricultural branches of MADR: CAD, ARFI, and APIA, all of which have been responsible for managing CAP programmes.

The small farmers were selected randomly with the help of Romanian lavender farmers' associations from the country's four macroregions, who distributed information about our research intentions on specialised social media groups (Facebook, WhatsApp). Afterwards, associations facilitated authors to contact the participants directly by phone to collect their answers in the first phase of February–March 2020.

Authors used the e-mails from official websites to contact civil servants with agricultural expertise from the three prominent institutions (CAD, ARFI, and APIA) located in each of the 41 counties and the capital city of Bucharest to establish details about the most convenient interviewing channel. Bucharest and two counties (Ilfov and Covasna) were afterwards excluded from the sampling because they included no lavender farms. In this way, we conducted asynchronous interviews by e-mail between December 2019 and March 2020, primarily because of Romania's pandemic-related restrictions. Secondly, this was a more manageable (less time and money consuming) way also used in other studies [100], which was maintained in the second interviewing phase.

The second interviewing phase took place in March 2022 to achieve an optimal sampling size in relation to the study's analytic purposes and methods. This was performed by e-mail as a more accessible and preferred communication channel in the case of civil servants with agriculture expertise and by both phone and face-to-face interviews in the case of the small farmers. The latter interviews were possible in the context of pandemic restrictions relaxation. They took place at the Lavender Fair "Lavandişor" in Bucharest, one of the most significant events where small lavender farmers from all the country's regions converge annually. Both surveys were performed according to the academic requirements and EU legislation on managing personal data [101]. Each interview started with a presentation section detailing the aim of the study, authors' affiliation, details on the personal data confidentiality, and participants' agreement.

2.2. Methodological Tools and Sample Size

Surveys and interviewing methods targeting stakeholders represent a broadly utilised technique for obtaining data in studies that approach the topic of small farmers [37], their business practices, and market participation [102,103], especially when considering recently introduced crops, for which external governmental stimuli and start-up financial loans are vital [104,105]. In addition, interviewing methods are successfully applied in exploratory studies that provide empirical insights into the small farm household sector [106].

We aimed to fill a research gap about the business experience and perspectives of small lavender farmers in Romania, which has not previously been considered as far as we know from the existing scientific literature. This study had an exploratory approach that aimed to capture the complexity of this emerging sector in the autochthonous agricultural landscape.

In this respect the semi-structured qualitative survey, containing mainly open questions and some Likert scale questions, was the most appropriate method to identify the relevant issues, dimensions, and characteristics of our topic of interest [107] and corresponded to the rationale of our approach. This method allowed us first to obtain raw data from interview transcripts and then cover the complexity of the topic while avoiding the limitations imposed by a pre-structured or deductive type of qualitative survey [107]. In a second stage, data were coded and transformed mainly into nominal variables in SPSS v.23 (IBM), allowing us analytic generalisations and certain empirical quantitative approaches necessary for "systematic comparisons" [108], p. 7. Jansen [107] confirms the general practise of inserting quantitative approaches (e.g., cross-tabulation of data) into qualitative research in the current "paradigmatic situation," which increased permissiveness toward mixing methods [109], p. 2. Despite several quantitative techniques utilised to analyse nominal data in SPSS (e.g., cross tabs and Spearman coefficient—particularly suitable for nominal data), the current research approach was meant to explore the topic and depict the diversity of the sample target population and interviewed groups rather than emphasise "numerical distribution" [107], p. 3. Quotations from both small farmers and civil servants consistently complement the analyses to explain and further validate the research hypothesis. In order to keep the anonymity of the respondents, the quotes were coded "F" for farmers and "CsA" for civil servants with agriculture expertise. In addition, the code also includes the names of the county and macroregion of each respondent.

Besides the possibility to emphasise the opinion diversity of a sample group for a certain phenomenon—in our case of the small-scale lavender farming business in Romania—the coding process allowed us to adjust the sample to an optimal satisfactorily size through data saturation techniques. The sample dimension in qualitative research is debatable and varies among scientists [110–113].

Researchers consider data saturation, a relevant parameter, as achieved, thus allowing a sufficient confidence level in survey results when no new answers are offered. Moreover, particularly in qualitative research, "the goal of sampling is to collect data that either further develop or challenge existent hypotheses", as this study also attempted [114], p. 70.

Taking into consideration all the above, as well as the territorial extension of this agricultural crop in Romania and the scientific paradigm of our research [112], we adjusted our sample size for small-scale lavender farmers in Romania to a satisfactory level of 162 respondents. Officially, only 267 small and very small lavender farms are registered by the APIA statistics (Table 1) as they are currently receiving subsidies for this crop through this institution. Their number is, in fact, higher, but official data in this respect are missing creating a limitation for the present study.

In the case of civil servants with agricultural expertise, the sample numbered 47 valid answers that were considered to represent optimal institutional and territorial coverage. These 47 respondents came from all the following above-mentioned agricultural institutions: 39 respondents work within CAD (1 answer for each of the 39 counties—NUTS3—with lavender farms); 4 respondents work within AFIR; 4 respondents work within APIA (1 answer from each institution on a macroregion scale—NUTS1). We chose to focus on CAD employees because this institution is the one responsible for organising promotional and training programmes for farmers regarding the PAC measures and sub-measures mentioned in sub-chapter 1.1.3 EU agricultural policies (M 10., M 11., sM 6.1., sM 4.1., sM 4.2., sM 17.1.). AFIR and APIA are tasked with European fund absorption by implementing EU-funded projects for the first and granting subsidies in the case of the latter.

2.3. Research Hypotheses

The surveys for both target groups included addressing the three qualitative research hypotheses, namely, H1: Small farmers' participation in training programmes leads to their accessing structural funds; H2: The social and economic impacts of lavender farms on Romanian rural communities are low; H3: Lavender farming has mostly beneficial ecological effects.

Hypothesis 1 (H1). *Small farmers' participation in training programmes leads to their accessing structural funds.*

This is an intuitive hypothesis, as once a beneficiary participates in a training programme for accessing European funding, his or her ability to do so would improve. In this regard, the Rural National Development Programmes for 2015–2020 propose, through Pillar II, different activities such as those organised by authorised institutions oriented towards financial instruments. Other types of proposed activities are more applicable as follows: training by farmers for farmers, farmers' working groups where they can share know-how with other countries and vice versa, learning to use capital to provide equipment for certain kinds of competencies (e.g., processing the raw produce in-situ), and demonstration events [115].

Training programmes that target helping farmers access structural funds are a series of tools aimed at increasing the absorption rate of European funding. However, the implementation of these programmes differs among EU member countries. The participation rate of farmers in the training programmes and the degree to which they access European funds are closely related to the organisation and functioning of the institutions responsible for carrying out the training programmes (CAD, ARFI, and APIA). This can sometimes represent failure factors due to bureaucracy, lack of communication between public authorities at all levels, or sometimes expertise and knowledge gaps of some of the employees [116].

The willingness of farmers to participate in training programmes dedicated to financing and establishing new or agri-environmental crops does not always guarantee the success of the entrepreneurial initiative [117]. The reluctance regarding training programmes is determined by their theoretical specificity, the limited specialised field the farmer is interested in (vegetable grower, apple-grower, etc.), as well as the (im)possibility of learning about successful experiences by directly exchanging know-how with other farmers [118]. Participation in training programmes has had a positive impact on accessing different funding axes and developing initiatives in agriculture, with interest in participation being higher among young people than among other age groups [119]. EU statistics show that in 2016, about 19% of young farmers had received full agricultural training compared to 2.6% of farmers over 65 years of age, which led to an increase in the number of investments in small and medium farms set up by young people (i.e., 27.5% of EU farms) [120,121]. However, we can say that young farmers still face significant difficulties in obtaining funding, given the large investments and the low yield of a farm in its infancy/start-up phase [122].

Hypothesis 2 (H2). *The social and economic impacts of lavender farms on Romanian rural communities are low.*

This hypothesis was constructed to mirror the reality of Romanian agriculture, namely, the predominance of small plots, which will probably continue to manifest due to farmers' hesitance to associate. Hence the agriculture practised on a small scale would have a negligible impact. From an economic point of view, introducing lavender farming into rural areas is advantageous considering their high investment return. However, although lavender has the potential to increase their income, farmers' profits can be small if production planning does not correlate with market consumption [123]. Lavender farms' profitability must also weigh the general costs of labour wages in the region, renting agricultural equipment, and, last but not least, reinvested capital. This being said, low-profit margins pressure farmers to keep their traditional activities—animal husbandry and cultivating traditional crops—as a primary source of income. Their profits will also decrease if they do not benefit from any form of subsidies or if they are not part of a larger agricultural organisation that can develop a high capacity to process, promote, and commercialise their products [37].

Farmers who diversify their activities and include other crops besides the traditional ones also benefit from governmental support through specific financing mechanisms [124]. From a social point of view, lavender farming can revitalise an area by creating new jobs due to a diversification of the local economic activities, leading to increased incomes and the involvement of more community members in a shared venture, thus strengthening social cohesion. Apart from generating income from raw material processing, lavender fields also encourage tourism and trade entrepreneurship, which in cases such as France have proven to be very profitable [125]. Other possible positive social effects include decreasing the risk of demographic ageing by motivating young people to remain in rural areas as the new jobs are more suitable to their preferences and training (product designers, managers, tourism guides, promoters, internet content creators, etc.) [22,126].

Lavender tourism success rate directly depends on the region's annual tourism capacity, with its development becoming viable if it complements an already existing rural tourism market. Rural areas already known as tourism destinations that have introduced lavender fields into their portfolio were able to benefit from diversifying the range of their tourism offer with products such as breakfasts or tours in lavender gardens, lavender-scented village scenery combined with unique nature walks, photo shootings, as well as from commercialising lavender-based products (oil, scented water, honey, soap, jam, potpourri, pillows) [17].

In other words, tourism can contribute to the future increase in the living standards of rural communities' members if farmers take into account and incorporate the best examples of good practices in countries such as France, Turkey, Croatia, etc. on the tourist exploitation of lavender crops [127–129].

Hypothesis 3 (H3). *Lavender farming has mostly beneficial ecological effects.*

The hypothesis started from the results of previous studies that showed that lavender, in general, has had a positive effect on the environment as its farming does not cause significant damage to the quality of the air, water, and soil [130]. At the same time, lavender is a source of pollen and nectar for bees and is important for the local biodiversity, environmental balance, and the functioning of its ecosystem [131].

Examples of positive effects and reduced environmental impact of lavender farming which were identified by the specialised literature are linked to beekeeping and biodiversity—bio lavender crops, in particular, contribute to increased biodiversity as the plant is a magnet for many species of insects and birds (butterflies, bees, pheasants, or families of kestrels—*Falco tinnunculus*), which can find lavender fields as a perfect habitat [132–134]; natural insecticide—aromatic plants such as lavender could be an adequate alternative to chemical insecticides (they could be a healthier option to the chemical substances that plants are pulverised in order to eliminate pests) that would also target the share of the population that prefers to consume bioproducts (foods that have not been treated or treated with natural-based insecticides or pesticides) [135–139]. These fields also create a pleasing aesthetic landscape [127,140], with the added bonus of strong fragrances released by volatile oils, allowing for a natural aromatherapy session in the middle of lavender fields [36,137].

Authors such as Fayet [141] show in their study that landowners perceive EU CAP policy as a powerful tool that can strongly influence farmers to stop land abandonment and stimulate opportunities for redevelopment of agricultural land. Lavender is not a very demanding shrub; it adapts easily to degraded, eroded, or abandoned lands and positively influences the environment [53].

2.4. Survey Content and Sample Characteristics

The surveys for both target groups were semi-structured and started with questions referring to socio-demographic data (Tables 3 and 4).

Table 3. Socio-demographic profiles of farms owners (small farmers).

Variable	Item	%	Variable	Item	%
Gender	Male	54.3	Occupation	Labourers (plumbers, tractor drivers, car mechanics, couriers)	46.3
	Female	45.7		Farmers	19.8
Age groups M = 43.98	25–34	21.6		Legislators and higher officials (public administration officials or lawyers)	6.7
	35–44	32.1		Other professionals (engineers, economists, IT technicians, doctors)	27.2
	45–54	30.3			
	55–64	16.0			
Marital status	Married	87.7			
	Not married	12.3			
Education	Primary	3.7	Source of funds for starting the farm	Private savings	80.3
	Highschool	40.7		Private savings/Bank loan and SF (Structural Funds)	12.9
	University or higher	55.6		Private savings and SF (Structural Funds)	6.8
Agricultural education	No	71.6			
	Yes	28.4			
Participated in a training activity related to CAP/PNDR	No	71.6	Regions	MR 1 (Transylvania)	34.6
	M 11.	17.3		MR 2 (Moldova-Dobrogea)	30.3
	sM 6.1 (former M 112)	10.5		MR 3 (Muntenia)	14.2
	sM4.1	0.6		MR 4 (Banat-Olt)	20.9

Table 4. Socio-demographic profiles of civil servants with agricultural expertise.

Variables	Item	Frequency (%)	Variable	Item	Frequency (%)
Gender	Male	44.7	Last attended programme	MAKIS-2009 (Modernizing Agricultural Knowledge and Information Systems)	14.9
	Female	55.3		No attendance	19.1
Education	Highschool	8.5		CAP 2014–2020	38.3
	University or higher	91.5		PNDR 2014–2020	23.4
Institution	CAD	83.0		ARFI training activities	4.3
	ARFI	8.5	NUTS	MR 1 (Transylvania)	25.5
	APIA	8.5		MR 2 (Moldova-Dobrogea)	34.1
Age groups	25–34	8.5		MR 3 (Muntenia)	19.1
	35–44	17.0		MR 4 (Banat-Olt)	21.3
	45–54	25.6			
	55–64	48.9			

For the first hypothesis, questions for farmers focused on their business experience, financing outlets used by farmers to set up their lavender business (private savings, structural funds (SF), bank loans), and their attendance at information dissemination or training programmes organised by the authorities about various CAP measures and sub-measures.

Open-ended questions meant to present both the economic and social impact of lavender farming were further coded into *family subsistence and tourism development*, respectively. *New jobs* and *new business trends* were formulated to test the second hypothesis.

The Likert Scale (1—very dissatisfied, 5—very satisfied) was used to measure farmers' perception of their satisfaction level towards this particular business. Closed questions that were afterwards coded into ordinal variables addressed business development (e.g., lavender farming, processing, selling) and business perspectives.

Open-ended questions regarding the respondents' opinions on the ecological effects of lavender farming were included in order to verify the third hypothesis.

Questions for civil servants with agriculture expertise mainly followed the same structure, dictated by the three main research hypotheses, but some questions were adapted for this group of respondents (e.g., if their organisation had implemented farmers' training programmes under CAP measures; Likert scale questions on their perception of the success of lavender business in Romania, which are explicitly the success factors for such a business; to name both the positive and negative ecological environmental effects of these cultures; to exemplify possible agricultural measures necessary in farming this niche crop).

Both questionnaires end with two questions addressing barriers and solutions for developing lavender business farming in Romania (see Supplementary Material).

3. Results

3.1. Farmers Attendance to Training Programmes on CAP/PNDR Measures

Of the farmers that attended the training and information programmes regarding structural funds, more than half participated in meetings on M 11. (Ecological agriculture) and more than a third on sM 6.1. (Setting up new farmers).

Crosstabulation between the type of financing and training and information programme attendance shows that 18.2% of respondents who accessed structural funds had participated in training and information programmes organised on CAP measures and sub-measures. The chi-square test result for this table rejects the null hypothesis regarding the association between these two variables. The results for the ($X^2(2) = 7.735$, $p = 0.021$) test show there is enough evidence to suggest an association between the type of financing and programme attendance as p is below the considered significance level of ($a = 0.05$).

The association between education level and programme attendance is not statistically significant, and the crosstabulation percentages do not prove that the attendance level increases with their education level. Therefore, it seems that participation in a training or information dissemination programme is mainly influenced by their personal motivation to have a successful lavender business and not by their education level.

This aspect is also confirmed by the Spearman correlation coefficient, which showed a positive statistical relationship between the level of attendance at these programmes and the satisfaction level of small farmers regarding the development of their lavender business. The two-tailed *p*-value rejects the null hypothesis, and the correlation coefficient has a value of 0.325 (Table 5).

Table 5. Spearman correlation coefficient between programme attendance and satisfaction level.

			Programme Attendance	**Satisfaction**
Spearman's rho	Programme attendance	Correlation Coefficient	1.000	0.325 **
		Sig. (2-tailed)	.	0.000
		N	162	162
	Satisfaction	Correlation Coefficient	0.325 **	1.000
		Sig. (2-tailed)	0.000	.
		N	162	162

** Correlation is significant at the 0.01 level (2-tailed).

The civil servants with agricultural expertise unanimously declared that their institutions organised CAP-related information dissemination and training programmes. Furthermore, approximately two out of five respondents confirmed knowing small farmers who had applied for structural funds as a result of attending such programmes, further confirming our hypothesis. A significant percentage of 21.27% of this target group declared, however, that they did not know or had no information regarding either the attendance level of small farmers in training programmes or the degree to which small farmers' attendance had led to them accessing structural funds. This proves the flawed communication between civil servants and small farmers or the lack of frequent monitoring of the results obtained after attending training and information programmes. These kinds of issues were also brought forward by studies in other countries [116]. The lack of awareness from the civil servants with agricultural expertise could result from personal factors, expertise, or specific knowledge of certain employees.

The degree to which authorities (e.g., CAD, AFIR, APIA) organise and follow up on these events influences the level of farmers' participation in the training programmes and their applying for and obtaining European funds. However, many times, research studies have noted the critical failure of this relationship due to bureaucracy, crucial delays, and lack of communication among public authorities at national, regional, and local levels, and the low level of expertise and knowledge of particular employees or authorities, especially at regional and local levels [116].

The low rate of participation in these programmes could be a factor that leads to low information levels and, thus, to farmers not accessing structural funds. Only a small part of the interviewed small farmers financed their business through structural funds but combined with their private savings (6.79%).

3.2. The Social and Economic Impacts of Small-Scale Lavender Farming on Romanian Rural Communities

The economic impact of lavender farms was coded as a source for family subsistence and the potential for tourism development.

Most interviewed small farmers (63.58%) declared that their lavender business is primarily a source of income for their family, showing the low economic impact of these small and very small lavender farms, known in Romania as semi-subsistence and subsistence farms.

The crosstabulation between the economic impact of lavender farming and the satisfaction level regarding their lavender business shows that most of both categories (those who identify the economic impact of lavender farming as "family subsistence" or "tourism development") are partly satisfied or satisfied (more than 30% of each category) (Figure 2).

This proves that the presence of lavender farms generates income primarily for family members who are engaged in the day-to-day agricultural work, thus having a social impact associated with the economic impact, which is explicitly connected but unfortunately limited to each lavender farm business.

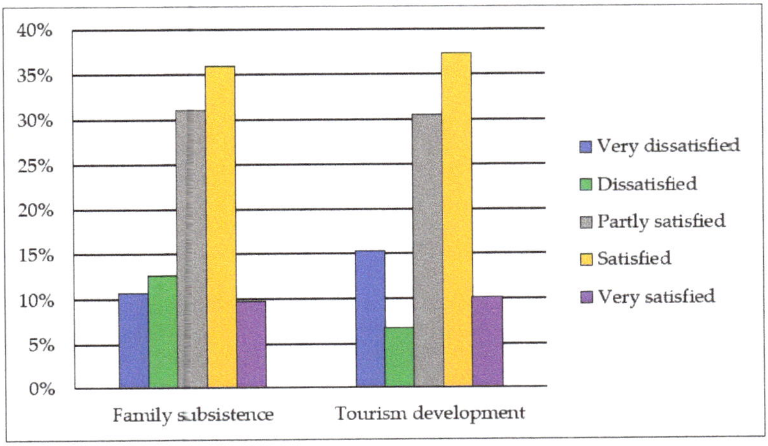

Figure 2. Crosstabulation between economic impact and business satisfaction level.

Lavender farms diversify economic activities at the local community level, and their importance was perceived differently depending on small farmers' education level. Thus, small farmers with primary and high school education levels overwhelmingly declared that the farms ensure their subsistence-level incomes. In contrast, those with university or higher levels of education identified the potential for entrepreneurial development of lavender farms in domains such as tourism (Table 6).

Table 6. Crosstabulation between economic impact and education level.

	Economic Impact		Total
Education Level	**Family Subsistence**	**Tourism Development**	
Primary	83.30%	16.70%	100.00%
Highschool	80.30%	19.70%	100.00%
University or higher	50.00%	50.00%	100.00%
Total	63.60%	36.40%	100.00%

The weight of answers naming tourism development, in other words, a diversification of the economic profile of lavender businesses, represents a superior level of their economic impact and is influenced by small farmers' education level and occupation. This is proven by the Spearman correlation coefficient, with a value of 0.313 showing a significant statistical association between education level and economic impact (Table 7).

Table 7. Spearman's rank correlation coefficient between farmers' education level and farms' economic impact.

			Education Level	Economic Impact
Spearman's rho	Education level	Correlation Coefficient	1.000	0.313 **
		Sig. (2-tailed)	.	0.000
		N	162	162
	Economic impact	Correlation Coefficient	0.313 **	1.000
		Sig. (2-tailed)	0.000	.
		N	162	162

** Correlation is significant at the 0.01 level (2-tailed).

Their farms' economic impact is also viewed differently according to small farmers' occupation, which is directly correlated to their education level. Labourers (whose main economic profession is something other than farmers, such as plumbers, tractor drivers, car mechanics, and couriers) consider that the economic impact of their farms consists mainly of ensuring their families' subsistence (61.3%), while farmers declare the same thing but with an even higher percentage of over 84%. When the small farmers' main occupation requires a professional level of training/education, such as legislators and higher officials (public administration officials or lawyers) or others (engineers, economists, IT technicians, doctors), they identified tourism as a viable complementarity to the agricultural development of their lavender business more often (Figure 3). The higher the occupational standing, the more small farmers are interested in diversifying the economic impact of lavender farms towards other domains such as tourism.

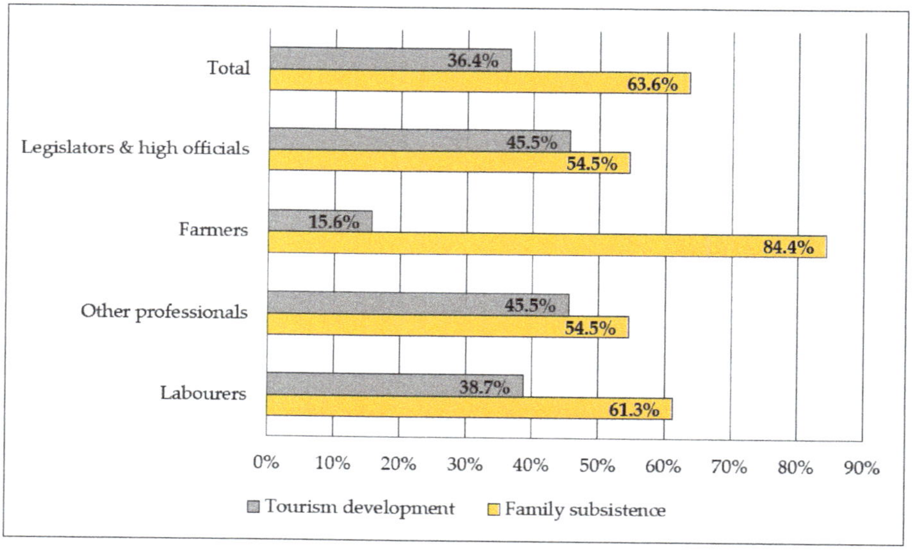

Figure 3. Correlation between lavender farms' economic impact and farmers' occupation.

Including tourism-related activities in their business stems from the following widely popular (especially among tourists) feature of lavender fields: their aesthetics [127,142]. In Romania, at this point, this phenomenon is just in its infancy, with small festivals (i.e., "open gates" events) organised inside lavender fields during their flowering period in June. *"We organise a small festival in June in Mădăraș, exactly in the fields; we have a 2 ha plot, where we usually receive those who love the purple gold"* F, Mureș County—MR1; *"in June when the fragrance is so strong that it covers the entire 5 ha of the farm, we wait for people to come to our festival, so they can enjoy nature and relax in a lovely place; we took inspiration from what people are doing in Provence"* F, Brăila County—MR2). In addition to such events, farmers organise small brunches, picnics, or photo sessions, as follows: *"we organise picnics straight in the middle of the lavender fields, in Bonțida, where people can taste locally made lavender flavoured products—cookies, ice cream, lemonade"* F, Cluj County—MR1; *"for June we made unique decorations, we have a lavender farm in Fibiș where we organise professional photo sessions"* F, Timiș County—MR4. Interviews showed that these events have the added benefit of being an opportunity for small farmers to promote and commercialise their products, such as the following: essential oils, floral water, diffusers, creams, soaps, bath bombs, etc.

Small farmers with older lavender farms that have capitalised on their businesses by including them in other related domains, such as tourism, are also more satisfied with their development. The Spearman correlation coefficient shows a positive and statistically significant relationship (Sig. (2-tailed)—0.000) between duration (the length of time of their lavender farms/business) and satisfaction level. This strengthens the rationale that lavender farming can be a suitable solution for improving the economic and social environment of rural communities in Romania (Table 8).

Table 8. Spearman's rank correlation coefficient between duration and satisfaction level.

			Duration	Satisfaction
Spearman's rho	Duration	Correlation Coefficient	1.000	0.287 **
		Sig. (2-tailed)	.	0.000
		N	162	162
	Satisfaction	Correlation Coefficient	0.287 **	1.000
		Sig. (2-tailed)	0.000	.
		N	162	162

** Correlation is significant at the 0.01 level (2-tailed).

Regardless of how old their lavender farm is (1, 2, 3, 4, 5, 5–10 years old), more than half of the interviewed farmers do their best to commercialise lavender products, not just the raw material. As they gain more experience and seniority in this business, the percentage of those who just cultivate lavender decreases and the diversity of ways in which the farms are capitalised increases. For example, they sell lavender cuttings alongside traditional products (handmade cosmetics, handmade decorative products, etc.). In total, 33.3% of those active for 4 years and 45.5% of those active between 5 and 10 years in the lavender business declared they engage in more than one type of commercial activity, which increases the farms' overall economic impact. The Chi-square test for the crosstab between duration and business development had a statistically significant *p*-value (Asimp. Sig. (2-sided)—0.027). Moreover, experienced farmers support new farmers with the knowledge they earned by ensuring free coaching to those who buy cuttings and want to start a lavender farm or by offering paid counselling to those who wish to start a business, as follows: *"I offer free counselling for the first 3 years to those to buy cuttings from me, but they can also contact me after that in case they face different problems"* F, Brăila County—MR2; *"when I deliver lavender cuttings, I also include a detailed technical chart with everything they need to know about cultivating lavender, about its planting and maintenance"* F, Timiș County—MR4).

Creating new jobs and new business trends are social elements that can improve a community's living conditions by increasing its income levels (and economic status). They are also elements that the small farmers identified as applicable to lavender farming. Respondents classified this social impact into different degrees of importance according to their education level and occupation (Table 9).

Table 9. Social impact according to education level and occupation.

		Social Impact		Tota
		New Jobs	**New Business Trend**	
Profession	Labourers	54.70%	45.30%	100.00%
	Other professionals	56.80%	43.20%	100.00%
	Farmers	71.90%	28.10%	100.00%
	Legislators and high officials	45.50%	54.50%	100.00%
TOTAL		58.00%	42.00%	100.00%
Education level	Primary	66.70%	33.30%	100.00%
	Highschool	68.20%	31.80%	100.00%
	University or higher	50.00%	50.00%	100.00%
TOTAL		58.00%	42.00%	100.00%

Full-time farmers overwhelmingly (71%) indicated that the main social impact generated by lavender farming is the creation of new jobs. They mentioned a new business trend as a secondary effect and to a lesser degree than the other occupational categories.

Similar to the economic impact, the Spearman correlation coefficient shows a positive, but weaker, relationship between education level and social impact (Sig. (2-tailed)—0.023; correlation coefficient—0.178).

Identifying social opportunities correlates with the extent of their lavender business (i.e., duration). The hierarchical importance of the following two types of social impact differs: with start-up level farmers declaring that the most important social impact is creating a new business trend (62.5%), while farmers who have been active for more than 2 years place the creation of new jobs first (Figure 4).

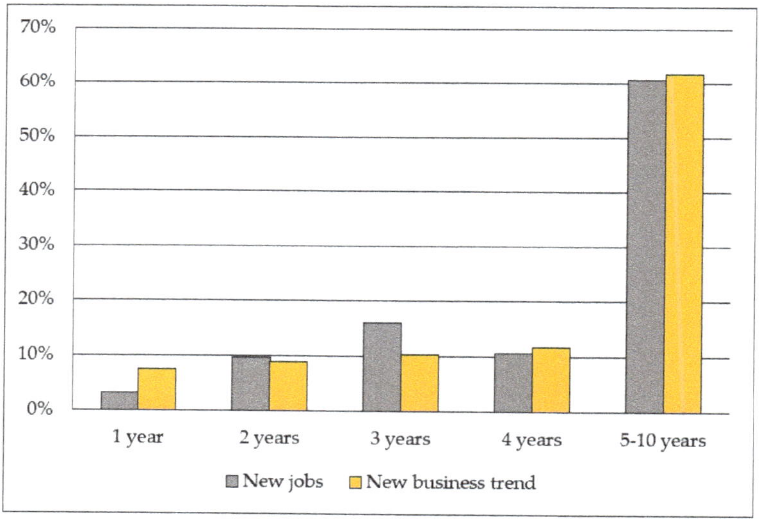

Figure 4. Relationship between the identified social impact of lavender farming and business duration.

Creating new jobs translates in reality into involving family members in the agricultural activities for farms younger than 2 years (*"we started to grow lavender two years ago, so far we are doing everything ourselves, family members trim the crops, take out the weeds, gather the flowers"* F, Teleorman County—MR3), and creating new seasonal or permanent jobs for community members for farms older than 4 years (*"our business is varied: we cultivate and maintain the fields, process the raw product, and commercialise various things from cuttings to*